MW01517713

Developments in Primatology:
Progress and Prospects

Series Editor
Russell H. Tuttle
Department of Anthropology
The University of Chicago

For further volumes, go to
http://www.springer.com/series/5852

Sharon Gursky-Doyen · Jatna Supriatna
Editors

Indonesian Primates

 Springer

Editors
Sharon Gursky-Doyen
Department of Anthropology
Texas A&M University
College Station, TX
USA
gursky@tamu.edu

Jatna Supriatna
Conservation International Indonesia
University of Indonesia
Jakarta
Indonesia
j.supriatna@conservation.org

ISBN 978-1-4419-1559-7   e-ISBN 978-1-4419-1560-3
DOI 10.1007/978-1-4419-1560-3
Springer New York Dordrecht Heidelberg London

Library of Congress Control Number: 2009942275

Printed on acid-free paper

Springer is part of Springer Science+Business Media (www.springer.com)

*S.L. Gursky-Doyen dedicates this volume
to her parents, Ronnie Bender and Burt
Gursky, who after all these years still do not
really know what she does, but they proudly
display her books on their coffee table; and
to her husband Jimmie who taught her what
love is.*

*Jatna Supriatna dedicates this volume to all
Indonesian primatologists who have been
struggling to save the most diverse primates
on earth. Also to his family and to
colleagues in the Department of Biology
at the University of Indonesia and
Conservation International Indonesia who
have supported me throughout my tenure.*

Acknowledgments

The contributions and comments of many people have been critical to the development of this edited volume and we would like to acknowledge their pivotal role. Our profound thanks to all the authors who contributed to this volume. We would also like to thank the many anonymous reviewers for the time and effort they put into critiquing the papers presented in this volume. Thanks to Janet Slobodien, the editor at Springer (and her assistants), and Dr. Russell Tuttle for allowing this volume to be a part of the Developments in Primatology series.

S.L. Gursky-Doyen would like to acknowledge a number of organizations that have been exceedingly generous with their logistic and financial support. Texas A&M University has provided both financial support for Gursky-Doyen's research as well as logistic support in the form of time off from teaching responsibilities. The Indonesian Institute of Sciences (LIPI) and the Indonesian Department of Forestry (PHPA) regularly granted Gursky-Doyen permission to conduct research in protected forest. The Fulbright Foundation, Primate Conservation Inc., Conservation International Primate Action Fund, LSB Leakey Foundation, Wenner–Gren Foundation for Anthropological Research, the National Science Foundation and the National Geographic Society have all provided Gursky-Doyen with funds at various stages of her fieldwork in Indonesia.

Jatna Supriatna would like to thank several individuals and organizations that have supported his work in Indonesia. Dr. Russell Mittermeier and Dr. Anthony Rylands from Conservation International and IUCN-SSC-Primate Specialist Group, respectively, have each enriched his understanding of primate conservation throughout the world. While in Indonesia, Dr. Sukarya Somadikarta, Dr. Indrawati Ganjar, Dr. Noviar Andayani, Mr. Jarot Arisona Aji from the University of Indonesia helped him develop a fuller understanding of Indonesian biodiversity. Both Conservation International Indonesia and the Department of Biology at the University of Indonesia have supported his work either by providing funds or students to work with him. In addition, the Indonesian Primatologist Association, the Southeast Asia Primatologist Association and IUCN-SSC-Primate Specialist Group have also worked with Jatna to develop primate action plans and workshops to save Indonesia's primates.

Contents

Contributors

Asep S. Adhikerana
Fauna and Flora International Indonesia Program, Jalan Harsono,
RM. No.1 Ragunan, Jakarta 12550, Indonesia

Noviar Andayani
Wildlife Conservation Society Indonesia Program Jl. Burangrang 18,
Bogor, Indonesia
Department of Biology, University of Indonesia, Depok, Indonesia
nandayani@wcs.org

James A. Bailey
Department of Political Science and Law Enforcement, Minnesota State
University, Mankato, MN, USA

M. Bismark
Forest and Nature Conservation Research and Development Center,
Bogor, Indonesia
bismark_forda@yahoo.com

Gil Brogdon
Department of Radiology, University of Southern Alabama Medical Center,
Mobile, AL, USA

Susan M. Cheyne
Wildlife Conservation Research Unit, Department of Zoology,
Oxford University, Tubney, Abingdon Road, OX13 5QL, United Kingdom
Orang-utan Tropical Peatland Project, Center for the International
Cooperation in Management of Tropical Peatlands (CIMTROP),
University of Palangka Raya, Indonesia
susan.cheyne@zoo.ox.ac.uk

David Chivers
Wildlife Research Group, The Anatomy School, University of Cambridge,
Downing Street, Cambridge CB2 3DY, UK
Djc7@cam.ac.uk

Roberto A. Delgado Jr.
Integrative & Evolutionary Biology, University of Southern California,
Los Angeles CA 90089, USA
radelgad@usc.edu

Gregory A. Engel
Swedish/Cherry Hill Family Practice Residency, Seattle, WA, USA
Washington National Primate Research Center, University of Washington,
Seattle, WA, USA

Agustin Fuentes
Department of Anthropology, University of Notre Dame, Notre Dame, IN, USA
afuentes@nd.edu

Sharon Gursky
Department of Anthropology, Texas A&M University, MS 4352,
College Station, TX 77843, USA
gursky@tamu.edu

Michael D. Gumert
Assistant Professor, Division of Psychology, School of Humanities and Social
Sciences, Nanyang Technological University, Singapore 639798
gumert@ntu.edu.sg

Hope Hollocher
University of Notre Dame, Department of Biology, Notre Dame,
IN 46556-5611, USA
Hope.Hollocher.1@nd.edu

Lisa Jones-Engel
Washington National Primate Research Center, University of Washington,
Seattle, WA, USA
jonesengel@bart.rprc.washington.edu

Kelly Lane
Department of Biology, University of Notre Dame, Notre Dame,
IN 46556-5611, USA
klane3@nd.edu

Susan Lappan
Department of Anthropology, Appalachian State University, Boone, NC 28608, USA
lappan@nyu.edu

Michelle Lute
Department of Biology, University of Notre Dame, Notre Dame, IN 46556-5611
mkulaga@nd.edu

Andrew J. Marshall
Department of Anthropology, University of California, Davis, CA, USA
ajmarshall@ucdavis.edu

Dr. Vicky Melfi
Whitley Wildlife Conservation Trust, Field Conservation and Research
Department, Paignton Zoo Environmental Park, Totnes Road, Paignton,
Devon, TQ4 7EU, UK
vicky.melfi@paigntonzoo.org.uk

Stefan Merker
Institute of Anthropology, Johannes-Gutenberg University Mainz,
Colonel-Kleinmann-Weg 2, 55099 Mainz, Germany
merker@uni-mainz.de, tarsius@gmx.net

Alan R. Mootnick
Gibbon Conservation Center, P.O. Box 800249, Santa Clarita, CA 91380, USA
alan@gibboncenter.org

Rachel Munds
Nocturnal Primate Research Group, School of Social Sciences and Law,
Oxford Brookes University, Oxford OX3 0BP, United Kingdom
rmunds27@gmail.com

K.A.I. Nekaris
Nocturnal Primate Research Group, School of Social Sciences and Law,
Oxford Brookes University, Oxford OX3 0BP, United Kingdom
anekaris@brookes.ac.uk

Vincent Nijman
Department of Anthropology and Geography, School of Social Sciences
and Law, Oxford Brookes University, OX3 0BP, Oxford, United Kingdom
vnijman@brookes.ac.uk

Lisa M. Paciulli
Department of Anthropology, University of West Georgia, Carrollton,
GA 30118, USA
lpaciull@westga.edu

Arta Putra
Fakultas Kedokteran Hewan, Pusat Kajian Primata, Udayana University,
Bali, Indonesia
artaputra@telkom.net

Erin P. Riley
Department of Anthropology, San Diego State University, San Diego,
CA 92182-6040, USA
epriley@mail.sdsu.edu

Aida Rompis
Fakultas Kedokteran Hewan, Pusat Kajian Primata, Udayana University,
Bali Indonesia
Agungida@indo.net.id

Michael A. Schillaci
Department of Social Sciences, University of Toronto Scarborough,
1265 Military Trail, Toronto, Ontario, M1C 1A4 ,Canada
schillaci@utsc.utoronto.ca

Elisabeth H. M. Sterck
Behavioral Biology, Utrecht University, PO Box 80086, 3508 TB,
Utrecht, The Netherlands
Ethology Research, Biomedical Primate Research Center, PO Box 3306 ,2280
GH Rijswijk, The Netherlands
E.H.M.Sterck@uu.nl

Jito Sugardjito
Research Center for Biology, The Indonesian Institute of Sciences,
Jalan Raya Jakarta-Cibinong Km.48, Cibinong 16911, Indonesia
jitos@cbn.net.id

Jatna Supriatna
Conservation International Indonesia, Jl. Pejaten Barat 16 A,
Jakarta 12559, Indonesia, jsupriatna@conservation.org
Department of Biology, University of Indonesia, Depok, Indonesia

Sri Suci Utami Atmoko
Faculty of Biology, Universitas Nasional, Jakarta, Indonesia
suci_azwar@yahoo.co.id

Carel van Schaik
Anthropological Institute and Museum, University of Zurich, Switzerland
vschaik@aim.unizh.ch

I. Nengah Wandia
Fakultas Kedokteran Hewan, Udayana University, Bali, Indonesia
ngh.wandia@plasa.com

Serge A. Wich
Great Ape Trust of Iowa, 4200 SE 44th Avenue, Des Moines, IA 50320, USA
swich@greatapetrust.org

Achmad Yanuar
Conservation International Indonesia, Jl. Pejaten Barat 16 A,
Jakarta 12559, Indonesia
ayanuar@conservation.org

Jessica L. Yorzinski
Animal Behavior Graduate Group, University of California, Davis, CA 95616, USA
jyorzinski@ucdavis.edu

Chapter 1
Introduction

Nanda Grow, Sharon Gursky-Doyen, and Jatna Supriatna

Indonesia's primates are remarkable in their rich diversity and number of taxa. Indonesia belongs to the Sundaland Biodiversity hotspot in terms of both flora and mammal species (Supriatna et al. 2001), and this collection of over 17,000 islands may be home to the most diverse collection of primates in the world. Conservation is an important issue to consider for all primates, but the Indonesian primates are especially at risk. According to the IUCN/SSC Primate Specialist Group (IUCN/SCC 2008), over 70% of Asian primate species are threatened by extinction, and 84% of the over 40 Indonesian primate species are threatened. A report compiled by Primate Specialist Group of IUCN's Species Survival Commission (SSC), the International Primatological Society (IPS), and Conservation International (CI), identifies three of the world's twenty-five most endangered primates as species endemic to Indonesia, including the Siau Island tarsier (*Tarsius tumpara*), the pig-tailed langur (*Simia concolor*), and the Sumatran orangutan (*Pongo abelii*) (Mittermeier et al. 2007). Nearly half of the Indonesian primate species are endemic, a percentage second only to one country, Madagascar. This high proportion of endemic primates makes Indonesia a particularly significant place to study the evolution of variation in primate taxa.

In conjunction with its considerable biodiversity, Indonesia is also one of the most populated countries in the world. Human population growth and industrialization, combined with forest resource exploitation and the lack of a centralized protective infrastructure, threaten the survival of the Indonesian primates. Recent shifts in Indonesian national park policy, including transferring authority over forests to the community level, have contributed to rapid deforestation and the encroachment of villages into national parks. While conservation efforts are more necessary than ever before, it is difficult to develop plans that consider national law, local economies and attitudes, as well as the habitat needs of specific populations. In developing a conservation plan, researchers of primate behavior are essential to determining the habitat needs of specific populations.

N. Grow (✉)
Texas A&M University, Department of Anthropology, College Station, TX 77843-4352
e-mail: nanda.grow@gmail.com

S. Gursky-Doyen and J. Supriatna (eds.), *Indonesian Primates*,
Developments in Primatology: Progress and Prospects,
DOI 10.1007/978-1-4419-1560-3_1, © Springer Science+Business Media, LLC 2010

Primatologists serve an important role in conservation by determining which environmental variables are important to species survival, as well as by exploring how local human activity impacts primate populations. Researchers can identify the major threats to primate communities, observe patterns of change, predict how human activities might affect primates in the future, measure primate densities, and assess the validity of different conservation strategies (Chapman and Peres 2001). Primatologists also explore the variation in primate species by detecting cryptic species or providing confirmation that rare species still exist in the wild. For example, the Siau Island tarsier has only recently been discovered in Indonesia (Shekelle et al. 2008), and the pygmy tarsier (*Tarsius pumilus*) of Sulawesi was recently observed in its natural setting (Gursky-Doyen and Grow 2009). Both of these finds illustrate how the diversity of Indonesian primates is by no means fully understood. Without the protection of primates, much of this valuable diversity may be lost before researchers have the chance to even recognize it.

This volume synthesizes current research on the primates of Indonesia, which include apes, monkeys, and prosimians. These chapters demonstrate the diversity in Indonesian primates, ranging from Nekaris and Munds' study of using slow loris facial variation to distinguish species to Delgado's discussion of communication patterns among orangutans. In this volume, the common thread of diversity is inextricably linked to the theme of conservation.

Part One, Indonesia's Apes, contains chapters on the endangered orangutans and gibbons. The orangutan is the only Asian great ape and includes the Sumatran form (*Pongo abelii*) and the Bornean form (*Pongo pygmaeus*). As Sugardito and Adhikerana note, nest densities of the Bornean orangutan are severely threatened by logging activity, especially at national park boundaries. The authors assess the effectiveness of community-based patrolling as a means to protect the Bornean orangutan, as programs that include community members can reduce conflict between local residents and forest rangers. They find that this type of system is becoming an increasingly effective means of protecting orangutans.

Delgado discusses conservation of orangutans as well as explores the variation in their communication behavior and sociality. Although orangutans may not have cohesive groups based on proximity, they have social networks of individuals that associate and communicate. In order to answer questions concerning orangutan communication and social behavior, Delgado emphasizes the need for conservation of orangutans to allow for observational studies to continue.

Atmoko and van Schaik similarly stress the importance of protecting critically endangered Sumatran orangutans for both their survival and the continued study on their behavior. As the authors discuss, long-term studies of these long-lived primates are crucial to understanding their life history. Sumatran orangutans have extremely slow life histories, with long interbirth intervals, low mortality, and long lives. Long-term protection of their habitat is therefore necessary to answer questions about their life history and behavior.

The gibbons of Indonesia are also at risk. Supriatna et al. present the threats to the critically endangered Javan gibbon (*Hylobates moloch*). These threats include severe habitat loss from agricultural encroachment and logging, as well as capture for the illegal pet trade. The authors note that these threats were exacerbated after the Indonesian government decentralized forest management in 2001, where local authorities may choose the short-term gains of allowing forest resources to be exploited instead of a sustainable management plan. The authors also highlight the importance of genetic data when relocating primate groups to new habitats, underscoring the need to acquire information on the ecological needs of each specific protected animal. Scientists take a primary role in this aspect of conservation. For example, research like Lappan's study of gibbon feeding behavior and food availability can contribute to habitat restoration plans. Lappan examines spatial and temporal variation between groups of siamang (*Symphalangus syndactylus*) in Sumatra. Groups were found to have as much dietary variation from year to year as they differed from each other within each year, showing that temporal fluctuations in food availability has a larger effect on siamang activity patterns than spatial variation.

Other studies of the Indonesian gibbons investigate how human-induced changes in forest composition directly result in changes in primate behavior. Yanuar and Chivers explore how forest fragmentation from human development affects ranging and diet in the folivorous siamang and the frugivorous agile gibbons (*Hylobates agilis*). Both siamang and gibbons had shorter travel distances and smaller home ranges in forest fragments than in continuous forest, reflecting a negative relationship between food availability and day range.

Cheyne explores the habitat requirements of white-bearded gibbons (*Hylobates albibarbis*) in disturbed peat-swamp forest of Borneo, a type of forest that has received little conservation attention. Studying the effects of human disturbance on primate behavior is important for all types of forest. It is also important to compare the different habitat requirements for different types of primates. Along these lines, Marshall investigates how habitat quality affects behavior in both Indonesian apes (white-bearded gibbons, *Presbytis rubicunda rubida*) and red leaf monkeys. The study is particularly interesting because it compares the monogamous and frugivorous gibbons, with a slower life history characteristic of apes, to the polygynous seed and leaf specialist monkeys. Results indicate that certain types of resources determine habitat quality and not overall food availability, with different resources limiting the density of different species, possibly relating back to differences in social system and life history.

Part Two, Indonesia's Monkeys, investigates the diversity of monkeys in Indonesia. All authors express the importance of conservation. Yorzinski explores how isolated island primates react to predation pressure with the critically endangered pig-tailed langur (*Simias concolor*) of the Indonesian Mentawai islands. The langurs were found to be afraid of novel vocalizations, confirming other studies that suggest primates fear novel stimuli. The langurs did not treat felid predator vocalizations differently than the other novel vocalizations, however, suggesting they do not retain specific predator recognition for specific mainland predators. While the Indonesian islands provide a good opportunity to study predator recognition among naïve isolated primates, conservation of small island populations is necessary to compare to mainland populations.

Paciulli examines the relationship between vegetation and primate densities for all four primates of the Mentawai Islands, including Kloss's gibbon (*Hylobates klossii*), the pig-tailed macaque (*Macaca pagensis*), the Mentawai Island leaf langur (*Presbytis potenziani*), and the pig-tailed langur. Results interestingly show that vegetation and forest structure do not consistently correlate with primate density, where availability of certain plant foods was a stronger predictor of density. Determining more precise relationships between primate densities and specific vegetation variables contribute to conservation plans that include selective logging, where the most valuable trees to primate populations are spared.

Knowledge of a species' habitat requirements is especially crucial for conservation of species that are sensitive to declines in habitat quality, like the proboscis monkey (*Nasalis larvatus*) of Borneo. Bismark reviews the ecology and conservation situation of these monkeys, a species that requires quality habitat but is experiencing increasing habitat degradation. Group size and density decrease in disturbed habitat compared to higher quality wetland habitats, such as mangrove forest. While behavioral data for this species is limited, Bismark observed that in disturbed habitat the monkeys modify their behavior by choosing foods with higher mineral content as well as foraging in smaller groups.

Understanding the human relationship to wild primates is important when considering how to protect primates that have a history of close human interactions. Ethnoprimatological research is thus an essential component of conservation plans. Lane et al. illustrate the importance of ethnoprimatology in their exploration of the relationship between long-tailed macaques (*Macaca fascicularis*) and the culture of human residents of Bali. Human attitudes towards nature, such as the Balinese Hindu belief that macaques should be respected, can lead to protection of the macaques. However, this belief does not ubiquitously translate to protective behaviors. In the following chapter, Schilaci et al. notes that despite their sacred status, the monkeys are often the target of air rifle pellets as they are economic liabilities to farmers. Tolerance of their presence may also be due to their appeal to the tourism industry rather than a sacred standing. Macaque behavior in Bali is clearly intertwined with human activity. The behaviors of both humans and monkeys must both be considered in order to understand how they influence one another.

Part Two of this volume includes further interesting research on the behavior of Indonesian monkeys. Riley reports Sulawesian Tonkean macaques (*Macaca tonkeana*) males prefer to affiliate with adult females rather than other males. Male tolerance of other males appears related to sex ratio. With a nearly even sex ratio, male-male agonism is low and grooming interactions between males occurred, but male–male agonism is higher for groups with skewed sex ratios. Nijman reviews the ecology and conservation of the little-studied Bornean grey-backed langurs, *Presbytis hosei, P. (h.) canicrus*, and *P. (h.) sabana*. While these langurs are sympatric with other langurs, they may be subject to competitive exclusion with the other species. Habitat loss from logging and fire as well as hunting threatens these endangered langurs. Wich and Sterck explore the ecology and sexual conflict of Sumatran Thomas langurs (*Presbytis thomasi*). The authors find that food competition does not appear to limit group size, and females in larger groups actually have more surviving offspring. Finally, Gumert tests Seyfarth's social grooming model (1977), which predicts that females will preferentially groom higher ranked females to develop social alliances, among long-tailed macaques (*Macaca fascicularis*) in Kalimantan. This study is an important test of a traditional model on an Indonesian primate species. It assesses whether this general model can apply to a specific island primate society. The social grooming model was supported in that rank and competition affected grooming, but specific predictions of model were not strongly supported. Results showed most female grooming was not reciprocal, imbalanced because of the effects of rank where higher ranked females receive more grooming than they give to lower-ranked partners.

While there are only three chapters about prosimians in Part Three (Indonesia's Prosimians), there is a much greater diversity of prosimian primates in Indonesia than is represented in this volume. Despite this great diversity, there is currently a shortage of researchers exploring the variation in Indonesian prosimians. Two chapters in this section contribute to our knowledge of the Sulawesian tarsiers. Gursky-Doyen discusses the role of scent marking in territory defense among the Sulawesian spectral tarsiers (*Tarsius spectrum*). Like many nocturnal prosimians, these tarsiers use scent marks to demarcate territorial boundaries. Merker investigates the behavioral ecology of Dian's tarsier (*Tarsius dentatus* or *T. dianae*), and discusses which factors affect population densities. Nekaris and Munds explore less well-known Indonesian prosimians, the Indonesian slow loris species (*Nycticebus* spp.). Indonesian lorises are understudied, with virtually no data on the distribution, density, or habitat requirements of these rare primates (Nekaris et al. 2008). In this study, the authors explore how facial markings may serve as species recognition devices and further divide the Indonesian slow lorises into additional species or subspecies. While the facial markings distinguish the three known species, specific characteristics of facial markings (such as color) may correspond to additional taxonomic variation. The full diversity of Indonesian lorises is thus currently unknown, and there may be additional cryptic taxa, as has been suggested among tarsiers (Shekelle and Salim 2009). With new endemic species being discovered in Indonesia, conservation of this rich diversity is important, especially in our quest to understand speciation and variation.

All contributions to the volume underscore the importance of conservation of these threatened primates. Habitat degradation clearly impacts the behavior and density of primates. Conservation of Indonesian primate habitats is imperative to continue to explore the diversity of the primate order. The primary threats to Indonesian primates are habitat loss from deforestation, small population density combined with high human population density, hunting and the wildlife trade, and the lack of enforceable conservation plans. Primates in the Indonesian islands are further threatened by constrained habitat area. This volume illustrates both how these threats may impede future primatological research and how scientific research can contribute to conservation, especially with regard to determining specific habitat requirements. The relationship of each primate species to its specific environment must be understood, as one conservation plan is not sufficient to encompass the diversity of the Indonesian primates.

References

Chapman CA, Peres CA (2001) Primate conservation in the new millennium: the role of scientists. Evol Anthropol 10:16–33

Gursky-Doyen S, Grow NB (2009) Elusive highland Pygmy tarsier rediscovered in Sulawesi, Indonesia. Oryx 43(2):173–174

IUCN/SCC (2008) IUCN Red list of threatened species. Species survival commission, International union for the conservation of nature and natural resources. Gland. IUCN Publications, Switzerland and Cambridge, UK

Mittermeier RA, Ratsimbazafy J, Rylands AB, Williamson L, Oates JF, Mbora D, Ganzhorn JU, Rodríguez-Luna E, Palacios E, Heymann EW, Kierulff MCM, Yongcheng L, Supriatna J, Roos C, Walker S, Aguiar JM (2007) Primates in peril: the world's 25 most endangered primates, 2006–2008. Primate Conserv 22:1–40

Nekaris KAI, Blackham GV, Nijman V (2008) Conservation implications of low encounter rates of five nocturnal primate species (Nycticebus spp.) in Asia. Biodivers Conserv 17:733–747

Seyfarth RM (1977) A model of social grooming among adult female monkeys. J Theor Biol 63:671–698

Shekelle M, Groves C, Merker S, Supriatna J (2008) Tarsius tumpara: a new tarsier species from Siau ISLAND, North Sulawesi. Primate Conserv 23:55–64

Shekelle M, Salim A (2009) An acute conservation threat to two tarsier species in the Sangihe Island chain, North Sulawesi, Indonesia. Oryx 43(3):419–426

Supriatna J, Manansan J, Tumbelaka L, Andayani N, Indrawan M, Darmawan L, Leksono SM, Djuwantoko SU, Byers O (2001) Conservation assessment and management plan for the primates of Indonesia: final report. IUCN/SSC Conservation Breeding Specialist Group, Apple Valley, MN

Part I
Indonesia's Apes

Chapter 2
Measuring Performance of Orangutan Protection and Monitoring Unit: Implications for Species Conservation

Jito Sugardjito and Asep S. Adhikerana

Introduction

The orangutan is the only great ape species that inhabits Asia. During the Pleistocene, they occurred throughout Southeast Asia, from Southern China in the North to Java in the South (Hooijer 1948; von Koeningswald 1981). Current distribution of this species is limited to the northern part of Sumatra and fragmented forest areas in Borneo (Reijksen and Meijaard 1999). Recently, experts have suggested that the orangutan populations on each island represent unique species (*Pongo abelii* on Sumatra and *P. pygmaeus* on Borneo; Groves 2001; Warren et al. 2001). The Bornean species is generally divided into three subspecies: *P. p. pygmaeus*, *Pongo p. wurmbii*, and *P. p. morio* (Groves 2001).

The Bornean orangutan is categorized on the 2002 IUCN Red List as Endangered (IUCN 2002), and it has been estimated that approximately 17,000–30,000 individuals reside within the protected areas (Sugardjito and van Schaik 1992). The 90,000 hectares of Gunung Palung National Park (GPNP) contains large patches of continuous forest that are capable of supporting a healthy orangutan population. Holding some 2,500 orangutans with great habitat diversity from peat swamp up to the hill forests, the GPNP forms one of the most important refuges for this population (IUCN 2002). Orangutan density in the GPNP is higher than the other protected areas in Kalimantan, ranging from 4.1 in peat forests and 2.4 in montane forests (Johnson et al. 2005). Despite the high conservation value of forest habitat in the GPNP, the adjacent forest corridors are threatened by conversion to oil palm plantations and transmigration sites. Recently, density estimate of Bornean orangutans living in both outside and inside protected areas is about 60,000 individuals, with the largest population located in the south-western of Borneo (IUCN 2004).

J. Sugardjito (✉)
Research Centre for Biology, The Indonesian Institute of Sciences,
Jalan Raya Jakarta-Cibinong Km.48, Cibinong, 16911, Indonesia
e-mail: jitos@cbn.net.id

S. Gursky-Doyen and J. Supriatna (eds.), *Indonesian Primates*,
Developments in Primatology: Progress and Prospects,
DOI 10.1007/978-1-4419-1560-3_2, © Springer Science+Business Media, LLC 2010

The main threats to the survival of the orangutans residing outside the protected areas are habitat loss and illegal trade (Reijksen and Meijaard 1999).

Previous studies have shown that logged forest reduces nest densities, an index used for population densities, in swamp forest to 21% and in evergreen lowland forest from 30% to local extinction (Felton et al. 2003; Rao and van Schaik 1997; Aveling 1982). A more recent survey of orangutan population density in GPNP area indicated that the density has declined, especially in the area near the park boundary where logging were active recently (Prasetyo and Sugardjito 2007). The threat to immediate survival prospects of the orangutan has been recognized by the experts meeting in Balikpapan, East Kalimantan (IUCN 2002). During the workshop, it was recommended that in order to strengthen law enforcement for habitat and orangutan protection, it is necessary to develop a community-based patrol specializing in orangutan protection. In particular, it was suggested that these patrols follow similar programs previously developed for rhino and tiger protection in Sumatra. These programs place a high priority on preventing illegal capture, habitat protection from logging, and conversion through law enforcement and community awareness programs. In responding to this need, the Orangutan Protection and Monitoring Unit (OPMU) was established in 2003. OPMU regularly patrol forest habitat in order to prevent forest crime. The inclusion of community rangers in patrolling has significantly reduced the potential for conflicts between community and forest rangers when combating illegal loggers and other intruders. To further reduce remaining forest crimes from local communities, OPMU has also conducted community outreach programs by facilitating the development of buffer zone village forest protection regulations and agreements, securing community access to forests surroundings the National Park. These programs have been conducted from 2004 to 2007.

In this paper, we use data from 2004 to 2007 to measure the effectiveness of OPMU activities. This includes description of findings as well as the actions taken to address the findings in the field. Further, we discuss the importance of the community-based patrol system and the implication for species conservation.

Methods

Orangutan Protection and Monitoring Unit (OPMU)

During 2004–2006, each OPMU team consisted of three community members and one forest ranger as a leader. The three team leaders were chosen from among the forest park rangers, all of which possess a license for prosecution. The nine other team members were selected from the local communities surrounding the park. Because of the extension of area covered in the beginning of 2007, the OPMU was restructured in accordance with the jurisdiction of conservation agencies in the region. Two units in the park, each was led by a forest ranger, and the other one unit operates outside the park was led by personnel from the Province Nature Conservation Agency. Both members and the leaders of OPMU were selected following standard recruitment procedures. Prior to its operations, the OPMU teams received training including jungle survival, intelligence, ground checking, forest fire control, search and rescue techniques, and primate monitoring technique.

The OPMU teams undertake the operation in a shift schedule scheme. When two teams are operating, the other team is stationed at the OPMU base. They then shift, so that all teams have equal time operating in the field as well as at the base. The OPMU is primarily concerned with four issues related to protection: (1) patrolling; (2) monitoring of wildlife; (3) forest crime investigation; and (4) conservation awareness.

Patrolling

The forest crimes encountered in the field by the OPMU primarily fall into four main types: (a) illegal logging; (b) forest encroachment; (c) illegal gravel mining; and (d) animal hunting.

Illegal Logging

Groups of people cutting down trees with the purpose of selling the wood constitutes illegal logging. During observations of illegal logging, the following data were collected: location coordinates using GPS, date and time of encounter, condition of the crime site (old or new), type of logs (log or processed timber), tree species when possible, and the number of illegal loggers. When the loggers are found in the field, an investigation is undertaken to obtain information on the purpose, financial support, market, and transportation methods of the wood. The OPMU team then undertakes confiscation, issues a warning letter, and promotes conservation awareness in the area.

Forest Encroachment

When people cut down trees individually, with the purpose of conversion of forest land to agricultural land, it constitutes forest encroachment. The data recorded for forest encroachment include: the location coordinates using GPS, date and time of encounter, condition of the crime site (old or new), type of planting system ("ladang" or plantation), and the number of illegal farmers. When the illegal farmers are found in the field, an investigation is undertaken to obtain information on the reason behind the forest encroachment. The OPMU team then confiscates the tools, issues a warning letter, and promotes conservation awareness in the area.

Gravel Mining

When people extract minerals from the land for commercial purpose, it constitutes gravel mining. The data recorded for gravel mining include: location coordinates using GPS, date and time of encounter, condition of the crime site (old or new), type of mining, and the number of illegal miners when directly encountered. When illegal miners are found in the field, an investigation is undertaken to obtain information on the purpose. The team then confiscates the minerals, presents the miners with a warning letter and promotes conservation awareness in the area.

Animal Hunting

When people seek either live or dead animals for consumption or future sale, it constitutes animal hunting. The data for animal hunting include: location coordinates using GPS, date and time of encounter, condition of the crime site (old or new), equipment used for hunting animals, and number of hunters. When the hunters are found in the field, an investigation is undertaken to obtain information on the purpose, financial support, and market. The team then confiscates the hunted animals, issues a warning letter whilst promoting conservation awareness in the area.

Wildlife Monitoring

During operation, the team also recorded the wildlife species encountered in the field, with special attention to the presence of the orangutan. Records are commonly made on the geoposition, date and time of encounter, species encountered, and whether the species encountered could be seen or heard. When it was seen, the number of individuals was noted. As for the orangutan, the number of nests observed was also recorded.

Crime Investigation

The data consist of all information that can be collected from the perpetrators of the crime. Specific information, such as incident location, the people who are financially supporting the crime, the market destinations, the size of log/timber being collected, the current price of logs or timbers, and the transportation system for transferring the logs/timber out of the forest, can also be collected.

Promoting Awareness

Implementation of a village meeting with either formal or informal leaders in the village adjacent to the orangutan's habitat. During patrolling in the field, consultation meetings with farmers bordering the park are also undertaken.

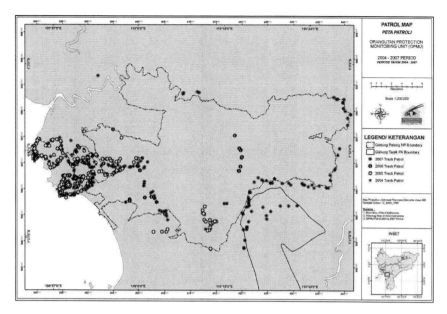

Fig. 2.1 Sites in which the OPMU patrolled in Gunung Palung National Park during the period of 2004 up to 2007

The data are recorded in the form of journal from the beginning of field work at 7.00 until 17.00. Later at the field station, the journal is transferred into a spread-sheet and converted into a database format suitable for GIS mapping purposed. At the early stage, OPMU patrolled inside the GPNP, and later, it was expanded to nearly all critical habitats of orangutan (Fig. 2.1). The same sites could be revisited during the operations, but the records will cover with the changes from the previous conditions.

Data Analysis

The data used for analysis consist of all the illegal activities observed between 2004 and 2007 as well as the frequency the OPMU team encountered wildlife. For the analytical purpose only, those new encounters are included in the analysis, i.e., those observed as new illegal logging activity, new encroachment area and/or activity, new mining activity, or new animal hunting events. Otherwise, all records will become an accumulation of all records during the operations. For wildlife, the records covered all noticeable and identifiable species, but the analysis only addressed specific species, such as hornbill, orangutan, agile gibbon, and maroonleaf monkey, all of which are very much dependent on forest habitat.

We used statistical analysis described in Sokal and Rohlf (1969). The nature of analysis for this paper is more descriptive rather than analytical. This approach was applied since the aim is describing all efforts that have been accomplished for effectiveness of patrolling and monitoring systems.

Although the OPMU has been operating since 2004, the early data are not appropriate for any analysis. This was due to the fact that in the beginning of its operation, the OPMU has focused on the training for its members in patrolling system. Consequently, only 6 patrols comprising 20 patrol days for 2004 were included in this paper. Patrol day ratio was calculated from the number of patrol days in a year divided by total patrol days for 4 years. Average patrol day was calculated from the total patrol days divided by the number of patrols. The staff ratio was obtained from the number of staff in a year divided by total number of staff for 4 years. The OPMU performance denotes the annual staff performance, which also represents the effectivenes of a patrol man-day. The OPMU performance is calculated by dividing the total patrol day ratio with the average number of staff ratio. The patrol cost was obtained from the total annual cost divided by patrol days of corresponding year, whereas the patrol cost-day was calculated from the patrol cost divided by average patrol day. The patrol cost day was converted to log-based number in order to measure the efficiency. The OPMU efficiency was calculated from patrol performance divided by patrol cost-day, whereas the OPMU efforts is the number of points observed during the operation, and only the percentages of efforts undertaken "inside" and "outside" of the Gunung Palung National Park are utilized for the analysis. Encounter rate is calculated from the number of encounter divided by the efforts.

Results and Discussion

Table 2.1 summarizes characteristics of OPMU related activities from 2004 to 2007. At its early stage, OPMU was only operating inside the national park. However, it was estimated that about 75% of orangutan populations occur outside the protected areas (Reijksen and Meijaard 1999). Therefore, a collaboration was developed with the Province Nature Conservation Agency (BKSDA) in 2006 to respond the critical orangutans' habitat outside the Park. As a consequence, efforts to observe the areas outside the park were gradually increased beginning 2006 (Fig. 2.2). It shows that such a need is later justified by the findings revealing that forest crimes are more prevalent outside the park boundaries. Figure 2.3 shows that both the OPMU operational performance and its operational efficiency have been improving since its early establishment. This means that the OPMU has a better understanding on how to perform well with efficient financial management. When the OPMU is confronted with forest crimes, such as illegal logging, encroachment, gravel mining, and animal hunting, OPMU always undertakes firm actions, including seizure of the illegal materials, legal notification, destruction of confiscated materials (equipments such as chainsaw, axes or machetes; huts; and logs), and

Table 2.1 OPMU performance and its operational efficiency

Year	2004	2005	2006	2007	Total
Patrol days	20	89	89	146	344
Number of patrol	6	16	11	18	
Number of staff	78	208	143	216	645
Average number of staff	13	13	13	12	
Average patrol days	3.33	5.56	8.09	8.11	
Patrol day ratio = patrol day in a year: total patrol days for 4 years	0.06	0.26	0.26	0.42	
Number of staff ratio = number of staff in a year : total number of staff for 4 years	0.12	0.32	0.22	0.33	
Patrol performance = Effective patrol man-day = patrol day ratio: number of staff ratio	0.48	0.80	1.17	1.27	
Total annual cost (IDR)	139,275,600	174,568,773	174,400,225	57,801,250	418,376,548
Average monthly budget (IDR)	11,606,300	14,547,398	14,533,352	14,450,313	
Average monthly budget (USD)	1,290	1,616	1,615	1,606	
Patrol cost = total annual cost: patrol days	580,315	1,961,447	1,959,553	395,899	
Patrol cost-day = patrol cost: average patrol day	174,095	352,620	242,192	48,809	
Patrol cost-day in USD (USD 1 = IDR 9000)	19.34	39.18	26.91	5.42	
Patrol cost-day (log-based number of patrol cost-day)	1.29	1.59	1.43	0.73	
OPMU Efficiency = performance: patrol cost-day	0.37	0.50	0.82	1.73	

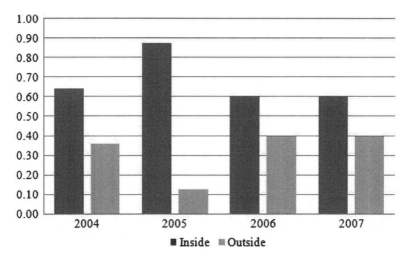

Fig. 2.2 OPMU patrolling efforts inside and outside of the Gunung Palung National Park

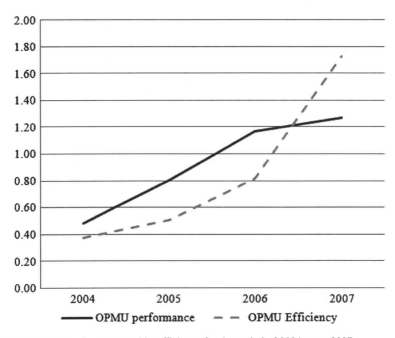

Fig. 2.3 OPMU performance and its efficiency for the period of 2004 up to 2007

keeping out the intruders from the areas. Such "on-the-spot" prosecution actions have significantly deterred the offenders to enter illegally into the park, and directly reduced forest crimes.

Table 2.2 shows the number of locations observed during OPMU operations, whereas Table 2.3 indicates the encounter rate with illegal activities during OPMU

Table 2.2 Number of points observed (efforts) during OPMU operations in the Gunung Paalung National Park in 2004–2007

Locations	2004	2005	2006	2007	
Inside	41	385	149	193	768
Outside	23	56	98	126	303
Total	64	441	247	319	1,071

Table 2.3 Encounters with forest crime and wildlife during OPMU operations

Encounter rate = number of encounter: efforts		Number of encounters				Encounter rate			
		2004	2005	2006	2007	2004	2005	2006	2007
Illegal logging	Inside	4	13	4	2	0.10	0.03	0.03	0.01
	Outside	3	11	6	12	0.13	0.20	0.06	0.10
	Total	7	24	10	14	0.11	0.05	0.04	0.04
Encroachment	Inside	9	46	0	3	0.22	0.12	0.00	0.02
	Outside	3	13	3	12	0.13	0.23	0.03	0.10
	Total	12	59	3	15	0.19	0.13	0.01	0.05
Gravel mining	Inside	0	5	0	0	0.00	0.01	0.00	0.00
	Outside	5	3	2	0	0.22	0.05	0.02	0.00
	Total	5	8	2	0	0.08	0.02	0.01	0.00
Animal hunting	Inside	4	5	1	1	0.10	0.01	0.01	0.01
	Outside	2	3	3	3	0.09	0.05	0.03	0.02
	Total	6	8	4	4	0.09	0.02	0.02	0.01

operations. It shows that illegal logging activity inside the park drastically decreased in 2005, but increased outside the park, particularly in 2007. Such an increase may not be attributed to a "real" increase, but may represent the increased efforts of the OPMU to observe the areas outside the park. The illegal logging might have been occurring outside the park for a long time, but it was just considerably recorded since OPMU expanded the operation outside the Park in 2007. The correlation between patrol performance, total efforts, and illegal activities can be seen in Table 2.4. The negative correlation means that the performance imposes a positive impact on the forest crimes. More specifically, increased performance by the OPMU results in a noticeable decrease in forest crimes.

Forest encroachment in the park has also drastically decreased since 2005 if it is compared with outside the park (Table 2.3). However, in 2006, it was the opposite, whereby forest encroachment outside the park was higher than inside the park. Once again,this might be due to the results of expansion of observation areas outside the park. This might also show that the "on-the-spot" prosecution effectively deters the loggers.

Both gravel mining and animal hunting inside the park were significantly decreased when compared with those outside the park in 2005, whereas in 2006, animal hunting was higher outside the park than inside the park. On the other hand, no more gravel mining has been observed since 2006. Table 2.4 shows that the OPMU operation has a significant impact on the findings of these illegal activities

Table 2.4 The direction of performance, which impacts on the encounters of illegal activities in the field. Coefficient correlation between Patrol Performance and total efforts, and encounter rates of forest crimes for the period of 2004–2007

| | Correlation between patrol performance | | |
	Total	Outside	Inside
Total efforts	0.48	0.99	0.22
Forest Crimes:			
Illegal logging	−0.44	−0.03	−0.63
Forest encroachment	−0.47	−0.17	−0.52
Gravel mining	−0.94	−0.96	−0.24
Animal hunting	−0.98	−0.98	−0.98

Note: The number on each cell is coefficient correlation between patrol performance, total efforts, and forest crimes encountered by OPMU during 2004–2007.

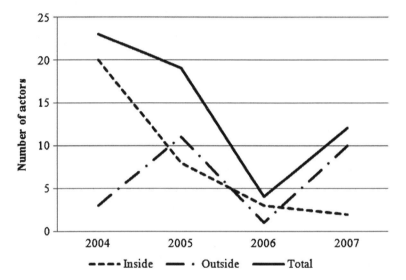

Fig. 2.4 Offenders detected both inside and outside bordering the park during 2004 up to 2007

in the field. This also indicates that the OPMU operation has imposed a deterrent impact on those offenders. The number of offenders (i.e., loggers, farmers, miners, and hunters) operating inside the national park has been declined drastically, although it was increased outside the park in 2007 (Fig. 2.4). The extent of encroachment areas has also been reduced inside the park (Fig. 2.5).

Wildlife Encounters

The nature of wildlife observation during the OPMU operation is supplemental to forest crime patrol and monitoring. Observational notes are always made for certain

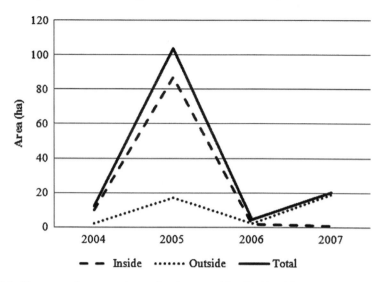

Fig. 2.5 The encroachment or pioneer farming areas identified by OPMU during the period of 2004 up to 2007

wildlife species encountered in the field, especially those that can be identified by the OPMU team. There are, for example, a number of hornbill species in the operation areas, but the team has no capacity in identifying them into the species level. This paper only deals with all hornbills that were recorded by the team during the operations. On the other hand, the team can easily identify orangutan and its nest since the team has been provided with a training on identifying primates of the areas, such as agile gibbon (*Hylobates agilis*) and red-leaf monkey (*Presbytis rubicunda*). The encounters with wildlife during OPMU operations are presented in Table 2.5. When all wildlife encounters are lumped together, it shows that the wildlife encounter rate tends to decrease in 2007. This could actually be attributed to the low encounter with wildlife outside the national park area as well as increasing efforts to observe the areas outside the park where most of the forest has been disturbed. As a consequence, it has affected the records of orangutans and their sleeping nests when their encounters are low in the period of 2007.

Large animals such as the orangutan need large areas with adequate productive fruit tree densities, and are therefore threatened by habitat loss and fragmentation. Furthermore, capturing orangutans can provide considerable financial gains. These factors have lead to the poaching of baby orangutans while killing the mother, further reducing their population densities and increasing their dependency on protected areas for survival. This phenomenon could be supported by the results showing that the orangutan and its sleeping nests are more common inside the national park areas than outside the GPNP. The same figure also shows that hornbills, agile-gibbon, and red-leaf monkey are more easily found inside than outside the park (Table 2.5).

The size of the protected area is particularly important because "edge effect" is pronounced for large arboreal animals such as orangutan (Woodroffe and Ginsberg 1988). However, large protected areas are also more costly to maintain and many

Table 2.5 Encounters with wildlife during OPMU operations

Encounter rate = number of encounter: efforts		Number of encounters				Encounter rate			
		2004	2005	2006	2007	2004	2005	2006	2007
Hornbill	Inside	2	22	5	16	0.05	0.06	0.03	0.08
	Outside	0	0	12	3	0.00	0.00	0.12	0.02
	Total	2	22	17	19	0.03	0.05	0.07	0.06
Agile gibbon	Inside	3	24	9	13	0.07	0.06	0.06	0.07
	Outside	0	2	12	8	0.00	0.04	0.12	0.06
	Total	3	26	21	21	0.05	0.06	0.09	0.07
Orangutan	Inside	6	183	104	61	0.15	0.48	0.70	0.32
	Outside	0	9	49	21	0.00	0.16	0.50	0.17
	Total	6	192	153	82	0.09	0.44	0.62	0.26
Number of nests	Inside	4	202	101	59	0.10	0.52	0.68	0.31
	Outside	0	9	49	19	0.00	0.16	0.50	0.15
	Total	4	211	150	78	0.06	0.48	0.61	0.24
Red-leaf monkey	Inside	3	18	10	12	0.07	0.05	0.07	0.06
	Outside	1	5	11	6	0.04	0.09	0.11	0.05
	Total	4	23	21	18	0.06	0.05	0.09	0.06

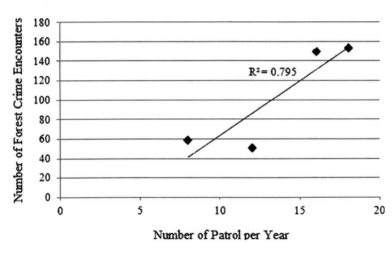

Fig. 2.6 Mean successful number of forest crimes detected and processed per year by OPMU during patrol

are underfunded (James et al. 1999). Therefore, concentrating OPMU patrols in key areas, with a mandate to detect and destroy all traps and deter poachers or offenders with force, if appropriate, may circumvent the need to protect the whole area. During OPMU patrols, poaching equipments such as air-rifles and traps were found, and they have been either confiscated or disabled (Fig. 2.7). The more frequent patrol conducted by OPMU per year the more the incidences encountered (Fig. 2.6).

Orangutans were recorded in all habitat types in GPNP but most frequently within lowland hill and peat swamp forests (Johnson et al. 2005). This emphasizes

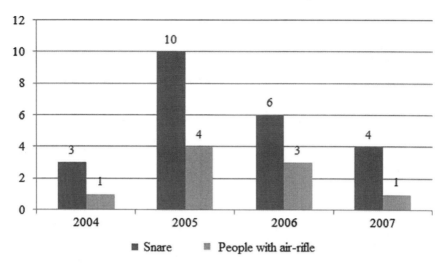

Fig. 2.7 Poaching efforts on wildlife in Gunung Palung National Park as indicated by the number of animal traps found, and the air-rifles which have been confiscated during the patrols

the need to protect this habitat, which occurs at the borders of GPNP (Prasetyo and Sugardjito 2007). However, these lower elevation forests experience the greatest human population pressure, with palm oil plantation, forest fires, commercial and illegal logging, mining operations, and pioneer farming, all resulting in loss of orangutan habitat.

Conclusions and Recommendations

Along the way, the OPMU team has been able to improve its performance and its operational efficiency. This provides evidence that continuous support to the OPMU has had a phenomenal impact on its operation and that surely further support will not only enhance its performance and efficiency but also its capacity in dealing with forest crime and wildlife identification. The OPMU structure, which consists of a team of forest rangers and community members, provides an opportunity of transparency in dealing with processing forest crime cases. Indeed, the OPMU team could only promptly react to forest crime in the field by an "on-the-spot" prosecution. However, such an action against the forest crime seems to be effective in deterring the offenders to disturb further the forest habitat. The offenders then realise that there is a rule of protection enforced in the field. The OPMU is also needed for the protection of orangutan habitat that is not situated in the conservation area. More than 50% of orangutan population in Kalimantan inhabit forest outside the conservation area (Reijksen and Meijaard 1999). The concerned areas could be orangutan habitat in Ketapang and Kapuas Hulu districts, where peat swamp and lowland forests being converted to industrial forest for oil palm plantations.

Although the main goal of OPMU is to protect orangutan populations and their habitat, the monitoring functions particularly well in detecting changes due to their intensive appearance in the field. Many incidences on forest habitat could be detected early. Frequently, the OPMU team destroyed wildlife traps or bird nets that were set by the hunters. Despite the effectiveness of OPMU for species conservation and habitat protection, they need to be refreshed regularly in order to avoid fatigue. A refreshment program such as physical training or outbound is needed for every personnel who works in securing natural wildlife habitat.

Acknowledgments We thank USFWS and KNCF for providing support to FFI for the implementation of OPMU. We are grateful to the management and staff of GPNP, especially Mr. Prabani Setyoherdianto for assisting field operations of OPMU. We have a debt of gratitude to the unwavering commitment and diligence of the OPMU staff.

References

Aveling RJ (1982) Orang-utan conservation in Sumatra, by habitat preservation and conservation education. In: de Boer LEM (ed)Orang-utan, its biology and conservation. The Hague, pp 299–316

Felton AM, Engstrom LM, Felton A, Knott CD (2003) Orangutan population density, forest structure and fruit availability in hand-logged and unlogged peat swamp forests in West Kalimantan. Indonesia Biol Cons 114:91–101

Groves CP (2001) Primatew taxonomy. Smithsonian Institution Press, Washington, DC

Hooijer DA (1948) Prehistoric teeth of man and the orangutan from central of Sumatra with notes on the fossil orangutan from Java and Southern China. Zoologische Mededeelingen, 1–175. Reijksmuseum, Leiden

IUCN/SSC (2002) Orangutan conservation and reintroduction workshop 19–22 June 2002. Palangka Raya, Kalimantan, Indonesia. Final Report

IUCN/SSC (2004) Orangutan population and habitat viability assessment 15–18 January 2004. Jakarta, Indonesia. Final Report

James AN, Gaston KJ, Balmford A (1999) Balancing the Earth's accounts. Nature 401:323–324

Johnson AE, Knott CD, Pamungkas B, Pasaribu M, Marshall AJ (2005) A survey of the orangutan (*Pongo pygmaeus pygmaeus*) population in and around Gunung Palung National Park, West Kalimantan, Indonesia based on nest count. Biol Conserv 121:495–507

Prasetyo D, Sugardjito J (2007) Orangutan survey in Gunung Palung National Park and its surroundings. Final Report to Defra

Rao M, van Schaik (1997) The behavioural ecology of Sumatran orangutans in logged and unlogged forest. Trop Biodivers 4:173–185

Reijksen HD, Meijaard E, (1999) Our vinishing relative: the status of wild orangutan at the close of the twentieth century. Kluwer Academic Publishers, Dordrecht, The Netherlands

Sokal RR, Rohlf FJ (1969) Biometry. Freeman, San Francisco

Sugardjito J, van Schaik CP (1992) Orangutans: current populations status, threats, and conservation measures. In: Proceedings of the Great Apes Conference (Jakarta, Pangkalan Buun) 1991. Jakarta, pp 142–152

von Koeningswald GHR (1981) Are there still orangutans in China (in Germany). Nat Mus 111(8):260–261

Warren KS, Verschoor EJ, Langenhuijzen S, Heriyanto, Swan RA, Vigilant L, Heeney JL (2001) Speciation and intraspecific variation of Bornean orangutans, *Pongo pygmaeus pygmaeus*. Mol Biol Evol 18:472–480

Woodroffe R, Ginsberg JR (1988) Edge effects and the extinction of populations inside protected areas. Science 280:2126–2128

Chapter 3
Communication, Culture and Conservation in Orangutans

Roberto A. Delgado

Introduction

Several fragmented tropical forests within Southeast Asia, namely on northern Sumatra and across Borneo, are home to remnant and declining populations of wild orangutans (*Pongo spp*), the only extant nonhuman great ape found in Asia. These populations and other sympatric fauna are increasingly threatened by the alteration and destruction of their habitats. The latest available assessments from the International Union for the Conservation of Nature (IUCN) recognize Bornean orangutans as an endangered species, whereas their Sumatran counterparts, found at precipitously falling population numbers (Wich et al. 2003; Singleton et al. 2004; Wich et al., 2008b), are identified as critically endangered (IUCN 2008). While the value of preserving species such as orangutans has previously been touted as serving important biological functions, particularly from a community ecology perspective, a more recent emphasis has been on strengthening ties between the goals of biological conservation and socioeconomic development among the impoverished communities that are most likely to face direct human-wildlife conflicts related to local land use practices.

One reflection of this shift in conservation approaches is the sponsorship and involvement of such high-level organizations as the United States Agency for International Development (USAID) and Development Alternatives, Inc. (DAI), in addition to local and international nongovernmental organizations (NGOs). At the time of writing, DAI, a development consulting firm that leads an Orangutan Conservation Services Program (OCSP) consortium comprising Orangutan Foundation International (OFI) and The Nature Conservancy (TNC) is in the middle of a multi-year project, the primary aim of which is to maximize the protection and viability of wild orangutan populations. The specific mechanism to help reach

R.A. Delgado (✉)
Integrative and Evolutionary Biology, University of Southern California,
Los Angeles, CA 90089, USA
e-mail: radelgad@usc.edu

S. Gursky-Doyen and J. Supriatna (eds.), *Indonesian Primates*,
Developments in Primatology: Progress and Prospects,
DOI 10.1007/978-1-4419-1560-3_3, © Springer Science+Business Media, LLC 2010

this goal is a series of activities that include but are not limited to implementation of policy reform, increased law enforcement, expanded public outreach, and site-based conservation measures at critical locations throughout Borneo and Sumatra. Clearly, conservation biology, at least as it pertains to wild orangutans, appears to be converging with business as well as development models and may have political ramifications at national, regional, and local levels. As this trend becomes more important across diverse taxa often noted as flagship, keystone or umbrella species, and spanning wide geographic localities, we must not forget the intrinsic biological value of populations and the key insights we can learn related to ecological and evolutionary processes.

Recent years have witnessed considerable discussion and contention surrounding the origins and evolution of (material) culture among nonhuman primates and the implications for understanding the role of social learning and its relationship, if any, with intelligence or other measurable cognitive abilities. Given the close behavioral, genetic, and morphological affinities to humans, studying great ape populations in their natural environments provides an excellent model for reconstructing ancestral adaptations, including the origins and evolution of language and culture. The observed and reported geographic variation in orangutan behaviors consisting of subsistence and comfort skills, as well as signals, lends support to the premise that the origin of cultures among great apes and man can be traced back to at least 14 million years ago (MYA), a time period purportedly coincident with the last common ancestor of humans, chimpanzees, gorillas, and orangutans (van Schaik et al. 2003; van Schaik 2004).

Yet, given the nature of imminent threats to conservation, habitat loss and disturbance may have profound effects on orangutan cultures. For example, van Schaik (2002) cites the loss of local traditions, reduced opportunities for social learning and innovation, reduced diffusion, and the loss of centers of traditions and sources of diffusion – not to mention attendant socioecological effects – as potential consequences resulting from such detrimental activities as habitat conversion, hunting, fragmentation, and the loss of habitat. With respect to communication, Delgado and colleagues (2006, 2007, 2009) report important differences (including geographic variation) in the acoustic structure and vocal behavior of adult male orangutans. However, central predictions derived from several explanatory hypotheses addressing ecological and genetic factors meant to account for the observed variance have not yet been tested rigorously though field studies are ongoing and planned. Hence, if orangutan populations and their habitats are not protected, allowing for continued observational studies and experimental research, then we may never understand the underlying factors leading to "dialectical" variation and other differences in cultural behaviors.

Field researchers are already well aware of considerable diversity in the social structure, opportunities for female mate choice and the nature of vocal signaling between orangutans and the African apes. For example, the African apes have either a single-male, multi-female group composition (typically seen in gorillas) or a dynamic fission-fusion community (more common in chimpanzees and bonobos), limited opportunities for female mate choice, and vocal signaling generally occurs most frequently within and between groups or parties. In contrast, orangutans are semisolitary to varying degrees, females at certain sites have the prospect of choosing a preferred mate – particularly in Sumatra (e.g., Fox 1998) – and vocal communication is principally between individuals. Hence, a closer examination of orangutan socio-ecology and the observed variance can assist us in developing more nuanced behavioral models for early humans and their ancestors.

Communication and Social Organization

By definition, communication is a social behavior because it requires both an actor to transmit a signal and at least one receiver to perceive and respond (cf. Bradbury and Vehrencamp, 1998). Although animals use a wide array of modalities to communicate with one another, including gestures and olfactory signals, the emphasis here will be on vocalizations because of the data that are available to date. Vocalizations are relatively conspicuous and more easily quantified under natural field conditions than other forms of communication, and, in particular, the adult male long call has been relatively well studied and, given the nature of dispersed societies at most sites, is considered to play an important role in social organization including reproductive strategies (Delgado 2003). However, before delving more deeply into the vocal repertoire of the orangutan, a brief overview addressing what is known about the social structure and mating system of orangutan societies will help to provide a framework for developing insights into the signal content and function of acoustic signals such as the long call.

Adult females tend to have highly overlapping home ranges of up to 900 hectares (ha) with the most overlap occurring in areas of high density (Horr 1975; Rijksen 1978; Galdikas 1985, 1995; Singleton and van Schaik 2001; Johnson et al., 2005; and for a review, see Singleton et al. 2009). More recent work at two Bornean sites (Gunung Palung, West Kalimantan, and Sabangau, Central Kalimantan) also confirms high range overlap. Specific home range estimates for adult females at Gunung Palung, based on three different methods – small grid, large grid, and polygon analyses – yield values between from as low as 280 ha to as high as ~ 800 ha (Knott et al. 2008); at Sabangau, home range estimates are between 200 and 500 ha for adult females and between 250 and 550 ha for unflanged males (Morrogh-Bernard unpublished data, cited in Harrison 2009). Knott et al. (2008) further suggest that adult female ranging patterns at Gunung Palung, specifically active avoidance, could reflect feeding competition and they report data consistent with range defense and exclusion as a means of establishing core areas. Flanged adult males do not actively defend territories, but appear to use long distance vocalizations in part as a site-independent spacing mechanism. In Sumatra, male home ranges are larger than those of females and may exceed 2,000–3,000 ha – though perhaps are lower for dominant individuals – and also have extremely high overlap (Singleton and van Schaik 2001); at Sabangau, home range estimates for flanged males are between 200 and 500 ha, similar to that of the adult females at this site (Morrogh–Bernard unpublished data, cited in Harrison 2009). Variation in male ranging patterns is probably tied to fluctuations in fruit availability, the availability of receptive females, and the presence of other, more dominant males (MacKinnon 1974; Rodman and Mitani 1987; Sugardjito et al. 1987; te Boekhorst et al. 1990; Mitani et al. 1991; van Schaik 1999; Delgado and van Schaik 2000; Singleton and van Schaik, 2002), but Singleton et al. (2009) also propose relationships between home range size and habitat heterogeneity, and perhaps local population density, as well as with subspecies membership; for the latter, specifically, a gradual increase in home range size from the easternmost populations in Borneo to the westernmost Sumatran populations. However, more data across multiple sites are needed to test predictions based on these and other factors.

Unlike all other diurnal anthropoids, orangutans do not have easily recognizable social units. Individuals are often solitary but associate in parties on a regular basis for social benefits although mean party size tends to remain small because of the high costs of feeding competition (Mitani et al. 1991; van Schaik and van Hooff 1996; van Schaik 1999). Behavioral observations such as preferential party associations further suggest individualized relationships between different adult animals (Galdikas 1984; van Schaik and van Hooff 1996; Singleton and van Schaik 2002). At some sites in northern Sumatra, clusters of presumably related females associate preferentially with one another (Singleton and van Schaik, 2002). Taken together, these studies suggest the most probable scenario for orangutan social organization is one in which animals form a network of socially distinct associations characteristic of individual-based fission-fusion societies (MacKinnon 1974; Sugardjito et al. 1987; van Schaik 1999; Delgado and van Schaik 2000). The network is most likely organized around the locally dominant flanged male (MacKinnon 1974), who tends to be the preferred mating partner of the area's females (Schürmann and van Hooff 1986; Utami and Mitra Setia 1995; Fox 1998). The other flanged males and probably all unflanged males visiting an area may form a separate class, covering several such loose communities (Singleton and van Schaik, 2002).

Adult male reproductive tactics vary between known Bornean and Sumatran populations, particularly among subordinate flanged and unflanged males. At several Bornean sites, adult males engage in short consortships, and both flanged and unflanged males force copulations with females although either small or low-ranking males (unflanged) do so with greater frequency (Table 3.1; Galdikas 1985a, b; Mitani 1985a). At Sumatran sites, flanged males hardly ever resort to using forced copulations when mating with females (Table 3.1; Schürmann and van Hooff 1986; Fox 1998). In contrast, subordinate flanged males in Sumatra rarely achieve matings, but both the dominant flanged male and unflanged males can maintain relatively long consortships with females (Fox 1998; Utami 2000). Interestingly enough, paternity analyses for one Sumatran population (Ketambe) indicate that unflanged males do conceive offspring at least some of the time (Utami et al. 2002) though it appears that the consortships leading to conceptions most often occurred during periods of male rank instability (Utami and Mitra Setia 1995; Utami et al. 2002). The reported island difference may be a consequence of systematic differences in habitat quality between the Bornean and Sumatran sites studied thus far. Bornean males, living in relatively poor habitats and potentially more limited by energetic constraints than Sumatran males, are only able to sustain brief consortships

Table 3.1 Frequency of forced and unforced mating committed by flanged and unflanged males across sites in Borneo and Sumatra

	Adult males	Unforced	Forced	N
Borneo				
Tanjung Puting	Flanged	96.7%	3.3%	30
	Unflanged	13.6%	86.4%	22
Kutai	Flanged	53.6%	46.4%	28
	Unflanged	4.6%	95.4%	151
Sumatra				
Ketambe	Flanged	96%	4%	50
	Unflanged	44.7%	55.3%	38
Suaq Balimbing	Flanged	100%	0%	36
	Unflanged	63.6%	36.4%	99

Borneo - Tanjung Puting: Galdikas (1985a, 1985b); Kutai: Mitani (1985a).
Sumatra - Ketambe: Schürmann and van Hooff (1986); Suaq Balimbing: Fox (1998).

(Mitani et al. 1991). Low habitat quality should also result in lower local population densities and less frequent rates of association. As a consequence, males probably have little or no knowledge about the females they encounter and are likely to be more aggressive in their mating attempts. In contrast, Sumatran males, generally living in richer habitats at higher densities, encounter females more often and are capable of maintaining longer consortships and sustained associations with females likely to be fecund (Schürmann 1981; Fox 1998; van Schaik 1999).

Using an alternative reproductive strategy, unflanged males on both islands actively seek and follow females and engage in consortships that often involve forced copulations (Galdikas 1985a; Mitani 1985a; Schürmann and van Hooff 1986; Utami and Mitra Setia 1995). A striking difference based on means between sites for each island (Kutai and Tanjung Puting on Borneo, Ketambe and Suaq Balimbing on Sumatra), however, is that 90% of the mating between unflanged males and adult females at the Bornean sites involve forced copulations (Table 3.1: Galdikas 1985a; Mitani 1985a), whereas forcing characterizes only about 45% of the mating at the two Sumatran sites (Table 3.1: Schürmann and van Hooff 1986; Fox 1998). On both islands, unflanged males might be more constrained in the length of their consortships by social factors such as the presence of more dominant flanged males (Galdikas 1985b; Mitani 1985a; Utami and Mitra Setia 1995; Fox 1998, 2002). In the Sumatran sites, it is possible that the difference reflects a greater degree of monopolization of females in their fecund period by the resident dominant flanged male. This restricted access is more likely in habitats with high productivity where preferred flanged males can maintain longer consortships, making forced mating less likely. One alternative explanation for the island difference would be that Sumatran females display a lower degree of resistance as a result of higher encounter frequencies or lower mating costs, but the appropriate data to test these alternatives are not yet available. Another possibility regarding the evolution of the orangutan mating system addresses changes in resource availability during the late Miocene and Pliocene that may have led to alternate reproductive strategies between flanged and unflanged males (Harrison and Chivers 2007). One unproven assumption, however, is that the ancestral populations leading to orangutans were more gregarious and characterized by single-male polygyny, a mating system where a single adult male defends access to two or more adult females (Harrison and Chivers 2007). Nonetheless, these authors expound on a provocative evolutionary scenario and their hypotheses warrant further testing if possible.

In orangutans, the absence of cohesive group structures does not preclude the presence of social networks. If social units can be described as a group of animals that associate frequently and more so with each other than with individuals of other such groups (Struhsaker 1969), then close and permanent spatial proximity is not a necessary requirement for social units. Distance communication by vocalizations, as well as regular interactions, can lead to individual assessment, mating preferences, spatial coordination and thereby to a loose network of socially distinct relationships (Delgado and van Schaik 2000; Mitra Setia et al. 2009). Minimum requirements for such a social system include a high degree of spatial overlap between individual home ranges and the temporal stability of associations. In addition, it should be demonstrated that affiliative social encounters and partner preferences occur on a regular basis. Both Galdikas (1984) and Singleton and van Schaik (2002) have observed preferential associations among the adult females within their respective long-term study populations.

This brief review of orangutan social organization suggests at least two possible models for social structure: a system characterized by roving male promiscuity or a network of loose associations within a greater, socially distinct and open community organized around resident flanged males or (related) female clusters. Female clusters are in evidence at most sites, though with varying mean size and frequencies of encounters. What is more likely to vary is the degree that there is networking, or spatial coordination, with the locally dominant male as well as the extent of long distance roaming by males, which is probably linked to local habitat productivity and optimal male mating strategies. An apparent flexibility in social organization, reflecting site differences, could well be an adaptation to deal with variation in local resource distribution and abundance. In habitats with relatively low productivity, social structure might be better characterized by roving male promiscuity, whereas a socially distinct network of loose association is more sustainable at high productivity sites (Mitra Setia et al. 2009). The likelihood and validity of a community model could be strengthened by observations demonstrating spatial coordination between specific individuals of both sex classes.

One important functional distinction between the two models places an emphasis on the basis of female mating preferences, where observed. Under the roving male promiscuity system, indirect benefits are accrued if females choose the highest quality males in their populations. The community model, however, assumes that females receive direct benefits in the form of protection against sexual coercion such as harassment or the threat of infanticide (van Schaik and van Hooff 1996; Delgado and van Schaik 2000). Additional data focusing on mating and male-female associations as a function of female reproductive state are needed to resolve this question.

An indirect, but crucial, prediction of the basis for female mating preferences expects differential responses to the playbacks of adult male long calls as a function of familiarity. Under a roving male promiscuity model, cycling females might approach the call of unfamiliar males, whereas noncycling females would likely ignore the call. On the other hand, a community model predicts that females with small, unweaned infants should strongly avoid the location of these calls and increase their association or spatial coordination with protector males if present. In general, association with her protector should mitigate any increased risk to the female or her infant as posed by the presence of unfamiliar flanged and unflanged males. At the same time, long calls are probably the best way to regulate such encounters and maintain spatial coordination among dispersed individuals.

Orangutan Vocal Communication

Much of the previous work in primate communication has focused on reporting both qualitative and quantitative data on the acoustic properties of vocalizations and in identifying potential markers for individual discrimination (Marler and Hobbett 1975; Cheney and Seyfarth 1982; Steenbeek and Assink 1996). However, there has been comparatively little research with respect to locale- or population-specific differences in vocal behavior. Although a few studies report acoustic differences between conspecific populations at different scales (e.g. Maeda and Masataka 1987; Hohmann and Vogl 1991; Fischer et al. 1998; Hafen et al. 1998), they do not provide adequate evidence for the causal factors underlying these differences.

The interactions between a species' evolutionary history, the nature of social encounters among individuals, and the local sound environment are thought to influence the structure and diversity of vocal signal repertoires in nonhuman primates (Range and Fischer 2004). For example, if acoustic features are anatomically constrained and vocal production is largely genetically determined, then phylogenetic history is expected to be important in accounting for observed differences between populations, though genetic differences are predicted to be low among subspecies or between closely related species. In similar fashion, the frequency, quality, and type of social interactions among individuals are also suggested to affect variation in vocal signaling (Elowson and Snowdon 1994; Maestripieri 1996; Snowdon and Elowson 1999), particularly when encounters are common and associated with other means of communication. Admittedly, rather than invoking vocal learning, these findings suggest that changes in the social situation may affect the subjects' internal states and consequently have an effect on the acoustic structure of calls. Furthermore, if vocal signals are integral to a species' evolved strategies, then it is reasonable to assume that acoustic features will be selected to eliminate, or minimize, signal degradation and maximize sound transmission properties within a particular habitat (Gish and Morton 1981; Brown et al. 1995). More specifically, Marler (1975, 1976) posited that either discrete or graded vocal signals would evolve as a function of habitat type and the absence or presence of other communicative cues. That is, continuous variation in calls should be more common in open habitats and in species with more frequent face-to-face encounters, whereas distinct vocal signals are expected to be more prevalent in forested environments and when vocalizations are used for long-distance communication among dispersed individuals. These predictions remain to be tested widely and, to date, relatively little information is available on the vocal repertoires of the extant hominoids (e.g. gorillas: Harcourt et al., 1993; orangutans: Hardus et al. 2009) or on the demographic and ecological determinants of geographic variation in their structure and acoustic features.

Adult male long calls in orangutans have already been well established as serving a spacing function between males based on relative rank (Galdikas 1983; Mitani 1985b, 1990; Galdikas and Insley 1988; van't Land 1990). That is, the locally dominant male approached calling subordinates, whereas subordinate males avoided the calling dominant (Galdikas 1983; Mitani 1985b; van't Land 1990). Although at least one important function had been identified, this finding did not preclude the possibility of there being other functions for adult male long calls. These additional functions may include but are not limited to attracting cycling females over long distances to facilitate mating (Rodman 1973; MacKinnon 1974; Horr 1975; Rijksen 1978) and coordinating dispersed individuals within a network of loose associations (MacKinnon 1974; van Schaik and van Hooff 1996; Delgado and Van Schaik 2000; Mitra Setia et al. 2009).

In contrast, evidence for a female-attraction function in orangutans has been inconsistent. Results of experimental playbacks conducted in the Kutai Game Reserve, East Kalimantan, revealed that sexually active females do not move toward long calls (Mitani 1985b). However, observational studies within the Gunung Leuser National Park in northern Sumatra document cycling females approaching and consorting with the locally dominant male, who made significantly more long calls than subordinate individuals (Utami and Mitra Setia 1995).

Proposed long call functions such as male-spacing and female-attraction rely on the assumption that identity and/or assessment cues are encoded in the vocal signals. For example, subordinate males might avoid the calling dominant for at least two reasons. First, subordinates may recognize the identity of the locally dominant male by his long call alone and can associate previous agonistic encounters with this male. In this scenario, vocal signals could provide indirect information about male quality if listeners can associate a signaler's identity with past performance (Rubenstein and Hack 1992). Conversely, properties inherent in the long call itself may convey information such as the signaling male's resource-holding potential or fighting ability, his willingness to escalate in aggressive interactions or his current condition. Similarly, these same factors could influence behavioral responses by females. In other words, adult male long calls in orangutans may provide a basis for direct or indirect evaluation, depending on whether there are assessment or identity cues, respectively, encoded in the vocal signals. In addition, there is no reason to believe that one type of cue (e.g. identity) should be present to the exclusion of the other (e.g. assessment). For instance, identity markers could be embedded within acoustic features, whereas assessment criteria are possibly reflected in calling patterns such as the speed and duration of each vocalization.

If, indeed, adult male orangutan long calls can be used for assessment, then energetic costs to the signaler are expected to be associated with vocal behavior (Clutton-Brock and Albon 1979; Zahavi 1982; Hauser 1993; Andersson 1994; Fitch and Hauser 1995; Gouzoules and Gouzoules 2002). The available data suggest that adult male orangutans face low energetic costs to vocal signaling (Delgado and van Schaik unpublished data); but, since there are differences in the rates of calling behavior related to male status at other sites (Tanjung Puting: Galdikas 1983; Galdikas and Insley 1988; Mentoko: Mitani 1985a; Cabang Panti: Knott unpublished data; Ketambe: Utami and Mitra Setia 1995), this result may be restricted to high-productivity sites such as Suaq Balimbing (and perhaps Ketambe) in northern Sumatra. Nonetheless, it is reasonable to suspect that adult males, especially subordinates, may face social costs. That is, subordinate males face the risk of being chased or attacked by more dominant individuals, especially if the distance separating the two is less than 400 m (Galdikas 1983; Mitani 1985a; van't Land 1990; Delgado pers. obs). But this cost is likely to be met only a proportion of the time since flanged males have very large home ranges and are usually found at low densities.

There is limited evidence to suggest that variation in the speed and duration of adult male long calls evokes differential responses by orangutan subjects (Delgado 2003, 2006). Fast calls of long duration were often responded to as if more threatening than slow calls of short duration (Delgado 2003); hence, long call speed and duration possibly give listeners a good idea about a signaler's current condition, especially for familiar individuals, if not about the caller's intrinsic (genetic) quality. In contrast, if adult male long calls in orangutans provide information about individual male identity, then variation in acoustic features between individuals is predicted, as found in other nonhuman primate species (Marler and Hobbett 1975; Waser 1977; Chapman and Weary 1990; Butynski et al. 1992; Zimmermann and Lerch, 1993). Indeed, orangutan long calls contain sufficient variation to identify individuals (Delgado 2007; Delgado et al. 2009) and experimental field playbacks demonstrate that orangutans can distinguish familiar males from unfamiliar males (Delgado 2003).

The capacity for individual recognition based on vocalizations alone is key if adult male long calls also function to coordinate dispersed individuals within a network of loose associations. Long-term behavioral data indicate that adult females coordinate their travel with calling flanged males, even when the females are not sexually active (Mitra Setia and van Schaik 2007; Mitra Setia et al. 2009; van Schaik unpublished data). This result suggests that females accrue some benefit to maintaining spatial proximity to flanged males. In fact, Fox (2002) demonstrates that females who maintain spatial association with flanged males, either through consortships or nonmating temporary parties, receive lower rates of sexual harassment from unflanged males. Close, permanent associations of females and flanged males are restricted by high feeding costs, but are observed more frequently during times of high fruit abundance, also coinciding with the highest incidence of sexual coercion by unflanged males (Fox 2002).

Females have the opportunity of tracking flanged males by homing in on the male's conspicuous long calls. Other subordinate males, those most likely to engage in sexual coercion, are generally kept away by these calls (Galdikas 1983; Mitani 1985b; van't Land 1990). Consequently, a signaling male creates a protective sphere for females in association with him. Observational accounts report that females will travel quickly toward long-calling males when being harassed by unflanged males (Fox 1998, 2002; van Schaik unpublished data) or when faced with a potential infanticidal threat (Delgado 2003). These observations imply that females recognize individual males only based on their long calls and sometimes seek associations with flanged males for protection against sexual coercion by nonpreferred males. However, females typically show no overt response upon hearing long calls. Thus, females selectively react only under motivational conditions that are associated with a real or perceived threat.

Orangutans interact and know conspecifics, but do not necessarily spend most of their time near one another (due to increased feeding constraints) unless benefits of association outweigh the costs. For both males and females, these advantages usually include access to mates; in addition, for females, there is an added benefit of protection against both conspecific threats and sexual coercion such as harassment or the risk of infanticide (van Schaik and Dunbar 1990; Brereton 1995; van Schaik and Kappeler 1997; Treves 1998). For females vulnerable to sexual harassment, long calls can facilitate localizing a flanged male for protection.

Behavioral responses to field playback experiments also demonstrate that females carrying newborn infants will avoid unfamiliar males (Delgado 2003); this finding is consistent with an infanticide threat. Further, unattended infants foraging independently from their mother also react fearfully and immediately approach their mothers upon hearing the playbacks of unfamiliar males (but not familiar males), suggesting that they perceive some potential risk to themselves.

Finally, what do adult male long calls reveal about orangutan social organization and reproductive strategies? Female mate choice does not appear to rely on long call production at most long-term study sites. However, observations from Ketambe (e.g. Utami and Mitra Setia 1995; Delgado 2006) indicate that calling rate is an important factor during brief and contested mating periods. Hence, long distance vocal signaling by flanged males and social interactions in orangutans appear to be related. Orangutans display a flexible social structure, possibly linked to local habitat productivity and population density. Results from Sumatran sites are consistent with a dispersed community model, whereas data from Borneo reflect a system characterized by roving male promiscuity. Under both scenarios, long calls facilitate encounters with flanged males in dispersed orangutan societies. Males announce their presence and relative location. Listeners may choose whether or not to respond, making the functions and consequences of adult male long calls in wild orangutans primarily receiver-dependent. Hence, adult male long calls play an important role in the complex social networks found within orangutan societies.

Culture and Social Learning

Two of mankind's closest living relatives, chimpanzees and orangutans, have increasingly well documented local traditions involving learned skillful behaviors, often involving tools, which vary from place to place and are maintained by social transmission (Whiten et al., 1999; van Schaik 2003). These local traditions are most likely the antecedents of human culture (van Schaik 2003, 2004). The best examples for a cultural interpretation have come from geographically isolated chimpanzee and orangutan populations demonstrating variation in the expression and forms of tool use in feeding (McGrew 1992, 2004; Wrangham et al. 1994; van Schaik and Knott 2001), but cultural interpretations can also be applied to other population-specific behaviors that are not explained by clear genetic or ecological differences. With respect to geographic variation in tool use among primates, researchers have argued that suitable conditions for social transmission are critical factors explaining the observed distribution (van Schaik et al. 1999; van Schaik and Knott 2001; van Schaik 2003, 2004). This assertion can be applied plausibly to other observed local traditions including the emergence of population-specific vocalizations or dialects that, in turn, imply vocal learning – a fundamental attribute of human language. Nonetheless, those factors leading to suitable conditions for social transmission remain to be identified.

Orangutans are a very good species in which to examine the determinants of local traditions because they differ in population density and the nature of social interactions across sites (Galdikas 1985a; Mitani et al. 1991; Sugardjito et al. 1987; van Schaik et al. 1999). Such diversity in gregariousness provides a natural experiment, allowing researchers to test predictions about the social conditions under which population-specific cultures can emerge. Previous studies have already demonstrated that orangutans show inter-site differences, including variation in tool manufacture and use, nest building and other behaviors (van Schaik and Knott 2001; van Schaik et al., 2003; Merrill 2004; Delgado 2007).

Site-specific variation in skilled behaviors such as tool use and manufacture may indicate population differences in genetic expression or local ecological factors, or patterns of innovation and the appropriate demographic conditions for social learning and diffusion. When comparing the feeding techniques and tool-using skills used by orangutans at one site each on Borneo and Sumatra, van Schaik and Knott (2001) found that neither genetic nor ecological differences were adequate reasons to explain the observed distribution of tool use. While it is exceedingly difficult to discriminate between population-specific trends in innovation or the extent of social learning as potential causal factors, the relatively high local population density and the social tolerance at the Sumatran site (i.e. Suaq Balimbing) suggests favorable conditions for the emergence and spread of tool using behavior (van Schaik et al. 1999; van Schaik and Knott 2001; van Schaik et al. 2003; van Schaik 2004). High densities of orangutan populations residing in other Sumatran field sites with reported tool use also provide evidence that suitable transmission conditions are an important factor determinant of local traditions we recognize and label as "cultural" (van Schaik and Knott 2001). Furthermore, a broader, multi-site comparison has found correlations between (1) geographic distance and cultural difference, and (2) the abundance of opportunities for social learning and the size of the local cultural repertoire, but no effect of ecological differences on the presence or absence of particular cultural variants in behavior (van Schaik et al. 1999).

Vocal Cultures?

Previous research has demonstrated geographic variation in the acoustic parameters of long distance vocalizations among diverse animal taxa including primates (Hunter and Krebs 1979; Maeda and Masataka 1987; Bradbury and Vehrencamp 1998; Fischer et al. 1998; Doutrelant et al. 1999; Peters et al. 2000; Wich et al., 2008a, b). For the closest living relative of anatomically modern humans, Mitani et al. (1999) have posited that factors such as vocal learning, habitat acoustics, the local sound environment, genetic variation and body size may underlie vocal variation between chimpanzee (genus *Pan*) communities, but testable predictions for these hypotheses have not yet been examined rigorously. Hence, specific causal factors explaining the observed variance among populations remain poorly understood. Nonetheless, through years of research, at least four explanations have emerged to account for the reported differences in the acoustic features of vocal signals. First, variance in body size, with corresponding differences in vocal tract length, may lead to differences in acoustic parameters such as fundamental and resonant frequencies (Davies and Halliday 1978; McComb 1991; Hauser 1993; Fitch 1997; Fischer et al. 2002; Pfefferle and Fischer 2006). Second, ecological factors such as landscape topography, habitat structure, climatic patterns (i.e. temperature and humidity), and background environmental noise may differ among populations, selecting for particular site-specific acoustic features that enhance the physical transmission properties of vocal signals within those localities (Marten et al. 1977; Waser and Waser 1977; Wiley and Richards 1978; Richards and Wiley 1980; Mitani and Stuht 1998). Third, genotypic differences could lead to phenotypic differences among populations either in anatomical structures related to vocal production or developmental processes and patterns of vocal behavior (Baker 1975; Ryan and Brenowitz 1985; Ryan 1986; Wycherley et al. 2002; Bernal et al. 2005). Finally, flexibility in population structure and social organization among populations can affect the frequency and nature of interactions and, thus, the opportunities for social transmission among community members within populations; as a result, local traditions in vocal culture, or dialects, may arise among different populations (Crockford et al. 2004). Such within-group or community convergence in vocalizations is expected in species where individuals chorus against individuals from other groups or communities, such as in chimpanzees and howling monkeys (genus *Allouatta*). Although it is not yet clear what benefits flanged adult male orangutans would derive from within-community convergence of long-distance vocalizations, they might also have the capacity for developmental modifiability in their long calls (e.g. Wich et al. 2009).

Whether one looks broadly (between populations) or more narrowly (within populations), there is clear evidence that certain behavioral variations depend on opportunities for social learning. The investigation of cultural variation across a diverse array of species reveals the extent of variability, both genetic and behavioral, that is at risk when habitat is destroyed or fragmented (van Schaik 2002; Merrill 2004). However, such approaches provide insights into the origins and evolution of human cultures, and a richer understanding of the nuances and subtle differences between humans and other species in behavioral flexibility and social transmission leading to geographic variation.

Conservation and Considerations

The litany of threats to orangutan populations and other tropical fauna include, but are not limited to, habitat loss and disturbance (e.g. conversion, fires, fragmentation, logging), hunting, and live capture for illegal trade. Orangutans are particularly vulnerable due to their specific habitat requirements, reliance on spatially and temporally dispersed resources, and very slow life history, making population recovery unviable even after only modest population losses. If viable, wild populations of orangutans in Borneo and Sumatra are to survive increasing threats to their natural habitats, change must occur at three levels. Nationally, policies must support habitat protection and establish appropriate incentives for conservation. Regionally, decentralized authorities and conservation programs must have the knowledge and wherewithal to implement these policies in a manner that meets economic development and conservation goals. Locally, government, businesses, and communities must reach compromises that avoid the type of conflict that threatens to derail many of the country's well-conceived conservation initiatives.

To achieve these ends, the USAID-sponsored Orangutan Conservation Services Program (OCSP) hopes to establish sustainable orangutan conservation programs in at least four priority locations, covering about 2 million hectares of prime habitat, where 80 percent of the critically endangered Sumatran species and at least 9,000 members of two Bornean subspecies will be protected. Optimistically, it is thought that threat levels will be reduced through implementation of site-specific conservation plans supported by improved law enforcement, outreach, and sustainable financing schemes.

But what is the role of research, if any, in conservation activities aimed at preserving orangutans (or other animals) and their habitats? First, effective conservation strategies require current and reliable information such as the geographic distribution, relative population abundance and resource use, as well as continued biological monitoring to detect demographic trends over time. Second, a stable research presence tends to deter illegal activities within protected areas and, when able to provide training and/or employment opportunities, often establishes a good rapport with local community members. Third, researchers themselves can become goodwill ambassadors by promoting conservation education and outreach, sharing information with the general public, and building capacity within host countries.

Unfortunately, given the extent of human population growth, there is no quick and easy solution for biological conservation and resource management. Biologists, conservationists, government, and local community members must continue to work together in an effort to develop, establish, and maintain effective land use practices that minimize human-wildlife conflicts while simultaneously balancing the competing interests of habitat preservation and protection with socioeconomic development.

References

Andersson M (1994) Sexual selection. Princeton University, Princeton

Baker MC (1975) Song dialects and genetic differences in white-crowned sparrows (*Zonotrichia leucophrys*). Evolution 29:226–241

Bernal X, Guarnizo C, Lüddecke H (2005) Geographic variation in advertisement call and genetic structure of *Colostethus palmatus* (Anura, Dendrobatidae) from the Colombian Andes. Herpetologica 61:395–408

Bradbury J, Vehrencamp S (1998) Principles of animal communication. Sinauer Associates, Inc., Sunderland, MA

Brereton AR (1995) Coercion-defence hypothesis: the evolution of primate sociality. Folia Primatol 64:207–214

Brown CH, Gomez R, Waser PM (1995) Old world monkey vocalizations: adaptation to the local habitat? Anim Behav 50:945–961

Butynski TM, Chapman CA, Chapman LJ, Weary DM (1992) Use of blue monkey "pyow" calls for long-term individual identification. Am J Primatol 28:183–189

Chapman CA, Weary DM (1990) Variability in spider monkeys' vocalizations may provide basis for individual recognition. Am J Primatol 22:279–284

Cheney D, Seyfarth R (1982) Recognition of individuals within and between groups of free ranging vervet monkeys. Am Zool 22:519–529

Clutton-Brock TH, Albon SD (1979) The roaring of red deer and the evolution of honest advertisement. Behaviour 69:145–169

Crockford C, Herbinger I, Vigilant L, Boesch C (2004) Wild chimpanzees produce group-specific calls: a case for vocal learning. Ethology 110:221–243

Davies NB, Halliday TR (1978) Deep croaks and fighting assessment in toads *Bufo bufo*. Nature 274:683–685

Delgado RA. (2003). The function of adult male long calls in wild orangutans. Ph.D. Thesis. Duke University, Durham NC

Delgado RA (2006) Sexual selection in the loud calls of male primates: Signal content and function. Int J Primatol 27(1):5–25

Delgado RA (2007) Geographic variation in the long calls of male orangutans (*Pongo spp.*). Ethology 113:487–498

Delgado RA, van Schaik CP (2000) The behavioral ecology and conservation of the orangutan: a tale of two islands. Evol Anthropol 9(5):201–218

Delgado RA, Lameira A, Davila Ross M, Husson S, Morrogh-Bernard H, Davila Ross M, Wich SA (2009) Geographical variation in orangutan long calls. In: Wich SA, Utami Atmoko SS, Mitra Setia T, van Schaik CP (eds) Orangutans: geographic variation in behavioral ecology and conservation. Oxford University Press, Oxford

Doutrelant C, Leitao A, Giorgi M, Lambrechts M (1999) Geographical variation in blue tit song, the result of an adjustment to vegetation type? Behaviour 136:481–493

Elowson AM, Snowdon CT (1994) Pygmy marmosets, *Cebuella pygmaea*, modify vocal structure in response to changed social environment. Anim Behav 47:1267–1277

Fischer J, Hammerschmidt K, Todt D (1998) Local variation in Barbary macaque shrill barks. Anim Behav 56:623–629

Fischer J, Hammerschmidt K, Cheney DL, Seyfarth RM (2002) Acoustic features of male baboon loud calls: influences of context, age and individuality. J Acoust Soc Am 111:1465–1474

Fitch WT (1997) Vocal tract length and formant frequency dispersion correlate with body size in rhesus macaques. J Acoust Soc Am 102(2):1213–1222

Fitch WT, Hauser MD (1995) Vocal production in nonhuman primates: acoustics, physiology, and functional constraints on "honest" advertisement. Am J Primatol 37:191–219

Fox EA (1998) The function of female mate choice in the Sumatran Orang-utan, (*Pongo pygmaeus abelii*). Duke Univesity, Durham, NC

Fox EA (2002) Female tactics to reduce sexual harassment in the Sumatran orangutan (*Pongo pygmaeus abelii*). Behav Ecol Sociobiol 52:93–101

Galdikas BMF (1983) The orangutan long call and snag crashing at Tanjung Puting reserve. Primates 24:371–384

Galdikas BMF (1984) Adult female sociality among wild orangutans at Tanjung Puting Reserve. In: Small MF (ed) Female Primates: Studies by Women Primatologists. Alan R. Liss, New York, pp 217–235

Galdikas BMF (1985a) Adult male sociality and reproductive tactics among orangutans at Tanjung Puting. Folia Primatol 45:9–24

Galdikas BMF (1985b) Subadult male orangutan sociality and reproductive behavior at Tanjung Puting. Am J Primatol 8:87–99

Galdikas BMF (1985c) Orangutan sociality at Tanjung Puting. Am J Primatol 9:101–119

Galdikas BMF, Insley SJ (1988) The fast call of the adult male orangutan. J Mammal 69:371–375

Galdikas BMF (1995) Social and reproductive behavior of wild adolescent female orangutans. In Nadler RD, Galdikas BMF, Sheeran LK, Rosen N (eds.) *The Neglected Ape*, pp 163–182. New York: Plenum Press

Gish SL, Morton ES (1981) Structural adaptations to local habitat acoustics in Carolina wren songs. Zeitschrift fuer Tierpsychologie 56:74–84

Gouzoules H, Gouzoules S (2002) Primate communication: by nature honest, or by experience wise? Int J Primatol 23:821–848

Hafen T, Neveu H, Rumpler Y, Wilden I, Zimmermann E (1998) Acoustically dimorphic advertisement calls separate morphologically and genetically homogeneous populations of the grey mouse lemur (*Microcebus murinus*). Folia Primatol 69:342–356

Harcourt AH, Stewart KJ, Hauser M (1993) Functions of wild gorilla 'close' calls I Repertoire, context, and Interspecific comparison. Behaviour 124(1–2):89–122

Hardus ME, Lameira AR, Singleton I, Morrogh-Bernard H, Knott CD, Ancrenaz M, Utami Atmoko SS, Wich SA (2009) A description of the orangutan vocal and sound repertoire: with a focus on geographical variation. In: Wich SA, Utami Atmoko SS, Mitra Setia T, van Schaik CP (eds) Orangutans: geographic variation in behavioral ecology and conservation. Oxford University Press, Oxford

Harrison ME (2009) Orang-utan feeding behaviour in Sabangau, Central Kalimantan. Ph.D. Thesis. University of Cambridge, Cambridge, UK

Harrison ME, Chivers DJ (2007) The orang-utan mating system and the unflanged male: a product of declining food availability during the late Miocene and Pliocene? J Hum Evol 52:275–293

Hauser MD (1993) Rhesus monkey copulation calls: honest signals for female choice? Proc Roy Soc Lond B 254:93–96

Hohmann G, Vogl L (1991) Loud calls of male Nilgiri langurs (Presbytis johnii): age, individual, and population-specific differences. Int J Primatol 12:503–524

Horr DA (1975) The Borneo orang-utan: population structure and dynamics in relationship to ecology and reproductive strategy. In: Rosenbaum LA (ed) Primate behavior. Academic Press, New York, pp 307–323

Hunter M, Krebs J (1979) Geographic variation in the song of the great tit in relation to ecological factors. J Anim Ecol 48:759–785

IUCN (2008) 2008 IUCN red list of threatened species. www.iucnredlist.org

Johnson AE, Knott CD, Pamungkas B, Pasaribu M, Marshall AJ (2005) A survey of the orangutan (*Pongo pygmaeus wurmbii*) population in and around Gunung Palung National Park, West Kalimantan, Indonesia based on nest counts. Biol Conserv 121:495–507

Knott C, Beaudrot L, Snaith T, White S, Tschauner, Planasky G (2008) Female-female competition in Bornean orangutans. Int J Primatol 29:975–997

MacKinnon J (1974) The behaviour and ecology of wild orang-utans (*Pongo pygmaeus*). Anim Behav 22:3–74

Maeda T, Masataka N (1987) Locale-specific vocal behavior of the tamarin (*Saguinus l. labiatus*). Ethology 75:25–30

Maestripieri D (1996) Primate cognition and the bared-teeth display: a reevaluation of the concept of formal dominance. J Comp Psychol 110:402–405

Marler P (1975) On the origin of speech from animal sounds. In: Kavanaugh JF, Cutting JE (eds) The role of speech in language. MIT Press, Cambridge, MA, pp 11–37

Marler P (1976) Social organization, communication and graded signals: the chimpanzee and the gorilla. In: Bateson PPG, Hinde RA (eds) Growing points in ethology. Cambridge University Press, Cambridge, pp 239–280

Marler P, Hobbett L (1975) Individuality in a long-range vocalization of wild chimpanzees. Z. Tierpsychologie 38:97–109

Marten K, Quine D, Marler P (1977) Sound transmission and its significance for animal vocalization. Behav Ecol Sociobiol 2:291–302

McComb K (1991) Female choice for high roaring rates in red deer, *Cervus elaphus*. Anim Behav 41:79–88

McGrew WC (1992) Chimpanzee material culture. Cambridge University Press, Cambridge

McGrew WC (2004) The cultured chimpanzee: reflections on cultural primatology. Cambridge University Press, Cambridge

Merrill MY (2004) Orangutan Cultures? Tool use, social transmission and population differences. Ph.D. Thesis Duke University, Durham NC

Mitani JC (1985a) Mating behaviour of male orangutans in the Kutai Game Reserve, Indonesia. Anim Behav 33:392–402

Mitani JC (1985b) Sexual selection and adult male orangutan long calls. Anim Behav 33:272–283

Mitani JC (1990) Experimental field studies of Asian ape social systems. *Int J Primatol* 11:103–126

Mitani JC, Stuht J (1998) The evolution of nonhuman primate loud calls: acoustic adaptation for long-distance transmission. Primates 39:171–182

Mitani JC, Grether GF, Rodman PS, Priatna D (1991) Association among wild orang-utans: sociality, passive aggregations or chance? Anim Behav 42:33–46

Mitani J, Hunley K, Murdoch M (1999) Geographic variation in the calls of wild chimpanzees: a reassessment. Am J Primatol 47:133–151

Mitra Setia T, van Schaik CP (2007) The response of adult orang-utans toflanged male long calls: inferences about their function. Folia Primatol 78:215–226

Mitra Setia T, Delgado RA, Utami Atmoko SS, Singleton IS, van Schaik CP (2009) Social organization and male-female relationships. In: Wich SA, Utami Atmoko SS, Mitra Setia T, van Schaik CP (eds) Orangutans: geographic variation in behavioral ecology and conservation. Oxford University Press, Oxford

Peters S, Searcy WA, Beecher M, Nowicki S (2000) Geographic variation in the organization of song sparrow repertoires. The Auk 117:936–942

Pfefferle D, Fischer J (2006) Sounds and size – identification of acoustic variables that reflect body size in Hamadryas baboons (*Papio hamadryas*). Anim Behav 72:43–51

Range F, Fischer J (2004) Vocal repertoire of Sooty Mangabeys (*Cercocebus torquatus atys*) in the Taï National Park. Ethology 110:301–321

Richards DG, Wiley RH (1980) Reverberations and amplitude fluctuations in the propagation of sound in a forest: Implications for animal communication. Am Nat 115:381–399

Rijksen HD (1978) A field study of Sumatran orang utans (*Pongo pygmaeus abelii* Lesson 1827): ecology, behaviour and conservation. Ph.D. Thesis. Wageningen, University of Wageningen.

Rodman PS (1973) Population composition and adaptive organisation among orang-utans of the Kutai Reserve. In: Michael RP (ed) Comparative ecology and behaviour of primates. Academic Press, London, pp 171–209

Rodman PS, Mitani JC (1987) Orangutans: sexual dimorphism in a solitary species. In: Smuts BB, Cheney DL, Seyfarth RM, Wrangham RW, Struhsaker TT (eds) Primate Societies. The University of Chicago Press, Chicago, pp 146–154

Rubenstein D, Hack M (1992) Horse signals: the sounds and scents of fury. Evol Ecol 6:254–260

Ryan M (1986) Factors influencing the evolution of acoustic communication: biological constraints. Brain Behav Evol 28:70–82

Ryan MJ, Brenowitz EA (1985) The role of body size, phylogeny, and ambient noise in the evolution of bird song. Am Nat 126:87–100

Schürmann CL (1981) Courtship and mating behavior of wild orangutans in Sumatra. In: Chiarelli AB, Corruccini RS (eds) Primate behavior and sociobiology. Springer-Verlag, Berlin, pp 130–135

Schürmann CL, van Hooff JARAM (1986) Reproductive strategies of the orang-utan: new data and a reconsideration of existing sociosexual models. Int J Primatol 7:265–287

Singleton I, van Schaik CP (2001) Orangutan home range size and its determinants in a Sumatran swamp forest. Int J Primatol 22(6):877–911

Singleton I, van Schaik CP (2002) The social organization of a population of Sumatran orang-utans. Folia Primatol 73:1–20

Singleton I, Wich S, Husson S, Stephens S, Utami Atmoko S, Leighton M, Rosen N, Traylor-Holzer T, Lacy R, Byers O (2004) Orangutan population and habitat viability assessment: final report. IUCN/SSC Conservation Breeding Specialist Group, Apple Valley, MN

Singleton I, Knott CD, Morrogh-Bernard HC, Wich SA, van Schaik CP (2009) Ranging behavior of orangutan females and social organization. In: Wich SA, Utami Atmoko SS, Mitra Setia T, van Schaik CP (eds) Orangutans: geographic variation in behavioral ecology and conservation. Oxford University Press, Oxford, 205–212

Snowdon CT, Elowson AM (1999) Pygmy marmosets modify call structure when paired. Ethology 105:893–908

Steenbeek R, Assink P (1996) Individual differences in long-distance calls of male wild Thomas langurs. Folia Primatol 69:77–80

Struhsaker TT (1969) Correlates of ecology and social organization among African cercopithecines. Folia Primatol 11:80–118

Sugardjito J, te Boekhorst IJA, van Hooff JARAM (1987) Ecological constraints on the grouping of wild orang-utans (Pongo pygmaeus) in the Gunung Leuser National Park, Sumatra, Indonesia. Int J Primatol 8:17–41

te Boekhorst IJA, Schürmann CL, Sugardjito J (1990) Residential status and seasonal movements of wild orang-utans in the Gunung Leuser Reserve (Sumatera, Indonesia). Anim Behav 39:1098–1109

Treves A (1998) Primate social systems: conspecific threat and coercion-defense hypothesis. Folia Primatol 69:81–88

Utami SS (2000) Bimaturism in orang-utan males. Ph.D. Thesis. Utrecht University, The Netherlands

Utami SS, Mitra Setia T (1995) Behavioral changes in wild male and female Sumatran orangutans (Pongo pygmaeus) during and following a resident male take-over. In: Nadler RD, Galdikas BMF, Sheeran LK, Rosen N (eds) The Neglected Ape. Plenum Press, New York, pp 183–190

Utami SS, Goosens B, Bruford MW, de Ruiter JR, van Hoof JARAM (2002) Male bimaturism and reproductive success in Sumatran orangutans. Behav Ecol 13(5):643–652

van Schaik CP (1999) The socioecology of fission-fusion sociality in orangutans. Primates 40(1):69–87

van Schaik CP (2002) Fragility of traditions: the disturbance hypothesis for the loss of local traditions in orangutans. Int J Primatol 23(3):527–538

van Schaik CP (2003) Local traditions in chimpanzees and orangutans: social learning and social tolerance. In: Fragaszy D, Perry S (eds) The biology of traditions: models and evidence. Cambridge University Press, Cambridge, pp 297–328

van Schaik CP (2004) Among orangutans: red apes and the rise of human culture. Harvard University Press, Cambridge, MA

van Schaik CP, Dunbar RIM (1990) The evolution of monogamy in large primates: a new hypothesis and some crucial tests. Behaviour 115(1–2):30–62

van Schaik CP, Kappeler PM (1997) Infanticide risk and the evolution of male-female association in primates. Proc R Soc Lond B 264:1687–1694 van Schaik CP, Knott CD (2001) Geographic variation in tool use on Neesia fruit in orangutans. Am J Phys Anthropol 114:331–342

van Schaik C, van Hooff JARAM (1996) Toward an understanding of the orangutan's social system. In: McGrew WC, Marchant LF, Nishida T (eds) Great ape societies. Cambridge University Press, Cambridge, pp 3–15

van Schaik CP, Deaner RO, Merrill MY (1999) The conditions for tool use in primates: implications for the evolution of material culture. J Hum Evol 36:719–741

van Schaik CP, Ancrenaz M, Borgen G, Galdikas B, Knott CD, Singleton I, Suzuki A, Utami SS, Merrill M (2003) Orangutan cultures and the evolution of material culture. Science 299:102–105

Van't Land J (1990) Who is calling there? The social context of adult male orang-utan long calls. Doctorandus, University of Utrecht, Utrecht

Waser PM (1977) Individual recognition, intragroup cohesion and intergroup spacing: evidence from sound playback to forest monkeys. Behaviour 60:28–74

Waser PM, Waser MS (1977) Experimental studies of primate vocalization: specializations for long-distance propagation. Z. Tierpsychologie 43:239–263

Whiten A, Goodall J, McGrew WC, Nishida T, Reynolds V, Sugiyama Y, Tutin CEG, Wrangham RW, Boesch C (1999) Cultures in chimpanzees. Nature 399:682–685

Wich SA, Singleton I, Utami Atmoko SS, Geurts ML, Rijksen HD, van Schaik CP (2003) The status of the Sumatran orang-utan Pongo abelli: an update. Oryx 37(1):49–54

Wich SA, Schel AM, de Vries H (2008a) Geographic variation in Thomas Langur (*Presbytis thomasi*) loud calls. Am J Primatol 70:566–574

Wich SA, Meijaard E, Marshall AJ, Husson S, Ancrenaz M, Lacy RC, van Schaik CP, Sugardjito J, Simorngkir T, Traylor-Holzer K, Doughty M, Supriatna J, Dennis R, Gumal G, Knott CD, Singleton I (2008b) Distribution and conservation status of the orangtuan (Pongo spp) on Borneo and Sumatra: how many remain? Oryx 42:329–339

Wich SA, Swartz KB, Hardus ME, Lameira AR, Stromberg E, Shumaker RW (2009) A case of spontaneous acquisition of a human sound by an orangutan. Primates. doi:10.1007/s10329-008-0117-y

Wiley RH, Richards DG (1978) Physical constraints on acoustic communication: implications for the evolution of animal vocalizations. Behav Ecol Sociobiol 3:69–94

Wrangham RW, McGrew WC, de Waal FB, Heltne P (1994) Chimpanzee Cultures. Harvard University Press, Cambridge, Massachusetts

Wycherley J, Doran S, Beebee TJC (2002) Male advertisement call characters as phylogeogrpahical indicators in European water frogs. Biol J Linn Soc 77:355–365

Zahavi A (1982) The pattern of vocal signals and the information they convey. Behaviour 80:1–8

Zimmermann E, Lerch C (1993) The complex acoustic design of an advertisement call in male mouse lemurs and sources of its variation. Ethology 93:211–224

Chapter 4
The Natural History of Sumatran Orangutan (*Pongo abelii*)

Sri Suci Utami Atmoko and Carel P. van Schaik

Introduction

The orangutan is the only great ape of Asia. Its present range is confined to dwindling areas on the islands of Sumatra and Borneo (Rijksen and Meijaard 1999). In contrast to its African relatives, the chimpanzee, bonobo (genus *Pan*), and gorilla (genus *Gorilla*), it is extremely arboreal (the Sumatra species more so than the Borneo as Sumatra still harbors tigers). In fact, it is the largest and heaviest of all predominantly arboreal mammals. Among the diurnal primates, it is, moreover, exceptional in that it is comparatively solitary.

Orangutans are now considered to represent two distinct species, the Sumatran orangutan, *Pongo abelii*, now occurring only in the northern part of Sumatra, and the Bornean orangutan, *Pongo pygmaeus*, still occurring in many scattered parts of Borneo with three subspecies (*P. p. pygmaeus*, *P. p. wurmbii*, and *P. p. morio*; Zhi et al. 1996; Groves 2001; Warren et al. 2001; Steiper 2006; Goossens et al. 2009). Recent work suggests that they are different enough from their Bornean congeners that extrapolation from the Bornean species is risky.

On the basis of two active field sites in the Gunung Leuser National Park, northern Sumatra (Aceh), Indonesia, where wild orangutans are being studied, this chapter focuses on the socio-ecology and behavior of Sumatran orangutans (*Pongo abelii*).

Field Sites

The Ketambe orangutan population is the longest-studied wild Sumatran orangutan population. It has been studied continuously since 1971. Together with Suaq Balimbing population that has been studied since 1991, the long-term data have yielded what we know today about Sumatran orangutan biology. Ketambe (3°41′N,

S.S.U. Atmoko (✉)
Faculty of Biology, Universitas Nasional, Jakarta, Indonesia
e-mail: suci_azwar@yahoo.co.id

S. Gursky-Doyen and J. Supriatna (eds.), *Indonesian Primates*,
Developments in Primatology: Progress and Prospects,
DOI 10.1007/978-1-4419-1560-3_4, © Springer Science+Business Media, LLC 2010

97°39′E) is located in the upper Alas valley (a rift valley inside the Barisan mountain range) at an altitude of 350 ± 500 m asl. This study area mainly consists of primary rain forest and was described in detail by Rijksen (1978) and van Schaik and Mirmanto (1985). Suaq Balimbing (3°04′N, 97°26′E) is located in the western coastal plain, some 70 km to the south-west, separated by mountains to over 2,000 m asl with Ketambe, and consists of a variety of floodplain and hill forest habitats.

Life History

Based from 32-years' data at Ketambe and 5.5-years' data at Suaq Balimbing, Wich et al. (2004a, 2009) reported that Sumatran interbirth intervals were longer than those reported for Bornean sites, but that age at first reproduction was similar at 15.5 years. For Ketambe, the mean interbirth interval has been estimated to be 9.3 years (Wich et al. 2004a), while for Suaq Balimbing, estimates are at least 8.2 years (van Noordwijk and van Schaik 2005). The first longevity estimates from the wild (Ketambe) indicate life spans of over 50 years, with no evidence for menopause. Mortality rates were very low for both males and females, with no clear sex difference. These estimates establish the Sumatran orangutan as the nonhuman primate with the slowest life history pace (Wich et al. 2004a).

One of the most unusual features of Sumatran orangutans is the remarkable individual variation in the age at which sexually mature males develop their sexual secondary characteristics (SSC), a phenomenon called bimaturism. This bimaturism leads to the coexistence of two adult, sexually mature morphs: flanged and unflanged males. In Sumatra, SSC development may be delayed 15–20 years after reaching sexual maturity (Utami Atmoko and van Hooff 2004). Although unflanged mature males lack SSCs; they are fertile, sexually active, and are able to sire offspring (Kingsley 1982; Maggioncalda et al. 1999, 2002; Utami Atmoko 2000; Utami et al. 2002; Goossens et al. 2006).

Population Distribution

During the Pleistocene, orangutans could be found from the south in Java up to the foothills of the Himalayas and the Tropic of Cancer in China. This distribution was prehistoric, and the degree to which it has been influenced by humans can be disputed. The reason for the continuous decline in orangutan numbers and distributions is that humans and ape favor the same habitat, namely alluvial plains, peat-swamp forests, and valleys. Now, their habitat has been limited to the island of Borneo and Sumatra (Fig. 4.1).

The Sumatran population is concentrated in the northern part of the island (Aceh and North Sumatra provinces) and is estimated to total ca 6,600 individuals in (yr) (citation). Of the 13 identified populations, only 7 contain more than 500 individuals, the minimum number needed to have some prospects for long-term viability (Soehartono et al. 2007; Wich et al. 2008).

Sumatran orangutan (*Pongo abelii*) (Photo by Jeff Oonk)

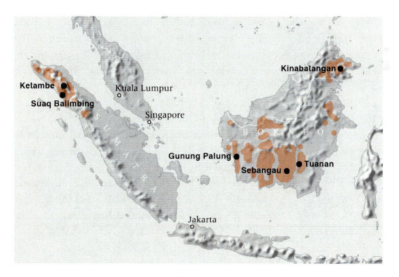

Fig. 4.1 Research study sites of Ketambe and Suaq Balimbing at Gunung Leuser National Park, Aceh, Indonesia (map by Perry van Duihoven)

Sumatran orangutans are found at higher densities (Ketambe 3–5 ind/km^2; Suaq Balimbing 7 ind/km^2) than most Bornean orangutans, although densities decline with increasing altitude in both species (Rijksen 1978; Djojosoedharmo and van Schaik 1992; van Schaik 1999; Rijksen and Meijaard 1999; Husson et al. 2009). Because the orangutan is a frugivore, studies suggest that orangutan densities are related to the proportion of soft pulp in a given area (Djojosoedharmo and van Schaik 1992; van Schaik et al. 1995; Buij et al. 2002) and the density of large strangling figs, at least on dry land (Wich et al. 2006). Lower densities at higher altitudes are probably a function of decreasing fruit availability.

Behavioral Ecology

Sumatra lies at the edge of a currently active subduction zone between two tectonic plates. The resulting recent and ongoing mountain building and volcanism lead to relatively intense erosion that provides continued mineral deposition in the lowland regions. As a result, a much higher proportion of Sumatran soils are productive and suitable for agriculture. In general, then, Sumatra is the product of recent geological processes, and, therefore, the soils of Sumatra tend to be higher in plant nutrients than those of Borneo (van Schaik et al. 2009c).

The implications of these island differences for fruit availability are probably the most important for orangutans. Fluctuations of food-fruit availability in Southeast Asian rain forest are particularly pronounced in lowland forests. As Wich et al. (2006) showed, there is very little systematic influence of fruit availability on Sumatran orangutan diets. At Ketambe, this might be due to the relatively high density of strangling figs and the general high productivity of the area. The figs' fruiting patterns are not strongly seasonal, which ensures that there are always a few huge fig trees in the area with abundant fruit in which orangutans can gather to feed (van Schaik 1986; Sugardjito et al. 1987; Utami et al. 1997).

As a result, Sumatran orangutans always have a high percentage of fruit in their diet; in this they differ from their cousins in Borneo, who must turn to leaves and bark as fallback food resources during low fruit availability. In Gunung Palung, on Borneo, orangutans suffered negative energy budgets during periods of prolonged low fruit availability after mast fruiting, as shown by ketones in their urine (Knott 1998; see Table 4.1). In Ketambe, however, we found no ketones in orangutans' urine and so no evidence of negative energy budget (Wich et al. 2006). These and other analyses of geographic variation in orangutans support the view that the Sumatran forests are generally better habitat for orangutans than Bornean forests (Marshall et al. 2009).

Activity Patterns

A general trend across sites shows that Sumatran orangutans divide their time among feeding (>50%), resting (22–42%), traveling (9–19%), with the remaining time spent on socializing and nest building (Table 4.2). Thus, orangutans in

Table 4.1 The comparison of fruit availability in Ketambe, Sumatra and Gn. Palung, Borneo (Wich et al. 2006; Knott 1998)

Condition	Ketambe (Sumatra)	Gn. Palung (Borneo)
Average monthly % of OU food trees fruiting	9.6 (sd = 3.6, range = 6.3–14.3)	6.1 (sd = 2.8, range = 2.5–12.5)
Minimum number of OU food trees fruiting/month	6.3%	2.5%
Minimum monthly % of fruit in the diet	50%	20%
Maximum monthly % constitute of cambium	5.3%	37%
Soil nutrients	More (volcanic origins)	Less

Table 4.2 Comparison of activity patterns in Ketambe and Suaq Balimbing (Wich et al. unpubl and van Schaik et al. unpubl cited in Morrogh-Bernard et al. 2009)

Site	Unflanged male			Flanged male			Sexually active female			Non-sexually active female		
	F	R	T	F	R	T	F	R	T	F	R	T
Ketambe	52.6	33.5	13.0	48.3	41.9	9.2	55.4	32.0	12.1	59.3	28.7	11.8
Suaq Balimbing	53.6	25.7	17.9	48.0	34.9	14.9	54.9	25.9	16.9	55.7	22.1	19.6

Sumatran forest have a high proportion of fruits in their diet year-round, which enables them to feed for more than 50% of their active period. They also restless and travel more than Bornean orangutans, at least Borneans that range in mixed-dipterocarp forest (Morrogh-Bernard et al. 2009).

In summary, non-sexually active females feed the longest and rest the least, flanged males rest the longest and feed/travel the least, and unflanged males travel the longest in Ketambe, while non-sexually active females travel the longest in Suaq. Because flanged males are larger than other sex-age classes, they can more easily satisfy their energy needs with foods that are harder and can, therefore, be harvested with less travel (Gaulin and Sailer 1985 in van Schaik et al. 2009b). Flanged males can also eat the same food items faster than other age-sex classes because of their larger body size, especially larger food items (e.g., in a giant strangler figs; see Utami et al. 1997).

Diet

Table 4.3 shows a comparison of diet composition, as measured by percentage of total feeding time. It appears that at both Ketambe and Suaq Balimbing, individuals show little variation in the time feeding on fruits, even though those at Ketambe have a slightly higher proportion of fruit in the diet. Wich et al. (2006) attributed this difference to the high density of large strangling fig trees at Ketambe. Strangler figs produce fruit year-round, thus are fed by orangutan constantly throughout the

Table 4.3 Orangutan diet composition at Ketambe and Suaq (percentages of total feeding time) (Morrogh-Bernard et al. 2009)

Site	Fruits	Leaves	Bark	Invertebrates	Other
Ketambe					
Mean	67.5	16.4	2.7	8.8	4.8 (inc. fl)
Low–high fruit	64.2–70.7	17.5–15.2	3.0–2.5	8.7–8.8	6.7–3.0
Suaq Balimbing					
Mean	66.2	15.5	1.1	13.4	3.8 (inc. fl)
Low–high fruit	62.7–69.6	18.3–12.7	0.8–1.4	14.6–12.2	3.6–4.1

year (Sugardjito et al. 1987). Figs are normally considered a fallback food that can be relied on when nonfig fruits are not available (Wich et al. 2006).

Meat Eating

Predation by orangutans on vertebrates is rare, considering the many hours of observation in the wild. However, occasionally at Ketambe and Suaq Balimbing, they catch slow loris, either by grabbing them directly or by quickly killing them by a bite to the head, after a short pursuit onto the forest floor. This qualifies as the stumble-upon-and-capture type of predation, and in this respect, it differs from the hunting that has been described for chimpanzees. So, capture of slow loris does not qualify as pursuit hunting (Utami Atmoko 1997; van Schaik et al. 2009b).

At Ketambe, no males have ever been seen to catch slow loris. Two females are loris capture specialists: their capture rate is higher than all other local orangutans (Utami and van Hooff 1997; Hardus et al. in preparation). At Suaq Balimbing, only three cases of loris capture have been observed, by three different individuals: two adult females, and one flanged male. There is therefore no evidence for a male bias in vertebrate capture among orangutans, as there is in chimpanzees (Boesch 1994a, b; Stanford et al. 1994a, b). If anything, the available data suggest a bias toward females, this could owe to the fact that catching loris typically takes place in the context of insect foraging (van Schaik et al. 2009c).

Tool Use

Orangutans at Suaq used tools in two main foraging contexts: extracting honey or social insects from nests hidden in tree holes and extracting the lipid-rich seeds from mechanically and chemically protected *Neesia* fruits. Use of seed extraction tools is somewhat biased toward females, but only because flanged males (and larger unflanged ones) are strong enough to open the fruits before they dehisce and before the protective stinging hairs have matured, so they can pick out the seeds with their fingers. Once fruits have opened, all individuals use tools at virtually all visits (see van Schaik and Knott 2001 for details; van Schaik et al. 2003, 2009b).

At Ketambe and Agusan (some 30 km to the North), orangutan use leaf gloves as tools to handle spiny fruits or spiny branches, or as seat cushions in trees with spines. Orangutan in Ketambe also use tools for sexual stimulation (auto-erotic), both female and male, and as young as two years of age (van Schaik et al. 2003, 2009b; Utami Atmoko et al. 2009a; Fox and bin Muhammad 2002).

Social Relationships

A local area contains a dominant flanged male, a number of unflanged males, adult females, often with offspring, and adolescent males and females, along with various males that pass through regularly. The flanged male is intolerant towards other fully flanged males that intrude into his vicinity, and is supposed to be the focal element around whom the other units are organized (van Schaik and van Hooff 1996; Utami Atmoko 2000; Utami Atmoko et al. 2009a).

Individual orangutans live in large home ranges (Table 4.4). Perhaps more on ranging in Sumatran OU's, e.g., larger ranges than in Borneo and why, ranging responses to seasonal fruit scarcities.

Table 4.4 shows in both sites that male home ranges are larger than those of females, even if no estimates were possible. This is consistent with the expectation that males competing for access to females maximize their access to females by ranging more widely. As a result, male home ranges overlap extensively (Utami Atmoko et al. 2009a). That local resident dominant males have smaller ranges may be because dominance allows them to monopolize the females in the area where they reside. Other males, instead, must always be looking for females not monopolized by local dominants, and this forces them to range much more widely.

The ranges of several individuals of both sexes overlap considerably. Females appear to be philopatric; among males, some flanged ones remain in a relatively small area (called "resident") while others range over greater regions (called "non-resident") (Singleton and van Schaik 2001, 2002; Goossens et al. 2006; Knott and Kahlenberg 2007). Patterns of male residency are not permanent, because resident males may be forced out by non-resident or resident challengers (Utami and Mitra Setia 1995).

Two major factors are known to affect the tendency to associate: food availability and mating opportunities (Sugardjito et al. 1987; te Boekhorst et al. 1990; van

Table 4.4 Home range estimates of Sumatran orangutans (Singleton and van Schaik 2001; Utami Atmoko et al. 2009a)

Sites	Study area size (ha)	Female HR (ha)	Flanged male HR (ha)	Unflanged male HR (ha)
Ketambe	450	300–400	>females	>females
Suaq Balimbing	500–2,000	≥850	≥2500[a]	≥2,500

[a]The locally dominant male ("resident") had a smaller home range than other flanged males, although it was still larger than the females' ranges.

Schaik and van Hooff 1996). Both the large size of orangutans and consequently, the high cost of their almost exclusively arboreal locomotion explain why they keep on their own. There is circumstantial evidence that living with others would impose to high costs in terms of the time and energy, budget because of the competition involved (te Boekhorst et al. 1990; van Hooff 1988; van Schaik and van Hooff 1996). This is indicated by the fact that on rare occasions, namely under conditions of eco-logical affluence, orangutans do congregate in groups and may even stay and travel together for several days (Sugardjito et al. 1987; Utami et al. 1997; Utami Atmoko 2000). However, this idea is not supported by the fact that the average party size at each site is not linked to the average fruit abundance at that site (van Schaik 1999; Utami Atmoko 2000; Wich et al 2006). It is likely that Ketambe and Suaq orang-utans, associate for rare benefits like mating and the transmission of social and for-aging skills, that was suggested to be important in orangutans (Wich et al. 2006; van Schaik and Knott 2001; van Schaik et al. 2003). It seems that orangutans in Sumatra are able to maintain relatively high mean party size without bearing much cost on fruit availability due to the high density of large strangling fig trees in Ketambe compared to other areas (Rijksen 1978; Wich et al. 2004b) and these figs are less seasonal in their fruiting patterns than other fruiting trees in Ketambe (van Schaik 1986). This also explains why Sumatran orangutans tend to be more often found in groups than their Bornean counterparts (see the following section).

Given these costs of association, it is clear that consort formation affects travel behavior in both classes of males; on average, flanged males traveled less than unf-langed males when traveling alone, but did not different significantly when they were in consort with a female (Utami Atmoko and van Hooff 2004). When consorting, flanged males had to increase their travel to keep up with female they accompany, whereas consorting unflanged males slowed down. The females' travel behavior did not change when they were consorting. In other words, the males adjusted to the females (van Schaik 1999; Utami Atmoko 2000; Utami Atmoko and van Hooff 2004).

Male–Male Relationships

Given the large size and high overlap of home ranges as well as the dense vegetation, it would seem impossible for flanged males to monopolize access to potentially reproductive females effectively, especially since orangutan females do not show visible signs of ovulation. Among orangutans, there is ample evidence for male-male contest competition for access to fertile females, as well as alternative male mating strategies driven by this contest, suggesting that multiple males can easily locate females (Utami Atmoko et al. 2009a).

Both in Ketambe (Utami Atmoko 2000; Utami Atmoko and van Hooff 2004; Wich et al. 2006) and in Suaq Balimbing (van Schaik et al. 2009a), the differences between the two male classes were as predicted based on their difference in social strategy. We found ecological differences between flanged and unflanged males that are direct expressions of the large difference in mobility (time spent moving, travel speed, day journey length). The greater mobility of unflanged males allowed them to feed more selectively, and thus have shorter feeding bouts. Flanged males may have supported their more sedentary life style by eating more fruit, in longer feeding bouts, and perhaps by spending more time on eating vegetable matter. Only the latter difference between the two male classes might also be linked to body size per se. To the extent that females differ significantly from any of these values, the sex difference is a product of multiple processes (reproduction, size, male socio-ecology) (van Schaik et al. 2009a).

Male–Female Relationships

Orangutans at the two Sumatran sites are more gregarious than Bornean orangutans (mean adult female party size 1.5–2.0 in Sumatra vs. 1.05–1.3 in Borneo; Mitra Setia et al. 2009). Sumatran orangutans occasionally congregate when they meet in large fruit trees. At Ketambe, for instance, large strangler figs often attract multiple adult orangutans simultaneously, and up to 14 individuals have been seen in or near a single tree (Mitra Setia et al. 2009).

Flanged males advertise their location by giving long calls. Long calls play a role in male spacing, by which relationships are communicated within the dispersed society, but their primary function in Sumatra is probably coordination of range use with adult females, and attraction of fertile females; unflanged males, in contrast, do not long call so they have to travel through an area to locate potentially fertile females (Delagado and van Schaik 2000; Mitra Setia et al. 2009; Utami Atmoko et al. 2009b). Females' responses to long calls suggest that they are trying to maintain earshot associations with the locally dominant male. The function of these loose associations is almost certainly that they allow females to seek refuge with the flanged males, especially dominant ones, if they are being harassed by other males. This evidence indicates the existence of loose communities in Sumatran orangutans, organized around dominant flanged males (MacKinnon 1974; Mitra Setia et al. 2009).

Female–Female Relationships

Among the Sumatran females, mean female party sizes at Ketambe and Suaq Balimbing are comparable in size, but associations at Ketambe involve fewer travel parties (van Schaik 1999). In Suaq, we found evidence for clusters of females, who may well be related, and whose ranges share similar boundaries with considerable overlap. Within these clusters they also show a tendency toward reproductive synchrony, as they have infants of similar ages, and preferential association with each other, even if home range overlap is taken into account (Singleton and van Schaik 2002).

Another example of female philopatry also found in Ketambe, where the daughters and granddaughters of the reintroduced rehabilitant Binjai all settled in the study area and they have more tolerant relationships, meet more often, show less aggression and feeding tolerance (Mitra Setia et al. 2009), except during lowest fruit availability (Utami Atmoko et al. in preparation). With mix female population (wild and ex-rehab) in Ketambe, displacements between adult females occurred even in large fig trees. They were unidirectional, with one exception. The females in the study area could be ordered in a nearly linear dominance hierarchy. Although the test for linear hierarchy showed only a trend, the displacement data had a high directional consistency, and it seems justified to claim that a hierarchy indeed exists. With the acquisition of more data, this could become significant (Utami et al. 1997).

Mating Strategies

Flanged and unflanged males differ in their mating strategies. Flanged males, and especially dominant individuals, often establish consortships with potentially reproductive females and are usually preferred by females (Utami Atmoko 2000). Unflanged males engage in consortships comparatively rarely, but often try to copulate with females, even when they resist and resulting forced copulation (Galdikas 1979; MacKinnon 1974; Mitani 1985; Rijksen 1978; Schurmann and van Hooff 1986).

Despite their semi solitary nature, behavioral and experimental evidence suggests that individualized sexual relationships exist in orangutans (van Schaik and van Hooff 1996; Delagado and van Schaik 2000). The majority of sexual interactions in Sumatra were cooperative and occurred during a consort relationship. Females select their sexual partners and choose when to consort and mate cooperatively, i.e., females show clear preferences for or aversions to particular males (Utami Atmoko 2000; Fox 2002; Utami Atmoko et al. 2009b). The reproductive success of flanged males is made possible by the females' preference for flanged over unflanged males. The dominant flanged male in an area may be able to exclude other flanged males from his immediate ranging area, but he certainly does not exclude all unflanged males. Female preference then sets the scene for male-male competition; if females had no preferences at all, unflanged males would be much more successful than flanged males due to their higher mobility (Utami Atmoko et al. 2009a). At least in Sumatran, subordinate flanged males do not seem to be successful at all (Utami Atmoko et al. 2009a; van Schaik 2004). Whether this is also true for Borneo needs to be assessed in future work.

On the basis of the long term behavioral data and genetic paternity studies at Ketambe, Utami et al. (2002) hypothesized that the two male morphs represent coexisting male reproductive strategies, "sitting, calling, and waiting" for flanged males vs. "going, searching, and finding" for unflanged males. The unflanged males travel faster and roam more widely, and can also endure longer associations because they lack SSCs and flanged males are more tolerant toward these smaller males. They are therefore better able to gain access to some potentially fertile females by following consort pairs closely and engaging in sneak mating sometimes by harassing the females and by engaging in voluntary consorts with nulliparous females who are less attractive to the flanged males (and with whom they may achieve their greatest siring success). Thus, only when a male is likely to achieve dominance in an area would the mating benefits of developing the full set of SSCs outweigh it costs. It is possible that a male may need to wait for a long time for such an opportunity to arise, which would explain the development arrest.

Conservation

Orangutans are of great scientific interest, representing a branch of great ape evolution distinct from the African great apes, and relevant to management of forest. They are regarded as "flagship" species that provide a symbol to raise conservation awareness to ensure survival of the forests that contain many other organisms.

The number of wild orangutans has declined continuously with the rapid loss of forest habitat, particularly in the lowland forests with their many timber trees species. The 2007 edition of the IUCN Red List (IUCN 2007) species recognized the Sumatran orangutan as Critically Endangered, whereas the Bornean orangutan has been listed as Endangered.

Orangutans everywhere cannot survive the conversion of forest into plantations, but on Sumatra, orangutans do not seem to cope well with selective logging. More research is needed to determine whether certain levels of extraction are compatible with orangutan conservation (Wich et al. 2008). Although the rate of forest loss in some areas remains high, in other areas there has been a decrease in forest loss rates and hence also a likely reduction in rates of orangutan decline. For instance, in recent years, annual forest loss in the Leuser Ecosystem in Sumatra decreased to 0.6% (Griffiths pers. com. in Wich et al. 2008). Recovery could be helped by retaining soft-pulp fruit bearing trees and climbers (especially *Ficus sp.* for Sumatran population) and strictly enforcing antipoaching laws (Rijksen and Meijaard 1999; Robertson and van Schaik 2001; Wich et al. 2008). However, habitat protection is most important as this is a key to orangutan survival.

Conclusions

Orangutans have been the subjects of long term field study at a number of different sites. Given their extended life span and slow development and the long-term cycles in affect their habitat, long-term studies like these are essential to document their behavior and life history. Only after more than 30 years of research are we beginning to understand orangutan life history to a certain extent, but we need much longer to complete the picture.

It is well appreciated that logging has a negative effect on orangutan density: on both islands orangutan density decrease after logging. Researchers have noted that the logging induced decrease in orangutan density seems less severe on Borneo than on Sumatra. Although this ability to endure habitat damage might enable Bornean orangutans to survive in the short-term, it is largely unknown how logging affects long-term survival of orangutan populations. As habitat protection has become the foremost issue in orangutan conservation, we should continue long-term studies in part to monitor the long-term effects of logging on Sumatran orangutans and to their long-term survival, as part of efforts to save Sumatran orangutans.

Acknowledgements We thank the Directorate General of Forest Conservation of the Ministry of Forestry for permission to work in the Gunung Leuser National Park; the Indonesian Institute of Sciences (LIPI) for permission to work in Indonesia; Utrecht University, Universitas Nasional, Paneco, and WOTRO for long term support of Ketambe Research Center; the Leuser Development Programme for logistic support; and Wildlife Conservation Society and Paneco for generous long-term support of the field site in Suaq Balimbing. We especially thank all the students (UNAS, UNSYIAH, STIK, UI, Utrecht University, Duke University), field assistants, and other researchers who have collected data at Ketambe and Suaq Balimbing. We would also like to thank Anne Russon for the comment and Dr. Jatna Supriatna for inviting us to be a part of the "Indonesian Primate" book.

References

Boesch C (1994a) Cooperative hunting in wild chimpanzees. Anim Behav 48:653–657

Boesch C (1994b) Hunting strategies of Gombe and Taï chimpanzees. In: Wrangham RW, McGrew WC, de Waal FBM, Heltne PG (eds) Chimpanzee cultures. Harvard University Press, Cambridge, pp 77–92

Buij R, Wich SA, Lubis AH, Sterck EHM (2002) Seasonal movements in the Sumatran orangutan (*Pongo pygmaeus abelii*) and consequences for conservation. Biol Conserv 107:83–87

Delagado R, van Schaik CP (2000) The behavioral ecology and conservation of the orangutan (*Pongo pygmaeus*): a tale of two islands. Evol Anthropol 9:201–218

Djojosoedharmo S, van Schaik CP (1992) Why orangutan so rare in the highlands? Altitudinal changes in a Sumatran forest. Trop Biodivers 1:11–22

Fox EA (2002) Female tactics to reduce sexual harassment in the Sumatran orangutans (*Pongo pygmaeus abelii*). Behav Ecol Sociobiol 52:93–101

Fox EA, bin Muhammad I (2002) New tool use by wild Sumatran orangutans (*Pongo pygmaeus abelii*). Am J Phys Anthropol 119:186–188

Galdikas BMF (1979) Orangutan adaptation at Tanjung putting reserve: mating and ecology. In: Hamburg DA, McCown ER (eds) The Great Apes. Benjamin/Cumings, Menlo Park, CA, pp 194–233

Gaulin SJC, Sailer L (1985) Are females the eco-logical sex? Am Anthropol 87:111–119

Goossens B, Setchell JM, James SS, Funk SM, Chikhi L, Abulani A, Ancrenaz M, Lackman-Ancrenaz I, Bruford MW (2006) Philopatry and reproductive success in Bornean orangutans (*Pongo pygmaeus*). Mol Ecol 15:2577–2588

Goossens B, Chikhi L, Jalil MF, James S, Ancrenaz M, Lackman-Ancrenaz I, Bruford MW (2009) Taxonomy, geographic variation and population genetics of Bornean and Sumatran orangutans. In: Wich SA, Utami Atmoko SS, Mitra Setia T, van Schaik C (eds) Orangutans: geographic variation in behavioral ecology and conservation. Oxford University Press, New York, pp 1–14

Groves CP (2001) Primate taxonomy. Smithsonian Institution Press, Washington, DC

Hardus M, Dominy NJ, Zulfa A, Wich SA (in preparation) Evolutionary and behavioral aspects of meat eating by orangutan (*Pongo abelii*)

Husson SJ, Wich SA, Marshall AJ, Dennis RD, Ancrenaz M, Brassey R, Gumal M, Hearn AJ, Meijaard E, Simorangkir T, Singleton I (2009) Orangutan distribution, density, abundance and impact of disturbance. In: Wich SA, Utami Atmoko SS, Mitra Setia T, van Schaik CP (eds) Orangutans: geographic variation in behavioral ecology and conservation. Oxford University Press, New York, pp 78–96

IUCN (2007) IUCN red list of threatened species. Species Survival Commission, International Union for the Conservation of Nature and Natural Resources. IUCN Publications, Gland, Switzerland and Cambridge, UK

Kingsley S (1982) Causes of non-breeding and the development of secondary sexual characteristics in the male orangutan: a hormonal study. In: de Boer LEM (ed) The orangutan, its biology and conservation. Den Haag, Junk, pp 215–229

Knott CD (1998) Changes in orangutan caloric intake, energy balance, and ketone response to fluctuating fruit availability. Int J Primatol 19:1061–1079

Knott CD, Kahlenberg SM (2007) Orangutans in perspective. Forced copulations and female mating strategies. In: Campbell CJ, Fuentes A, MacKinnon KC, Panger M, Bearder SK (eds) Primates in perspective. Oxford University Press, Oxford, pp 290–305

MacKinnon J (1974) The behavior and ecology of wild orangutans (*Pongo pygmaeus*). Anim Behav 22:3–74

Maggioncalda AN, Sapolsky RM, Czekala NM (1999) Reproductive hormone profiles in captive male orangutans: implications for understanding developmental arrest. Am J Phys Anthropol 109:19–32

Maggioncalda AN, Czekala NM, Sapolsky RM (2002) Male orangutan subadulthood: a new twist on the relationship between chronic stress and developmental arrest. Am J Phys Anthropol 118:25–32

Marshall AJ, Ancrenaz M, Brearley FQ, Fredriksson GM, Ghaffar N, Heydon M, Husson SJ, Leighton M, McConkey KR, Morrogh-Bernard HC, Proctor J, van Schaik CP, Yeager CP, Wich SA (2009) The effects of forest phenology and floristics on populations of Bornean and Sumatran orangutans. In: Wich SA, Utami Atmoko SS, Mitra Setia T, van Schaik CP (eds) Orangutans: geographic variation in behavioral ecology and conservation. Oxford University Press, New York, pp 97–117

Mitani JC (1985) Sexual selection and adult male orang-utan long calls. Anim Behav 33:272–283

Mitra Setia T, Delgado R, Utami Atmoko SS, Singleton I, van Schaik CP (2009) Social organization and male–female relationships. In: Wich SA, Utami Atmoko SS, Mitra Setia T, van Schaik CP (eds) Orangutans: geographic variation in behavioral ecology and conservation. Oxford University Press, New York, pp 245–254

Morrogh-Bernard HC, Husson SJ, Knott CD, Wich SA, van Schaik CP, van Noordwijk MA, Lackman-Ancrenaz I, Marshall AJ, Kanamori T, Kuze N, bin Sakong R (2009) Orangutan activity budgets and diet. In: Wich SA, UtamiAtmoko SS, MitraSetia T, van Schaik CP (eds) Orangutans: geographic variation in behavioral ecology and conservation. Oxford University Press, New York, pp 119–134

Rijksen HD (1978) A field study on Sumatran orangutans (*Pongo pygmaeus abelii*, Lesson 1827): Ecology, behavior, and conservation. H. Veenman and Zonen, Wageningen, The Netherlands

Rijksen HD, Meijaard E (1999) Our vanishing relative: the status of wild orangutans at the close of the twentieth century. Kluwer Academic Publishers, Dordrecht

Robertson JMY, van Schaik CP (2001) Causal factors underlying the dramatic decline of the Sumatran orang-utan. Oryx 35:26–38

Schurmann CL, van Hooff JARAM (1986) Reproductive strategies of the orangutan: new data and a reconsideration of existing socio sexual models. Int J Primatol 7:265–287

Singleton I, van Schaik CP (2001) Orangutan home range size and its determinants in a Sumatran swamp forest. Int J Primatol 22:877–911

Singleton I, van Schaik CP (2002) The social organisation of a population of Sumatran orangutans. Folia Primatol 73:1–20

Soehartono TR, Susilo HD, Andayani N, Utami Atmoko SS, Sihite J, Saleh C, Sutrisno A (2007) Strategi dan rencana aksi konservasi orang-utan Indonesia 2007–2017. Dirjen PHKA, Departemen Kehutanan, Indonesia

Stanford C, Wallis J, Matama H, Goodall J (1994a) Patterns of predation by chimpanzees on red colobus monkeys. Anim Behav 49:577–587

Stanford C, Wallis J, Matama H, Goodall J (1994b) Hunting decisions in wild chimpanzees. Behaviour 131:1–18

Steiper ME (2006) Population history, biogeography, and taxonomy of orangutans (Genus: *Pongo*) based on a population genetic meta-analysis of multiple loci. J Hum Evol 50:509–522

Sugardjito J, te Boekhorst IJA, van Hooff JARAM (1987) Ecological constraints on the grouping of wild orangutans (*Pongo pygmaeus*) in the Gunung Leuser National park, Sumatra, Indonesia. Intl J of Primatol 8:17–41

te Boekhorst IJA, Schurmann CL, Sugardjito J (1990) Residential status and seasonal movements of wild orangutans of the Gunung Leuser reserve (Sumatra, Indonesia). Anim Behav 39:1098–1109

Utami Atmoko SS (2000) Bimaturism in orang-utan males: Reproductive and ecology strategies. Ph.D. Thesis. Utrecht University, The Netherlands

Utami Atmoko SS, Mitra Setia T, Goossens B, James SS, Knott CD, Morrogh-Bernard HC, van Schaik CP, van Noordwijk MA (2009a) Orangutan mating behaviour and strategies. In: Wich SA, Utami Atmoko SS, Mitra Setia T, van Schaik C (eds) Orangutans: geographic variation in behavioral ecology and conservation. Oxford University Press, New York, pp 235–244

Utami Atmoko SS, Singleton I, van Noordwijk MA, van Schaik CP, Mitra Setia T (2009b) Male–male relationships in orangutan. In: Wich SA, Utami Atmoko SS, Mitra Setia T, van Schaik C (eds) Orangutans: geographic variation in behavioral ecology and conservation. Oxford University Press, New York, pp 225–234

Utami SS, Mitra Setia T (1995) Behavioral changes in wild male and female Sumatran orangutans (*Pongo pygmaeus*) during and following a resident male take-over. In: Nadler RD, Galdikas BMF, Sheeran LK, Rosen N (eds) The neglected ape. Plenum, New York, pp 183–190

Utami SS, van Hooff JARAM (1997) Meat-eating by adult female Sumatran orangutans (*Pongo pygmaeus abelii*). Am J Primatol 43:156–165

Utami SS, van Hooff JARAM (2004) Alternative male reproductive strategies: Male bimatursm in orangutans. In: Kappeler PM, van Schaik CP (eds) Sexual selection in primates: new and comparative perspectives. Cambridge University Press, Cambridge, pp 196–207

Utami SS, Wich SA, Sterck EHM, van Hooff JARAM (1997) Food competition between wild orangutans in large fig tress. Int J Primatol 18:909–927

Utami SS, Goossens B, Bruford MW, de Ruiter J, van Hooff JARAM (2002) Male bimaturism and reproductive success in Sumatran orangutans. Behav Ecol 13:643–652

van Hooff JARAM (1988) Sociality in primates a compromise of ecological and social adaptation strategies. In: Tartabini A, Genta ML (eds) Perspectives in the study of primates. Cozena, de Rose, pp 9–23

van Noordwijk MA, van Schaik CP (2005) Development of ecological competence in Sumatran orangutans. Am J Primatol 127:79–94

van Schaik CP (1986) Phenological changes in Sumatran rain forest. J Trop Ecol 2:327–347

van Schaik CP (1999) The socioecology of fission–fusion sociality in orangutans. Primates 40:69–86

van Schaik CP, Knott CD (2001) Geographic variation in tool use on Neesia fruit in orangutans. Am J Phys Anthropol 114:331–342

van Schaik CP, Mirmanto E (1985) Spatial variation in the structure and litterfall of a Sumatran rain forest. Biotropica 17:196–205

van Schaik CP, van Hooff JARAM (1996) Toward an understanding of the orangutan's social system. In: McGrew WC, Marchant LF, Nishida T (eds) Great ape societies. Cambridge University Press, Cambridge, pp 3–15 van Schaik CP, Azwar MS, Priatna D (1995) Population estimates and habitat preferences of orangutans based on line transect nests. In: Nadler RD, Galdikas BMF, Sheeran LK, Rosen N (eds) The neglected ape. Plenum, New York, pp 129–147

van Schaik CP, Ancrenaz M, Borgen G, Galdikas BMF, Knott CD, Singleton I, Suzuki A, Utami SS, Merril M (2003) Orangutan cultures and the evolution of material culture. Science 299:102–105

van Schaik CP, Ancrenaz M, Djojoasmoro R, Knott CD, Morrogh-Bernard HC, Nuzuar Odom K, Utami Atmoko SS, van Noordwijk MA (2009a) Orangutan cultures revisited. In: Wich SA, Utami Atmoko SS, Mitra Setia T, van Schaik C (eds) Orangutans: geographic variation in behavioral ecology and conservation. Oxford University Press, New York, pp 299–310

van Schaik CP, Marshall AJ, Wich SA (2009b) Geographic variation in orang-utan behavior and biology. In: Wich SA, Utami Atmoko SS, Mitra Setia T, van Schaik C (eds) Orangutans: geographic variation in behavioral ecology and conservation. Oxford University Press, New York, pp 351–362

van Schaik CP, van Noordwijk MA, Vogel ER (2009c) Ecological sex differences in wild orang-utans. In: Wich SA, Utami Atmoko SS, Mitra Setia T, van Schaik C (eds) Orangutans: geographic variation in behavioral ecology and conservation. Oxford University Press, New York, pp 255–268

Warren KS, Verschoor EJ, Langenhuijzen S, Heriyanto Swan RA, Vigilant L, Heeney JL (2001) Speciation and intrasubspecific variation of Bornean orangutans, *Pongo pygmaeus pygmaeus*. Mol Biol Evol 18:472–480

Wich SA, Buij R, van Schaik CP (2004a) Determinants of orangutan density in the dryland forests of the Leuser Ecosystem. Primates 45:177–182

Wich SA, Utami-Atmoko SS, Setia TM, Rijksen HD, Schürmann C, van Schaik CP (2004b) Life history of wild Sumatran orangutans (*Pongo abelii*). J Hum Evol 47:385–398

Wich SA, Utami Atmoko SS, Mitra Setia T, Djojosudharmo S, Geurts ML (2006) Dietary and ener-getic responses of *Pongo abelii* to fruit availability fluctuations. Int J Primatol 27:1535–1550

Wich SA, Meijaard E, Marshall AJ, Husson SJ, Ancrenaz M, Lacy RC, van Schaik CP, Sugardjito J, Simorangkir T, Traylor-Holzer K, Doughty M, Supriatna J, Dennis R, Gumal M, Knott CD, Singleton I (2008) Distribution and conservation status of the orangutan (*Pongo spp.*) on Borneo and Sumatra: how many remain? Oryx 42:329–339

Wich SA, de Vries H, Ancrenaz M, Perkins L, Shumaker RW, Suzuki A, van Schaik CP (2009) Orangutan life history variation. In: Wich SA, Utami Atmoko SS, Mitra Setia T, van Schaik C (eds) Orangutans: geographic variation in behavioral ecology and conservation. Oxford University Press, New York, pp 65–76

Zhi L, Karesh WB, Janczewski DN, Frazier-Taylor H, Sajuthi D, Gombek F, Andau M, Martenson JS, O'Brien SJ (1996) Genomic differentiation among natural populations of orangutan (*Pongo pygmaeus*). Curr Biol 6:326–336

Chapter 5
Javan Gibbon (*Hylobates moloch*): Population and Conservation

Jatna Supriatna, Alan Mootnick, and Noviar Andayani

Introduction

Java marks the most southwesterly limits for the range of the Asian primates. There is fossil evidence that the orangutan (*Pongo pygmaeus*) (Storm et al. 2005), siamang (*Symphalangus syndactylus*) (Hooijer 1960), and the pigtailed macaque (*Macaca nemestrina*) (Aimi 1981) once lived in Java. Those local extinctions are quite ancient, but other extinctions, such as the Javan tiger (*Panthera tigris javanicus*), date back only a few decades, reflecting the severity of environmental degradation in the island (Siedensticker 1987). There are five primates living in Java today, of which two, the Javan gibbon (*Hylobates moloch*) and the grizzled leaf monkey (*Presbytis comata*), are now categorized on the IUCN Red List as endangered (IUCN 1996). The remaining species, the Javan leaf monkey (*Trachypthecus auratus*) and Javan slow loris (*Nycticebus coucang javanicus*) are ranked as vulnerable, while the long-tailed macaque (*Macaca fascicularis*) is still relatively abundant (Supriatna and Wahyono 2000; Supriatna et al. 2001).

Current taxonomies indicate that there may be 14–16 gibbon species and four genera (Mootnick and Groves 2005; Mootnick 2006; Roos et al. 2007). There are seven gibbon species found in Indonesia: The agile gibbon (*Hylobates agilis*), siamang (*Symphalangus syndactylus*), and Sumatran lar gibbon (*Hylobates lar vestitus*) are native to Sumatra; the Kloss gibbon (*Hylobates klossii*) occurs only on the Mentawai Islands; the Javan gibbon is endemic to Java (Fig. 5.1), and the Mueller's gibbon (*Hylobates muelleri*) and the white-bearded gibbon (*Hylobates albibarbis*) are native to Kalimantan. Of these species, only the Javan gibbon has been listed on the IUCN Red List as critically endangered, having the highest risk of extinction due to habitat loss and hunting (Supriatna et al. 2001). The Javan gibbon is now only found in forest remnants of Western (*H. moloch moloch*) and Central Java

J. Supriatna (✉)
Conservation International Indonesia, Jl. Pejaten Barat 16 A, Jakarta, 12559, Indonesia
e-mail: jsupriatna@conservation.org

S. Gursky-Doyen and J. Supriatna (eds.), *Indonesian Primates*,
Developments in Primatology: Progress and Prospects,
DOI 10.1007/978-1-4419-1560-3_5, © Springer Science+Business Media, LLC 2010

Fig. 5.1 An adult male Javan gibbon *(Hylobates moloch)*, at the Javan Gibbon Center, Bogor, Indonesia

(*H. moloch pongoalsoni*) (Sody 1949). Four workshops have been carried out to examine the conservation status and discuss conservation measures for the species. A Population and Habitat Viability Analysis (PHVA) workshop held in 1994 was run by the IUCN/SSC Conservation Breeding Specialist Group (CBSG) (Supriatna et al. 1994), and the second workshop, organized by Conservation International, Indonesia, in collaboration with the University of Indonesia and the Nagao Environment Fund, Japan, in 1997, examined the rescue and rehabilitation programs for this species (Supriatna and Manullang 1999). The third workshop was part of the Indonesia gibbon conservation and management (Campbell et al. 2008), while the fourth was solely dedicated to review and develop a national action plan for 2008–2018. The last two were carried out by the Forestry Department and Indonesia Primatologists Association with technical support from CBSG-IUCN. These workshops, especially the second and fourth, resulted in intensified efforts on the part of experts, governmental agencies, and conservation organizations, in the protection of this species. The development of the first rescue and rehabilitation center for the Javan gibbon and the production of a government endorsed action plan are just two of the exemplary efforts made so far.

During the last two decades, much attention has been given to obtaining population estimates of the Javan gibbons that survive in the small patchy forests in West and Central Java (Asquith 1995; Asquith et al. 1995; Nijman and van Balen 1998; Supriatna et al. 1998; Djanubudiman et al. 2004; Nijman 2004; Atmoko et al.

2008). A number of students and scientists have carried out surveys in specific sites such as the Gunung Slamet Protected Forest (Supriatna et al. 1992), Ujung Kulon National Park (Gurmaya 1992; Wibisono 1995), Gunung Halimun National Park (Sugardjito et al. 1997; Sugardjito and Sinaga 1999; Suryanti 2006; Iskandar 2007), Gunung Gede Pangrango National Park (Purwanto 1996; Raharjo 2003), Gunung Simpang Protected Area (Subekti 2003), and Gunung Tilu Protected Area (Alrasyid 2003; Atmoko et al. 2008).

There have been numerous initiatives and campaigns to save the Javan gibbon. Notable among them was the media campaign and education program at the Bedogol Conservation Education Center in the Gunung Gede Pangrango National Park, set up and supported by the Gunung Gede National Park Management, Conservation International, the Alami Foundation, and the University of Indonesia. Every year, more than 5,000 people visit the site. The goal is to report and detail the plight of the Javan gibbon and promote an understanding of the link between conserving wildlife and the benefits to the people in securing their natural forests.

Threats to the Javan Gibbons

An island of about 130,000 km² (slightly larger than New York State), Java has been overcrowded for the last 200 years. Before Indonesia's independence in 1945, the Dutch colonial government had tried to relocate some of the human population to other islands in order to reduce the pressures on the environment. However, the population continues to grow and has, in fact, accelerated. From 1961 to 2000, the human population on Java nearly doubled from 63 million to more than 115 million (Whitten et al. 1996; Biro Pusat Statistik 2006). This burgeoning human population and the island's long history of farming, that dates back to at least 1,000 years ago has significantly reduced Java's forest cover. Whitten et al. (1996) estimated that more than 1.5 million ha had already been lost to farmland and teak plantations by 1,000 A.D. Prior to World War II, Java's forests had been reduced to 23% of their original size (Siedensticker 1987). By 1990, only 0.96 million ha of natural forest remained (FAO 1990). Today, most of the forest remnants are found in national parks or other variously effective forms of protected areas, including those for watershed protection. Large areas of "forest" cover on the island are tree plantations (teak, pine, and other trees), mixed community forests, or forest reserved for research purposes.

Java continues to lose its forests – significantly following the Indonesian government's decentralization of forest management to the regencies. In 2001, the central government adopted new laws on responsibilities for natural resource management and the allocation of the pertinent budgets. Forest management, except for conservation areas, have been given over to local governments, some of which focus on short-term economic gain from activities, such as logging, rather than the sustainable, longterm management of natural resources. One aspect which results in the persistence of these threats is that local people, including decision-makers, do not have adequate information concerning the importance of conservation, and the long-term benefits that local people can derive from these forests, such as watershed services. The major cause of natural forest loss today is not, however, industrial-scale logging, but encroachment and depredation by small holders – tree cutting for subsistence plots, collection of firewood, forest fires, and charcoal production.

The balance of 5 years of decentralization in the responsibilities for forest management adds to the loss of Javan forests. Satellite images from 1985, 1997, and 2003, respectively, show a reduction in forest cover not only in the watershed protection forests but also in protected areas (unpublished map of Planning Agency of Min. of Forestry, Indonesia). The forest of Gunung Simpang Protected Area lost almost 15% (from 15,000 ha) during this time, Ujung Kulon National Park lost 4% of its 76,100 ha, and Gunung Halimun National Park 2.5% of 42,000 ha (Director Conservation Area of Min of Forestry pers. comm. 2001) (Fig. 5.2).

The pet trade is another major problem for the Javan gibbon. Based on a recent survey by Atmoko et al. (2008), illegal animal traps to capture Javan gibbons, and other animals were still found. It is believed that nearly 300 individuals are illegally held in Indonesia, generally for the pet trade (Supriatna et al. 1994). The north coast of the island of Java is a major route for the trafficking of Indonesian nonhuman primates, which includes Java gibbons (Malone et al. 2004). As such, Javan gibbon hunters throughout the island are likely to be involved in the supply and sourcing of the illegal trade in primates and other wildlife. One of the biggest challenges in enforcing the regulations is the willingness of the authorities to become engaged in and carry through the required judicial procedures. Illegal logging, felling for firewood and local construction industries, encroachment of protected areas, and illegal trading in wildlife are widespread and yet unpunished.

Distribution and Key Populations

The first population survey of the Javan gibbon was carried out by Kappeler (1984). He identified 25 Javan gibbon populations in forest patches in West and Central Java. Asquith et al. (1995) resurveyed the populations located by Kappeler and identified additional populations in Western Java close to Gunung Simpang. The report on the 1994 Javan gibbon and Javan langur Population Habitat and Viability Assessment (PHVA) Workshop indicated that no more than 400 Javan gibbons lived in 30 protected areas, with an additional 386–1,957 living in 23 forest patches elsewhere (Supriatna et al. 1994). Asquith et al. (1995) estimated less than 3,000 individuals in Central and Western Java. A subsequent survey from 1994 to 1997 discovered a number of new sites and populations in Ujung Kulon and Gunung Halimun National Parks, which are currently major strongholds for two primate species' (Supriatna et al. 1998). Additional populations were brought to light by Nijman and his colleagues; one in a small forest area in West Java, and in three large significant forests in Central Java, on the southern slopes of Gunung Segara (Pembarisan Mountains), Gunung Cupu-Simembuat and Gunung Jaran (Nijman and Sözer 1995; Nijman and van Balen 1998; Nijman 2004). Nijman (2004) indicated the

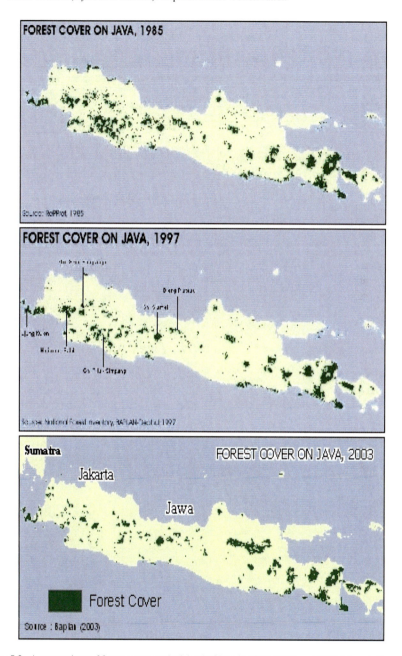

Fig. 5.2   A comparison of forest cover on the island of Java in 1985, 1997, and 2003 (above). *Sources*: RePPProt (1985); Min. of Forestry and World Bank (2000), and Min. of Environment (2007)

total number of wild gibbons in Java to be between 4,000 and 4,500. Following a year-long survey, Djanubudiman et al. (2004) estimated a population of between 2,600 and 5,304. Atmoko et al. (2008) reported on their extensive surveys in 63 sites

Table 5.1 Javan gibbon habitat areas, population with possible effective conservation

Protected Area	Habitat (km²)	Forest size (km²)	Forest type	Estimated population	Source
Ujung Kulon NP G. – Payung G. – Honje	3085	761	Lowland	300–560	Kappeler (1984), Gurmaya (1992), Wibisono (1995), Asquith et al. (1995), Iskandar (2007), Rinaldi (2003), Nijman (2004), and Djanubudiman et al. (2004)
Gunung. Halimun NP Gunung Salak	23576	400	Lowland, sub-montane, montane	900–1221	Supriatna (1998), Sugardjito and Sinaga (1999), and Djanubudiman et al. (2004)
Gunung. Gede Pangrango NP	50	140	Lowland, sub-montane, montane	447	Djanubudiman et al. (2004) and Suryanti (2006)
Gunung Papandayan PF	130		Sub-montane and montane	40–527	Djanubudiman et al. (2004) and Atmoko et al. (2008)
Telaga Warna PA		50	Sub-montane	476	Djanubudiman et al. (2004)
Gunung Simpang PA	110	150	Sub-montane	132	Asquith et al. (1995), Djanubudiman et al. (2004), and Atmoko et al. (2008)
Gunung Tilu PA	30	80	Sub-montane	20–196	Djanubudiman et al. (2004) and Atmoko et al. (2008)
Gunung Kendeng PF, Dieng Plateu	90		Sub-montane	492	Djanubudiman et al. (2004)
Gunung Slamet PF	38.6		Lowland, sub-montane, montane	96	Supriatna et al. (1992) and Djanubudiman et al. (2004)

NP national park; *PF* protection forest; *PA* protected area.

at the previously known areas of Javan gibbons, that some were declining, while other areas had more individuals based on direct sightings (Table 5.1).

Many of the forest patches maintaining Javan gibbons are minute and have less than ten individuals – a number well below the demographic and genetic thresholds for their mid to long-term persistence (Lande 1988). They are evidently at high risk of extinction unless subjected to intensive conservation efforts. Although conservation programs might best be focused primarily on core populations, such as those in the national parks of Gunung Halimun, Gunung Gede Pangrango, and Ujung

Kulon (Supriatna et al. 1994), consideration must be given to smaller populations functioning as critical stepping-stone populations, allowing for the maintenance of genetic diversity, genetic exchange, dispersal and colonization – processes vital for the long-term survival of this species.

The majority of the surviving Javan gibbons are now confined to small populations in isolated forest patches. With burgeoning human populations and the uncertain future of the already scarce and fragmented forests, there is a need to establish a wildlife sanctuary to allow the rescue and translocation of the scattered and isolated gibbons groups before their forests are destroyed. Although the translocation of wild animals is still fraught with difficulties, this strategy may be the only conservation option in this case, particularly when so much of the forest on Java is scheduled for imminent destruction. The translocation of rescued groups proved to be a highly successful component of the overall strategy for the conservation of the golden lion tamarin (*Leontopithecus rosalia*) in the Atlantic forest of Brazil. In the early 1990s, 42 golden lion tamarins in six groups, each isolated in tiny forest remnants, were captured and introduced to a secure forest. They thrived, and in May 2006 numbered more than 250 in about 25 groups, comprising about 18% of the entire population (1,400) in the wild (Kierulff et al. 2002; Kierulff pers. comm. 24 May 2006). Prolonged monitoring and in-depth studies of their demography, ecology and behavior need to accompany a program of this sort. Analyses are in progress to determine the extent and nature of genetic variability in the remnant populations and the degree of divergence among them. Such information will contribute to a decision as to whether such a strategy is necessary and justifiable, and if the answers are positive, this will determine which populations should be given highest conservation priority (Avise 1994).

Although estimates of remaining Javan gibbon numbers may vary, there can be no doubts as to the significant threats that all current populations are facing: principally from continuing habitat degradation and fragmentation. Today, almost all the remaining Javan gibbon habitats are sub-montane and montane forests (Gunung means mountain). The major exception is Ujung Kulon NP, which also has small portions of lowland forest plus small forest remnants in Gunung Halimun and Gunung Gede National Parks. Though the three national parks in West Java, Gunung Gede Pangrango, Gunung Halimun-Salak, and Ujung Kulon, plus several protected areas (Gunung Simpang, Gunung Tilu dan Telaga Warna) and protected forests for watershed (Gunung.Kendeng, Gunung Papandayan) are the only three areas that have the potential to maintain a population of more than 100 Javan gibbons.

Habitat and Remaining Forest

One of the most recent surveys by the YABSHI Foundation (supported by U. S. Fish and Wildlife Service) documented the disappearance of a number of forests over the last decade, notably Bojong Picung and Pasir Susuru, besides the imminent loss of gibbon habitat in Leuweung Sancang, Gunung Jayanti, Gunung Tangkuban

Perahu, and Telaga Warna, where only part of the remaining forests are within legally protected areas (Djanubudiman et al. 2004). Knowing the actual numbers of gibbons is important, but now the paramount is, where possible, the protection of those forests, avoiding their destruction and controlling hunting, and where gibbons and their forests are doomed, with some means to have them translocated or brought into captive breeding programs for future reintroduction.

To add to the Javan gibbon's problem, the majority of the earlier population predictions were based on extrapolating the density of existing forests, without a deep understanding of the forest structure and which did not give an accurate estimation of the Javan gibbons. Suryanti (2006) studied the suitability index and disturbance to Javan gibbons at Gunung Halimun National Park. Her conclusions were that the Javan gibbon was generally affected by two variables, which were the number of roads, and the patchiness or discontinuous forests, but was not so much from altitude, slopes, rainfall, or villages in the park. In the park, roads crisscross at the lowland area not only connecting the villages but is also used by illegal gold miners and off road vehicles from Jakarta, and therefore provide an easy access to hunters. The forest patchiness found in some areas was a result of vegetation transition between lowland and submountain, and human activities. Suryanti identified eight locations (Citorek, Ciparay, Cigudeg, Cisangku, Cikelat, Sampora and two small areas) that need to be intensively managed by the park authority because they are ecologically important for biodiversity conservation, but face high threat from humans, making these forests unsuitable for Javan gibbons.

Conservation Measures for the Javan Gibbon

Population and Habitat Viability Analysis Workshop

In May 1994, more than 50 primate specialists participated in a Population and Habitat Viability Analysis workshop for the Javan gibbon (Supriatna et al. 1994). The workshop established guidelines for a captive management program, not just as a hedge against extinction, but also to rationalize and facilitate the placing of confiscated animals. A public awareness campaign that focused on the threats to the Javan gibbon and its habitat was highly recommended. Follow-up workshops developed the criteria for site selection, guidelines for quarantine procedures and veterinary policies, with recommendations on enclosure design, nutrition, population sources, rehabilitation, and education and research programs. A plan to establish a Javan gibbon rescue and rehabilitation center was also proposed.

On the last day of the conference, a working group was established to lay out the guidelines for establishing a captive management program. Immediate recommendations included a survey of pets, the establishment of a Javan gibbon studbook, the preparation of a gibbon husbandry manual, and training in gibbon health and husbandry techniques for Indonesian Zoo Association (PKBSI) staff. Not all of these recommendations have been acted on, but nevertheless remain a priority. A survey of pets and gibbons held in Indonesian zoos was carried out in 1996 (Supriatna et al. 1998). Information was gathered from the Offices for Conservation and Natural Resources (BKSDA) of West Java, Central Java, and the city of Jakarta, in order to verify the number of Javan gibbons held in zoos and by pet owners. The number of registered pet Javan gibbons in West Java, Central Java, and Jakarta were 54, 41, and 36, respectively. Most pets were found in poor health; some were traded, or died from infectious disease, intestinal parasites or bacteria (Supriatna et al. 1998). During the Javan gibbon rescue and rehabilitation workshop in 1999, training in better gibbon health practices were recommended, which included the development of a gibbon husbandry manual. Eventually, an international Javan gibbon studbook was established in the Perth Zoo, Australia, which included captive populations in Java.

Rescue and Rehabilitation Program

Following the 1994 PHVA workshop, Conservation International, the University of Indonesia, and the Nagao Environment Fund (NEF), Japan hosted an international workshop on Javan Gibbon Rescue and Rehabilitation in August 1997 (Supriatna and Manullang 1999). Eight papers were presented on topics, such as population status in the wild (Gunung Halimun National Park), population genetics, *ex situ* conservation and cryopreservation, government policy on rehabilitation, management, nutrition, and enclosure design and protocols for their housing. A significant element of the workshop was the presentation of techniques, methods, and "lessons learned" by experts on the rehabilitation of gibbons in Thailand. Other aspects considered were the existing government policy on rehabilitation, the IUCN protocols, and the experiences of zoo personnel on gibbon enclosures and husbandry. Supriatna et al. (1998) also reported that the numbers of gibbon kept as pets were not entirely accurate because many individuals were misidentified.

The phylogenetic tree for hylobatids clearly shows the Javan gibbon to be a monophyletic group separated from other gibbon species (Takacs et al. 2005). DNA sequence data suggests the possible existence of two populations, a western population and an eastern population extending into Central Java (Supriatna et al. 1999; Andayani et al. 2001). Additional research on molecular genetic and vocalization may shed further light on this (Geissmann et al. 2002; Mootnick 2006). Morphological differences between these two gibbon populations are subtle. *Hylobates m. moloch* has a darker cap than the Javan gibbons of Eastern Java. The release of confiscated Javan gibbons into the wild must be conducted with caution until further genetic and vocalization investigations have concluded these findings. It is, therefore, essential that zoos properly identify the subspecies of their gibbons for their breeding program, if the ultimate goal would be to release future offspring into protected forests in their native habitat. The conclusion of this workshop resulted in a recommendation to the Government of Indonesia to establish a Rescue and Rehabilitation Center (Supriatna and Manullang 1999).

There was also a recommendation for a captive breeding program to preserve the genetic diversity of the species in captivity. It was debated if captive breeding programs would have a vital role in the survival of the Javan gibbon. There are a small number of Javan gibbons held in zoos outside Indonesia with only a few founders (Thompson and Cocks 2008), but one of the first steps would be to improve our understanding of the reproductive behavior and physiology of the species. Two graduate students from the University of Indonesia and Bogor Agriculture University are currently carrying out researches with respect to reproductive biology (Sjahfirdi et al. 2006a, b) and have already made significant inroads to understanding the menstrual cycle and the behavioral and physiological determination of the periovulatory phase. Hopefully, studies such as these, will contribute to greater understanding as to why in some cases the Javan gibbon has a lower captive reproductive rate.

Javan Gibbon Center

During the XVIII Congress of the International Primatological Society (IPS), held in Adelaide, Australia, in 2001, Conservation International (CI) and the Silvery Gibbon Project (SGP, Australia) agreed to collaborate to establish a Javan Gibbon Center (JGC) for the maintenance and rehabilitation of rescued and confiscated Javan gibbons. The JGC receives donated or confiscated Javan gibbons (generally pets) with the short-term goal of assessing their medical and behavioral status, and restoring their health. Because so few Javan gibbons remain in the wild, it is very important not to loose the genetic material of the illegally-held Javan gibbons, with the likelihood that some of these gibbons will be unreleasable. The JGC is working, therefore, to: (a) retrieve pet Javan gibbons; (b) manage an *ex situ* population; (c) conduct non-invasive research, including genome resource banking; and (d) provide public awareness and education programs, focusing on Javan gibbons and its imperiled status in the wild. This work is carried out in collaboration with the Indonesian Ministry of Forestry (Department of Forest Conservation and National Parks, the Provincial Natural Resources Agency, and the Forestry Research and Development Center), the Javan Gibbon Foundation, Conservation International Indonesia, Perth Zoo, Australia and the University of Indonesia.

The JGC formally opened in mid-2003 in Cigombong Lido Bogor, West Java, on land donated by a local ecotourism hotel. By June 2006, JGC housed six rescued Javan gibbons. Additional facilities were constructed, including a guard station, office, medical and quarantine facility, and individual, introduction, and socialization enclosures. The infrastructure and staffing of the JGC are still far from complete. More enclosures are needed to accommodate Javan gibbons that are currently turned away because of space and staff constraints. By 2008, the Javan Gibbon Center housed 27 Javan gibbons, including five established pairs, three newly introduced pairs, some housed singularly outdoors, or in quarantine. One of main struggling efforts at the Center is the introduction process of human-reared Javan gibbons. Copulation has been observed once the pair is compatible, with two pregnant females observed in 2008.

In parallel with the work at the JGC, there is an urgent need for educational outreach to local communities living in and around the Javan gibbon's remaining forests. Some efforts have been made but they are as yet incipient. Outreach is critical so that when gibbons are successfully rehabilitated and can be released, there will be ample support and understanding, and protection provided by the local communities involved.

Despite the gloomy assessment of most primate population, conservationists point to a notable success in helping targeted species recover. In Brazil, the black lion tamarin (*Leontopithecus chrysopygus*) was downlisted from critically endangered to endangered, as was the golden lion tamarin in 2003, as a result of three decades of conservation efforts involving numerous institutions. A similar pattern of downlisting the Javan gibbon from critically endangered to endangered, was not based on successful conservation mitigation, but the change of IUCN criteria 2004 toward evaluating the threatened species. Both Javan gibbon populations are now protected, but remain very small, creating an urgent need for reforestation to provide new habitats for their long-term survival.

Securing More Habitat via Corridor Development

As mentioned above, the molecular genetic study by Andayani et al. (2001) suggested the presence of possibly two populations of Javan gibbons. The western population is represented by the large population in Gunung Halimun, while the eastern population includes isolates around Cianjur–Sukabumi complex (possibly covering Gunung Masigit, Gunung Tilu, Gunung Ciremai, and Gunung Sawal) and Gunung Slamet in Central Java. The good news is that the Gunung Ciremai forest was declared as a new national park, and hopefully, the small Javan gibbon population will increase, as observed at the Gunung Halimun-Salak National Park. These findings have consequences for conservation policies: (1) gibbons in the Gunung Halimun complex should be managed as a separate and distinct conservation population – they should not be considered as a population to reinforce the threatened isolates of the eastern population; (2) the Cianjur–Sukabumi complex presents a second distinct population – gibbons from there can be moved among the different localities within this complex; (3) although the gibbons in Gunung Slamet are not evolutionarily distinct from populations in the west, they merit special attention as they might represent a case of peripheral isolation.

The forests and the gibbon population of Gunung Halimun are almost linked to the Gunung Salak Protected Forest and the Gunung Gede-Pangrango National Park. With approximately 1,800–2,000 individuals – almost half of the entire wild population – these three mountain ranges are the major stronghold for Javan gibbon populations today. These protected areas comprise an integrated conservation management system that protects the last remaining tropical forest remnants on Java, and guarantee water supplies for 35 million people in Jakarta, Indonesia's capital, and neighboring cities, besides numerous industries along the rivers that run north–south in Western Java. In 2003, the government agreed to create a corridor of these protected areas by incorporating Gunung Salak into Gunung Halimun National Park and enlarging the Gunung Gede Pangrango National Park. This decision, which increased the size of the two parks to 135,000 ha in total, more than doubled the amount of protected habitat for the Javan gibbon. The management of the Gunung Gede Pangrango National Park has created a buffer of vegetation to secure the new boundaries of this recent park expansion by developing a small community agroforestry and reforestation program (Conservation International Indonesia 2005).

Educating People to Save the Javan Gibbon

For more than 5 years (2000–2006), Conservation International Indonesia (CI) has led the GEDEPAHALA Consortium (Gede-Pangrango–Halimun-Salak), comprised of 17 nongovernmental organizations (NGOs), eight government institutions and research centers, four universities, and two private companies. The objective of the consortium is to raise the awareness of all stakeholders (including government, business enterprises and local communities) concerning the advantages of maintaining, protecting, and expanding, the two parks for human welfare, notably in the maintenance of a reliable long-term water supply, the generation of carbon sequestration benefits, and the protection of wildlife.

Approximately 1 h from Jakarta is the montane region of the Gunung Gede-Pangrango National Park, which is of major importance for tourism, and approximately 150 gibbons securely live. There are hundreds of hotels, restaurants and recreation areas, and for obvious reasons, the tourism industry must be a major target for awareness campaigns concerning the value of the forests, their wildlife, and the plight of Java's endemic ape. In 2001, the Alami Foundation, Conservation International Indonesia and the park authority created the Badogol Conservation Education Center to secure local support for the parks through an understanding of the behavior of wildlife and by generating direct and indirect benefits to the local communities. A Mobile Conservation Education Unit is used to take the conservation education program beyond the park's gates, visiting communities surrounding the Gunung Gede-Pangrango National Park, to encourage them to incorporate conservation concepts in their daily activities. The Mobile Conservation Education Unit uses the characters of "Moli," the Javan gibbon, and "Telsi," the Javan hawk-eagle, to deliver a conservation message, besides showing wildlife films, stimulating discussions, and playing interactive games, and making a small library accessible to local groups (Conservation International Indonesia 2005).

Other Conservation Measure Needs and Recommendations

There has been a dramatic loss of natural habitats throughout Indonesia. The massive destruction of its forests and the loss of the Javan tiger signal a clear extinction crisis in Java, as in so many other regions of the country. The last and richest habitats across

Java are now under the greatest pressure. Unprotected lowland forests of the island are likely to be completely cleared unless aggressive measures are taken by government officers and NGOs. The range of the Javan gibbon has been dramatically reduced by habitat loss and human encroachment. Of the 37 forests previously inhabited by this species and registered by Kappeler (1984), many were found to be severely degraded and no longer suitable to sustain viable populations 10 years later (Asquith et al. 1995). Djanubudiman et al. (2004) further emphasized that illegal poaching is another serious threat to the species. Specific recommendations for the conservation of the Javan gibbon include the need to encourage government officers to take action in curbing illegal trade in gibbons, to double their efforts to patrol the existing parks, to create programs to monitor populations both in and outside protected areas, and to discourage the pet trade by confiscating pets and placing them in a rehabilitation program.

The Indonesian forestry reform is moving rapidly, with a growing interest among stakeholders to seize this opportunity to promote greater sustainability in the forestry sectors, as well as to increase local community involvement in the management of their forest resources. There is a growing concern regarding the provision of effective long-term management for Indonesia's extraordinary system of conservation areas – comprising almost 90% of the island's remnant forests in Java. There is, consequently, an urgent need to implement a demonstrative program to earn public support for the potential direct and indirect benefits of the parks. The charm of the endangered Javan gibbon can be used to develop ecotourism programs, and generate income for all stakeholders in and around the protected areas where it occurs. Campbell et al. (2008) compiled recommendations for all gibbons in Indonesia, while the Javan Gibbon Action Plan urged for improvement of awareness and education of the general public, and to mainstream the action plan into regional planning of districts and provinces.

Legislation providing for regional autonomy, which went into effect in January 2001, is fundamentally reshaping the relationship between Jakarta and local authorities for all sectors, including forestry. Local governments are anxious to increase their revenues from natural resources, including efforts to levy taxes on private and state-controlled operations. District and provincial officers are now allowed to pass local regulations. These may have negative or positive implications for forest conservation and indigenous livelihoods. One positive implication is the increased facility and capacity for NGOs to lobby for local regulations that recognize indigenous rights to natural resources and promote the sustainable use of forests and their resources. A potential negative implication is that district administrators can now issue large numbers of permits for local companies to exploit their forests. This movement has to be anticipated by conservationists and government conservation officers, promote greater local participation in resource allocation decisions and demand a greater accountability on the part of regional governments.

The principle recommendation regarding the application of scientifically grounded conservation management of the Javan gibbon is the need for research on their population genetics. There is some genetic evidence that the Javan gibbon split, around 100,000 years ago, into two distinct lineages, Western and Central Java (Andayani et al. 2001). This finding must be considered when planning the relocation of groups from doomed habitats – a vital tactic for conservation of the genetic variability of the species. Genetic research on this species has, to date, been based on a limited number of samples and genetic analysis, and any plan for translocation should first be based on a more complete understanding of the demography and population genetics of the species in the various parts of its range. If we can still conserve the forests remaining today, and eliminate the hunting pressure, there might still be hope for the survival of the Javan gibbon.

Acknowledgments We thank our colleagues who helped us in gathering data and publications, and assisted us during fieldwork and literature studies on the Javan gibbon. We are most grateful to our colleagues at the Department of Biology, University of Indonesia for their support, especially the students who have been working with us in the study of the Javan gibbon. We also thank Dr. Didi Indrawan, Guritno Djanubudiman, R. M. Hidayat of Indonesia Biodiversity and Development Foundation (BIODEV), Anton Ario, Hendi Sumantri, Dr. Barita Manullang, William Marthy, Ermayanti, Iwan Wijayanto, and the entire staff at Conservation International Indonesia. Many thanks to our friends and supporters at the Javan Gibbon Foundation and the International Gibbon Foundation.

References

Aimi M (1981) Fossil *Macaca nemestrina* (Linnaeus, 1766) from Java, Indonesia. Primates 22(3):409–413

Alrasyid MH (2003) Populasi owa Jawa (*Hylobates moloch* Audebert, 1798) di Resort Wilayah Konservasi Mandala, Cagar Alam Gunung Tilu, Jawa Barat. BSc thesis, Bogor Agriculture University, Bogor, Indonesia

Andayani N, Morales JC, Forstner MRJ, Supriatna J, Melnick DJ (2001) Genetic variability in mtDNA of the silvery gibbon: implications for the conservation of a critically endangered species. Conserv Biol 15(3):770–775

Asquith NM (1995) Javan gibbon conservation: why habitat protection is crucial. Trop Biodivers 3(1):63–65

Asquith NM (2001) Misdirections in conservation biology. Conserv Biol 15(2):345–352

Asquith NM, Martarinza M, Sinaga RM (1995) The Javan gibbon (*Hylobates moloch*): status and conservation recommendations. Trop Biodivers 3(1):1–14

Atmoko SU, Wedana M, Oktavinalis H, Bukhorie E (2008) Preliminary report: survey update population and distribution estimates of the Javan gibbon. Gibbon's Voice 10(1):3–7

Avise JC (1994) Molecular markers, natural history and evolution. Chapman and Hall, New York

Biro Pusat Statistik (2006) BPS Statistics Indonesia: population profile. http://jateng.bps.go.id/2006/index06_eng.html, Yogyakarta: http://yogyakarta.bps.go.id/component/content/article/64-daftar-publikasi/147-daerah-istimewa-yogyakarta-dalam-angka-20062007-, Jawa Timur : http://jatim.bps.go.id/index.php?s=data+2006, Banten: http://www.bantenprov.go.id/home.php?link=dtl&id=1731, Jawa Barat : http://jabar.bps.go.id/web2006/index.html; DKI Jakarta: http://jakarta.bps.go.id/JDA/JDA2006/JDA2006.pdf , April 20, 2006, 10:15 am

Campbell C, Andayani N, Cheyne S, Pamungkas J, Usman F, Taylor-Holtzer K (2008) Indonesia Gibbon conservation and management workshop: final report. IUCN/SSC Conservation Breeding Specialist Group, Apple Valley, MN

Conservation International Indonesia (2005) Annual report. Conservation International, Jakarta, 35 pp

Djanubudiman G, Arisona J, Iqbal M, Wibisono F, Mulcahy G, Indrawan M, Hidayat RM (2004) Current distribution and conservation priorities for the Javan Gibbon (*Hylobates moloch*). Report to Great Ape Conservation Fund, US Fish and Wildlife Service, Washington, DC, Indonesian Foundation for Advance of Biological Sciences and Center for Biodiversity and Conservation Studies of University of Indonesia, Depok, 25 pp

FAO (1990) Situation and outlook of forestry sectors in Indonesia. Food and Agriculture Organisation (FAO), Jakarta

Geissmann T, Dallmann R, Pastorini J (2002) The Javan silvery gibbon (*Hylobatesmoloch*): are there several subspecies? In: Abstracts: XIXth congress of the international primatological society, 4–9 August 2002, Beijing, China. Mammalogical Society of China, Beijing (Abstract), pp 120–121

Groves CP (2005) Order Primates. In: Wilson DE, Reeder DM (eds) Mammal species of the world: a taxonomic and geographic reference, vol 1. Johns Hopkins University Press, Baltimore, pp 111–184

Gurmaya KJ (1992) Ecology and conservation of five species of Java's primates in Ujung Kulon National Park, West Java, Indonesia. Research report, Padjadjaran University, Bandung

Hooijer DA (1960) Quaternary gibbons from the Malay Archipelago. Zool Verhand Leiden 46:1–41

Iskandar E (2007) Habitat dan populasi owa jawa (*Hylobates moloch* Audebert, 1797) – di Taman Nasional Gunung Halimun-Salak Jawa Barat. Thesis Doktor, Universtas Indonesia, Bogor, p 141

IUCN (1996) 1996 IUCN red list of threatened animals. IUCN, Gland

Kappeler M (1984) The gibbon in Java. In: Preuschoft H, Chivers DJ, Brockelman W, Creel N (eds) The lesser apes: evolutionary and behavioral biology. Edinburgh University Press, Edinburgh, pp 19–31

Kierulff MCM, Procópio de Oliveira P, Beck BB, Martins A (2002) Reintroduction and translocation as conservation tools for golden lion tamarins. In: Kleiman DG, Rylands AB (eds) Lion tamarins: biology and conservation. Smithsonian Institution Press, Washington, DC, pp 271–282

Lande R (1988) Genetics and demography in biological conservation. Science 241:1455–1460

Malone NM, Fuentes A, Purnama AR, Adi Putra IMW (2004) Displaced hylobatids: biological, cultural, and economic aspects of the primate trade in Java and Bali. Indones Trop Biodivers 8(1):41–49

Ministry of Forestry and World Bank (2000) National forest inventory mapping. Jakarta

Mootnick AR (2006) Gibbon (Hylobatidae) species identification recommended for rescue or breeding centers. Primate Conserv 21:103–138

Mootnick AR, Groves C (2005) A new generic name for the hoolock gibbon (Hylobatidae). Int J Primatol 26(4):971–976

Nijman V (2004) Conservation of the Javan gibbon *Hylobates moloch*: population estimates, local extinctions, and conservation priorities. Raffles Bull Zool 52(1):271–280

Nijman V, Sözer R (1995) Recent observations of the grizzled leaf monkey (*Presbytis comata*) and extension of the range of the Javan gibbon (*Hylobates moloch*) in central Java. Trop Biodivers 3(1):45–48

Nijman V, van Balen B (1998) A faunal survey of the Dieng mountains, Central Java, Indonesia: status and distribution of endemic primate taxa. Oryx 32:145–146

Purwanto Y (1996) Studi habitat owa abu-abu (*Hylobates moloch* Audebert, 1798) di Taman Nasional Gunung Gede Pangrango Jawa Barat. BSc thesis, Jurusan Konservasi Sumber Daya Alam, Bogor Agriculture University, Bogor

Raharjo B (2003). Studi populasi dan analisis vegetasi habitat owa Jawa (*Hylobates moloch* Audebert, 1797) di Bedogol, Taman Nasional Gunung Gede-Pangrango, Jawa Barat. BSc thesis, Dept. Of Biology, University of Indonesia, Depok

RePPProt (1990) Land resources of Indonesia: a national overview. Overseas Development Administration (UK) and Dept. Of Transmigration, Jakarta

Rinaldi D (2003) The study of Javan Gibbon (*Hylobates moloch* Audebert) in Gunung Halimun National Park (distribution, population and behavior). In: Sakagushi N, ed. Research and conservation of biodiversity in Indonesia, vol XI: Research on endangered species in Gunung Halimun National Park. Bogor: JIKA Biodiversity Conservation pp 30–48

Roos C, Thanh Vu Ngoc, Walter L, Nadler T (2007) Molecular systematics of Indochinese primates. Vietnam J Primatol 1(1):41–53

Siedensticker J (1987) Bearing witness: observations on the extinction of *Panthera tigris balica* and *Panthera tigris sondaica*. In: Tilson RL, Seal US (eds) Tigers of the world: the biology, biopolitics, management, and conservation of an endangered species. Noyes, New Jersey, pp 1–8

Sjahfirdi L, Ramelan W, Yusuf TL, Supriatna J, Maheswari H, Astuti P, Sayuti D, Kyes R (2006a). Reproductive monitoring of captive-housed female Javan gibbon (*Hylobates moloch* Audebert,

1797) by serum hormone analyses. Proc. intl. assoc. Asia and Oceanic Society for comparative endocrinology, Bangkok, pp 365–370

Sjahfirdi L, Ramelan W, Yusuf TL, Supriatna J, Maheswari H, Astuti P, Sayuti D, & Kyes R. (2006b). Hormonal vaginal cytology of captive-housed female Javan gibbon (*Hylobates moloch* Audebert, 1797) by serum hormone analyses. Proc. intl. proc. intl. assoc. Asia and Oceanic Society for comparative endocrinology, Bangkok, pp 371–376

Sody HJV (1949) Notes on some primates, carnivora, and the babirusa from the Indo-Malayan and Indo-Australian regions. Treubia 20:121–126

Storm P, Aziz F, de Vos J, Kosasih D, Baskoro S, Ngallman, van den Hoek Ostende LW (2005) Late Pleistocene homo sapiens in a tropical rainforest fauna in East Java. J Hum Evol 49(4):536–545

Subekti I (2003) Populasi Owa jawa (*Hylobates moloch* Audebert, 1798) di Cagar Alam Gunung Simpang, Jawa Barat. Dept. of Biology, Padjadjaran University, Bandung

Sugardjito J, Sinaga MH (1999). Conservation status and population distribution of primates in Gunung Halimun National Park, West Java – Indonesia. In: Supriatna J, Manullang BO (eds) Proceedings of the international workshop on Javan Gibbon (*Hylobates moloch*): rescue and rehabilitation. Conservation International Indonesia and University of Indonesia, Jakarta, pp 6–12

Sugardjito J, Sinaga MH, Yoneda M (1997) Survey of the distribution and density of primates in Gunung Halimun National Park, West Java, Indonesia. In: Research and conservation of biodiversity in Indonesia, vol 2. The Inventory of Natural Resources in Gunung Halimun National Park, Bogor, pp 56–62

Supriatna J (2001) Primate conservation in the fragmented forest: case study in Indonesia. In: Supriatna J, Manansang J, Tumbelaka L, Andayani N, Seal U, Byers O (eds) Conservation assessment and management plan for the primates of Indonesia. Briefing book. Conservation Breeding Specialist Group (IUCN/SSC), Apple Valley, MN, p 516

Supriatna J, Hendras E (2000) Panduan lapangan primata Indonesia. Yayasan Obor Indonesia, Jakarta, p 332

Supriatna J, Manullang BO (1999) Proceedings of the international workshop on Javan Gibbon (*Hylobates moloch*): rescue and rehabilitation. Conservation International Indonesia and University of Indonesia, Jakarta

Supriatna J, Martarinza, Sudirman (1992) Sebaran kepadatan, dan habitat populasi lutung dan owa. Dept. Biologi, Universitas Indonesia, Depok

Supriatna J, Wahyono EH (2000) Panduan lapangan primata Indonesia. Yayasan Obor Indonesia, Jakarta

Supriatna J, Tilson RL, Gurmaya KJ, Manansang J, Wardojo W, Sriyanto A, Teare A, Castle K, Seal US (eds) (1994) Javan Gibbon and Javan Langur: population and habitat viability analysis report. IUCN/SSC Conservation Breeding Specialist Group (CBSG), Apple Valley, Minnesota

Supriatna J, Andayani N, Suryadi S, Leksono SM, Sutarman D, Buchori D (1998) Penerapan Genetika Molekular dalam upaya konservasi satwa langka: Studi kasus metapopulasi owa jawa (*Hylobates moloch*). Kantor Menteri Negara Riset dan Teknologi, Dewan Riset Nasional, Jakarta

Supriatna J, Andayani N, Forstner M, Melnick DJ (1999) A molecular approach to the conservation of the Javan gibbon (*Hylobates moloch*). In: Supriatna J, Manullang BO (eds) Proceedings of the international workshop on Javan Gibbon (*Hylobates moloch*): rescue and rehabilitation. Conservation International Indonesia and University of Indonesia, Jakarta, pp 25–31

Supriatna J, Manansang J, Tumbelaka L, Andayani N, Seal US, Byers O (2001) Conservation assessment and management plan for the primates of Indonesia. Briefing book. IUCN/SSC Conservation Breeding Specialist Group (CBSG), Apple Valley, MN

Suryanti T (2006) Ekologi Lansekap dalam manajemen dan konservasi habitat owa jawa (*Hylobates moloch* Audebert 1797) di Taman Nasional Gunung Halimun Jawa Barat. Thesis Doktor, Universtas Indonesia

Takacs Z, Morales JC, Geissmann T, Melnick DJ (2005) A complete species-level phylogeny of the Hylobatidae based on mitochondrial *ND3–ND4* gene sequences. Mol Phylogenet Evol 36:456–467

Thompson H, Cocks L (2008) International silvery gibbon studbook (*Hylobates moloch*)

Whitten T, Soeriaatmadja RE, Affif SA (1996) Ecology Java and Bali. Periplus Editions, Singapore

Wibisono HT (1995) Survei Populasi dan Ekologi Primata di Gunung Honje Taman Nasional Ujung Kulon. Yayasan Bina Sains Hayati Indonesia, Jakarta

Chapter 6
Siamang Socioecology in Spatiotemporally Heterogenous Landscapes: Do "Typical" Groups Exist?

Susan Lappan

Introduction

Tropical rainforest habitats are the most biologically diverse and spatially heterogeneous landscapes on earth, including a bewildering variety of life forms organized into a complex three-dimensional lattice. Sumatran forests contain enormous numbers of plant species, some of which are present at low densities, and the dominant tree species vary with elevation, soil structure, other landscape features, and geographic region (Whitten et al. 2000). Natural and anthropogenic disturbances produce a diversity of microhabitats within each general habitat type. Plant dispersal dynamics, interactions among plant species or individuals within rainforest habitats, and plant-animal interactions may also result in uneven distributions of plants across space (Condit et al. 2000; Silva and Tabarelli 2001). Therefore, each unit of area in a rainforest habitat contains a set of plants that differs at least slightly from that in the adjacent area. Accordingly, for territorial animal species that utilize plant resources, the actual and relative availability of specific plant foods in a given month may vary substantially between territories, even among neighboring groups.

Most primates consume parts of only a subset of the plant species found in their home ranges (Oates 1987; Ungar 1995). The availability of preferred plant foods in a habitat depends on the local densities of plant food species, but will also vary over time, as many plant parts important to herbivores (e.g., fruits, flowers, and new leaves) are produced only during some weeks or months of the year and may not be produced every year (Cannon et al. 2007). Plant phenological cycles differ among species and habitats. Some plant species produce fruits, flowers, and new leaves at fairly predictable intervals determined by local rainfall, day length, or temperature regimes, whereas others produce new leaves or reproductive parts at regular or irregular intervals that are not associated with seasonal climatic variation, and that may or may not be coordinated among individual plants within a species

S. Lappan (✉)
Department of Anthropology, Appalachian State University, Boone, NC, 28608, USA
e-mail: lappan@nyu.edu

S. Gursky-Doyen and J. Supriatna (eds.), *Indonesian Primates*,
Developments in Primatology: Progress and Prospects,
DOI 10.1007/978-1-4419-1560-3_6, © Springer Science+Business Media, LLC 2010

(Sakai et al. 1999; Whitten et al. 2000; Sakai 2001). Variation in overall food availability (i.e., the availability of all potential food species) in an area may follow roughly seasonal patterns in geographic areas displaying pronounced seasonal variation in rainfall, but in Sumatran and Bornean forests characterized by long periods of relatively high rainfall interspersed with periodic droughts associated with El Niño-Southern Oscillation (ENSO) events and mast fruiting events, interannual climatic variation may be more pronounced than seasonal variation, and temporal variation in food availability may be dramatic, yet unpredictable (Bawa 1983; Sakai et al. 1999; Wich and Van Schaik 2000; Kinnaird and O'Brien 2005; Cannon et al. 2007).

Natural selection should favor behavioral flexibility in primates living in complex, unpredictable landscapes. The diets of rainforest primates generally include many food species and often show substantial temporal and spatial variation (e.g., Cords 1986; Palombit 1997; Knott 1998; Chapman et al. 2002). The effects of spatial variation can be profound, even across relatively short distances. For example, Chapman et al. (2002) found that a group of red colobus in Kibale National Park had higher dietary overlap with a group of black-and-white colobus that shared their home range than with a neighboring red colobus group, even though the home ranges of the two red colobus groups had substantial overlap. Nonetheless, researchers often use studies of a single primate group to characterize the behavior of a species in a particular habitat type (e.g., Whitten 1982a; Kappeler 1984; Sangchantr 2004; Riley 2008; Silva and Ferrari 2008).

Gibbons (Hylobatidae) are generally described as monogamous, territorial, and frugivorous (Leighton 1987; Bartlett 2007), but as the number of gibbon groups studied has increased, it has become clear that gibbons are somewhat flexible in their grouping and mating patterns (Reichard 1995, 2009; Fuentes 1999, 2000, 2002; Jiang et al. 1999; Lappan 2007a, b; Malone 2007), territorial and ranging behavior (Whitten 1982b; Chivers 1984; Jiang et al. 1999; Bartlett 2007; Fan and Jiang 2008; Fan et al. 2008), and diets (Palombit 1997; Bartlett 2007; Fan et al. 2008; Elder 2009; Malone and Fuentes 2009). Even within a single neighborhood, home ranges may vary in size and quality (O'Brien et al. 2003; Savini et al. 2008). These differences matter. Variation in reproductive success and survivorship among gibbons of the same species living in different habitat types within a single landscape can be pronounced (O'Brien et al. 2003; Marshall 2009), and perhaps as a result, local gibbon densities often vary substantially between habitat types (O'Brien et al. 2004; Marshall 2009; Yanuar 2009). Nonetheless, for practical reasons, many studies of wild gibbon behavior have focused on only one or two groups of a given taxon, and some have followed each study group for 1 year or less (e.g., Raemaekers 1977; Gittins 1980; MacKinnon and MacKinnon 1980; Whitten 1982a; Kappeler 1984; Bartlett 1999; McConkey et al. 2002; Malone 2007; Fan and Jiang 2008; Fan et al. 2009). While these studies have offered remarkable insights about gibbon behavior, the limited sample sizes make it difficult to estimate the extent of behavioral variation within gibbon populations and species and to identify the underlying causes for this variation. To date, few studies have directly addressed the question of whether conspecific gibbon groups living in the same neighborhoods and habitat types display broadly similar activity patterns and diets, and therefore whether it is in fact appropriate to extrapolate from one or two groups to characterize an entire population or species.

In this chapter, I use behavioral data collected from five wild siamang (*Symphalangus syndactylus*) groups from the Way Canguk Research Area in Lampung, Sumatra, between October 2000 and August 2002 to examine patterns of temporal and spatial variation in gibbon behavior among a set of groups in a single local population. My central goal is to determine whether a study sampling the behavior of just one or two groups in this neighborhood could produce a meaningful estimate of the general patterns of behavior of the population of animals in the neighborhood as a whole. I also specifically examine temporal variation in siamang behavior by making longitudinal comparisons of the same groups across study months and years, and by comparing patterns of temporal variation in activity patterns and diets among neighboring groups. If seasonal variation in food availability associated with climatic cycles is the most important determinant of siamang diets, then all five groups should display broadly similar patterns of temporal variation in their diets and activity patterns. Conversely, if local or individual factors such as the densities and fruiting phenologies of rare tree species, group size or composition, female reproductive status, or the local densities of competitors have pronounced effects on the activity patterns of siamangs, then patterns of temporal variation in siamang diets may follow different trajectories in the neighboring groups.

Methods

Study Area and Siamang Population

The Way Canguk Research Station (5° 39′ 32″ S, 104° 24′ 21″ E) is located in the southern part of the Bukit Barisan Selatan National Park in Lampung Province, Sumatra, and is run by the Wildlife Conservation Society-Indonesia Program (WCS-IP) and the Indonesian Ministry of Forestry's Department of Forestry and Nature Conservation (PHKA). The trail system in the Research Area covers an area of 1,000 ha of lowland forest (elevation 50 m) and is bisected by the Canguk River. The study area consists of primary forest interspersed with forest damaged by drought, wind throws, earthquakes, illegal tree felling, and fire (O'Brien et al. 2003; Kinnaird and O'Brien 2005), and is contiguous with large areas of undisturbed primary forest as well as areas disturbed by human activities.

Annual rainfall at Way Canguk in most years is 3,000–4,000 mm/year, but rainfall as low as 1,600 mm/year has been recorded in drought years associated with ENSO events (Kinnaird and O'Brien 2005). Rainfall is weakly seasonal, with a short dry season typically occurring between June and September. Mean monthly rainfall from 1998 to 2002 was 287 mm/month, and mean monthly rainfall in the dry season was >60 mm/month (Kinnaird and O'Brien 2005). From October 2000–August 2002, when this study was conducted, mean monthly rainfall was

284 mm/month, and rainfall exceeded 100 mm every month (Kinnaird and O'Brien 2005). Tree species diversity in the study area is high: Kinnaird and O'Brien (2005) identified 365 distinct species of trees in 49 families in plots encompassing a total of 5 ha. The overall availability of fruits in the study area varies substantially over time, but Kinnaird and O'Brien (2005) did not find a strong pattern of seasonal variation in the availability of plant reproductive parts from 1997 to 2002, or a significant relationship between monthly rainfall and fruit availability in the study area.

In 1997, approximately 165 ha of the Research Area were damaged by fires of human origin during a drought associated with an ENSO event (Kinnaird and O'Brien 1998; O'Brien et al. 1998). While siamangs inhabited the fire-damaged areas prior to the fires, researchers from WCS-IP did not detect siamangs in the burned areas a month after the fires, suggesting that siamangs either fled the area or were killed (Kinnaird and O'Brien 1998; O'Brien et al. 1998). However, in subsequent years, some siamang groups have included fire-damaged areas in their home ranges (O'Brien et al. 2003), although no home range is comprised entirely of fire-damaged forest.

The siamang population at Way Canguk has been censused annually by WCS-IP since 1998 (O'Brien et al. 2003, 2008), generating an unparalleled demographic data set. The home ranges of 36–37 groups are found within or partially within the study area. Habituation of siamang groups in the study area was initiated in 1998 (Nurcahyo 1999), and at the time of writing, over ten siamang groups have been habituated to human observers.

I originally selected the five groups used for this study for a study of parenting behavior (Lappan 2008, 2009b); the feeding and ranging data used in this study were collected concurrently with parental care data. Thus, the study groups are more similar than would be expected in a random sample of groups in the study area because each group contained a young infant. Otherwise, the groups display a diversity of characteristics that appears to be representative of the groups in the study area. The compositions of the study groups and their approximate home ranges are shown in Fig. 6.1. The entire Way Canguk study area is embedded in siamang home ranges except in areas damaged by fire in 1997. Home ranges of four out of five groups consisted entirely of healthy forest, whereas one group (group C) had a home range that included fire-damaged areas as well as healthy forest. Groups B, C, F, and G have contiguous home ranges, while the home range of group A is approximately 1.5 km away, on the opposite side of the Canguk River (Fig. 6.1). However, the home range of group A is also surrounded by siamang home ranges, and is therefore comparable to those of the other study groups. The size of the study groups at the conclusion of the study ranged from three to six individuals, and four out of five groups contained two or more adult males during some part of the study period. Groups A, B, and C included two adult males that copulated with the adult female, and genetic data suggested that neither male was the offspring of the female, whereas both genetic and behavioral data suggested that group F was probably a socially monogamous group with a retained adult male offspring (Lappan 2007a).

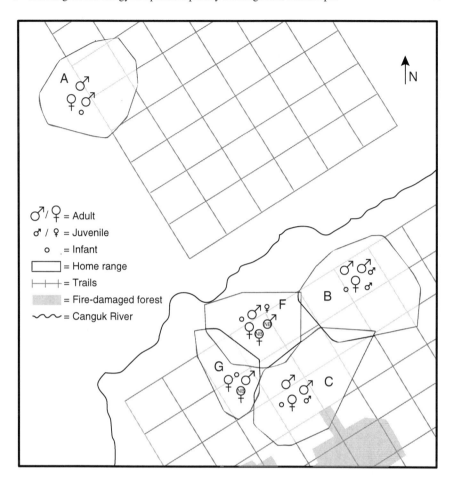

Fig. 6.1 Home ranges and compositions of groups A, B, C, F, and G. "NR" indicates adult-sized individuals that are believed to have been non-reproductive in their groups. Females were designated as "non-reproductive" based on nipple condition and group histories. Males were designated as "non-reproductive" if their mitochondrial DNA haplotype matched that of the group's adult female and they were never observed copulating (mtDNA data from Lappan 2007a)

Behavioral Data Collection

Behavioral data were collected by a team including three field assistants and myself during all-day follows of a focal adult. I selected adults as focal individuals on a rotating basis, and each group was followed until each adult had served as a focal individual for 2 days. The study groups were followed on a rotating basis. A team of two observers collected instantaneous samples of behavioral data at 5-min intervals from a focal adult, recording the focal animal's activity, food type (i.e., fig fruits, nonfig fruits, leaves, flowers, insects) and stage of ripeness (ripe or unripe fruits, new or mature leaves), and food species. Activities were classified as resting, feeding (reaching for, handling, chewing, or swallowing food), within-crown movement (movement within a feeding tree), travel (all other types of travel), social activities (e.g., vocalization, social grooming, aggression, social play, copulation), and other (e.g., urination, defecation, drinking). As all types of movement will incur energy costs, I subsequently grouped movement within a feeding tree (often referred to as "foraging") with other types of travel for analyses. One observer paced below the focal animal to estimate daily path lengths (DPL). We also recorded the location of the focal individual within the trail system opportunistically (i.e., when it did not interfere with other research activities) at 5-min intervals from October 2000 to December 2001, and systematically at 15-min intervals from January to August 2002. The approximate home range maps in Fig. 6.1 were derived from these data. We initiated behavioral data collection upon the departure of the focal adult from the sleeping site or upon the first encounter of the observers with the focal adult if sleeping site departure was not observed and collected data until the focal adult entered the subsequent sleeping site. Behavioral data were collected from group B starting in October 2000, from groups A and C from November 2000, from group F from February 2001, and from group G from May 2001. Behavioral data were collected from all groups until August 2002.

Data Analysis

I included data from all adult group members for all analyses. Percentages of time that groups spent engaging in specific behaviors or consuming specific food items were estimated as the percentage of instantaneous samples in which the relevant behavior was observed. Proportional data were arcsine square-root transformed prior to the use of parametric statistical tests (Zar 1996). For statistical analyses of individual activity data, I calculated daily means of hourly rates, excluding days from which fewer than 5 h of data were available, whereas for analyses of diet data, I simply calculated daily means and included all available data. Then, I grouped the data into 2-month blocks to maximize the comparability of the data among groups.

To examine variation in general activity patterns (i.e., the percentages of time spent in each activity and DPL) and plant parts eaten, I used repeated-measures General Linear Models (GLM) with mean values for the activity variable for each 2-month interval for which data from all groups were available as a repeated measure, and groups as subjects. The within-subjects effect was used to examine the effects of month-block on behavior, while the between-subjects effect was used to examine variation between study groups. As DPL and the time spent traveling on the same day were highly correlated (group A ($r_s = 0.651$, $N=44$, $p < 0.001$), B ($r_s = 0.432$, $N = 53$, $p = 0.001$), C ($r_s = 0.461$, $N = 58$, $p < 0.001$), F ($r_s = 0.639$, $N = 32$, $p < 0.001$), G ($r_s = 0.642$, $N = 39$, $p < 0.001$), I present only the analyses of DPL data. For the analysis of feeding time, I also conducted separate ANOVA analyses for each group to examine whether daily mean values varied significantly among 2-month blocks.

To assess variation in activity patterns between study years, I conducted separate analyses for each group using a multivariate General Linear Model with the monthly mean proportions of time spent feeding, resting, traveling, and socializing as variables, and study year (year 1 was October 2000–September 2001, year 2 was October 2001–August 2002) as a factor. Comparisons between years were only conducted for groups A, B, and C, from which almost two full years of data were available. Where the multivariate analysis indicated significant differences, separate analyses were conducted for each behavioral variable to identify variables contributing to the overall significance of the model.

Insects comprised <1% of the diet of each group, so insect-feeding time was not included in the dietary analyses. I calculated the annual diets for each group as the mean of means for each 2-month block of daily mean proportion of time spent feeding on each plant part or plant species between May 2001 and April 2002 for comparisons between groups, or between October 2000 and September 2001 (year 1) and October 2001 and August 2002 (year 2) for comparisons of the same groups in different years. To estimate the similarity in the diets of pairs of study groups, I calculated dietary overlap by summing the percentage overlap for each plant part and species (Holmes and Pitelka 1968). I calculated actual percentages of feeding time for plant parts from the 26 plant species that comprised at least 1% of the diet of at least one group, but grouped all nonfig lianas and treated them as a single taxon for each plant part, as I had difficulty in identifying many lianas. Other plant foods that could not be identified to the species or that comprised <1% of the diet of all five groups were grouped together into the category "other." "Other" foods comprised 2–10% of siamang diets. Therefore, this analysis will tend to slightly underestimate the actual differences in diet between pairs of groups or between years.

I tested the hypothesis that pairs of groups display similar patterns of temporal variation in activity patterns and diets for each pair of groups using Spearman correlation analysis of the mean percentage of time spent engaging in an activity, the mean DPL, or the mean percentage of feeding time spent eating a specific plant part or species in each 2-month block for each group. Analyses were one-tailed, as the hypothesis that seasonal variation in food availability is the primary predictor of siamang behavior predicts a positive correlation between the activity patterns of pairs of study groups. As I conducted ten separate pair-wise comparisons for each analysis, which substantially increases the risk of type I error, I considered sets of results in which two or fewer comparisons resulted in significant positive correlations as showing no evidence that the variation in the parameter was correlated among the study groups, whereas sets of results in which three or more comparisons resulted in significant positive correlations were considered to indicate that variation in the given parameter displayed a stronger correlation between groups than would be expected by chance.

To examine the effects of intergroup variation on calculations of siamang time budgets and the plant part compositions of siamang diets in samples including different numbers of groups, I compared estimates for each group based on annual means (of bimonthly means of daily means) with an estimate based on the annual grand mean of the mean values for each 2-month block for a sample including all five study groups. When the mean difference between each group and the whole sample for any variable was >3%, I made additional comparisons for sets of five random subsamples that included two or three of the study groups.

Results

Intergroup Variation in General Activity Patterns

There was a significant effect of group on the mean proportion of time spent feeding (between-subjects effect: $F_{1,4} = 1{,}130.324$, $p < 0.001$), resting ($F_{1,4} = 1{,}108.308$, $p < 0.001$), and socializing ($F_{1,4} = 700.581$, $p < 0.001$), but the effect of month-block did not reach statistical significance for any variable, although it approached significance for feeding (within-subjects effect: feeding: $F = 3.790$, adjusted df$=2.276$, $p = 0.060$; resting: $F = 1.169$, adjusted df$=1.675$, $p = 0.355$; socializing: $F = 1.364$, adjusted df$=2.105$, $p = 0.308$). In separate longitudinal analyses for each group, feeding time differed significantly among study months (in 2-month blocks) for most study groups (group A: $F_{9,66} = 1.970$, $p = 0.060$; group B: $F_{10,80} = 2.847$, $p = 0.005$; group C: $F_{10,76} = 3.166$, $p = 0.002$; group F: $F_{8,56} = 2.674$, $p = 0.016$; group G: $F_{7,41} = 1.206$; $p = 0.326$).

While variation in activity patterns between groups was statistically significant, the actual differences in annual means were very low (Fig. 6.2a). For example, the mean difference between a single group and the five groups combined was only 2% for feeding time, 3% for resting time, and 1% for social time.

The mean daily path length differed significantly between study months (within-subjects effect: $F_{1,4} = 14.107$, adjusted df$=1.515$, $p = 0.007$), and between groups (between-subjects effect: $F = 214.063$, df$=1$, $p < 0.001$). Mean daily path lengths were highest for groups A (mean$=1{,}288 \pm 78$ m) and C (mean$=1{,}227 \pm 44$ m), and lowest for groups F (mean$=1{,}088 \pm 64$ m), B (mean$=1{,}068 \pm 51$ m), and G (mean$=1{,}067 \pm 65$ m).

The proportion of time spent eating nonfig fruits and flowers differed between groups (between-subjects effect: nonfig fruits: $F_{1,4} = 72.198$, $p = 0.001$; flowers: $F_{1,4} = 133.236$, $p < 0.001$) and varied systematically among time periods (within-subjects effect: nonfig fruits: $F = 23.063$, adjusted df$=2.152$, $p < 0.001$; flowers: $F = 6.686$, adjusted df$=2.220$, $p = 0.015$). The proportion of time spent eating fig fruits and leaves also differed significantly between groups (fig fruits: $F_{1,4} = 48.108$, $p < 0.001$; leaves: $F_{1,4} = 2{,}154.360$, $p < 0.001$), but did not vary systematically among time periods (fig fruits: $F = 1.625$, adjusted df$=2.099$, $p = 0.254$; leaves: $F = 3.972$, adjusted df$=1.913$, $p = 0.067$).

Differences between groups in the proportion of time spent eating each plant part were relatively modest in most cases (Fig. 6.2b). The mean value for each group differed from the grand mean for all five groups by a mean of 7% for nonfig fruits, 5% for fig fruits, 4% for leaves, and 3% for flowers, whereas the grand mean for all five groups differed from the annual mean value for five randomly-chosen subsamples of two groups by a mean of 6% for nonfig fruits, 3% for fig fruits, 3% for leaves, and 2% for flowers, and from the annual mean value for five randomly chosen subsamples of three groups by 3% for nonfig fruits, 2% for fig fruits and leaves, and 1% for flowers.

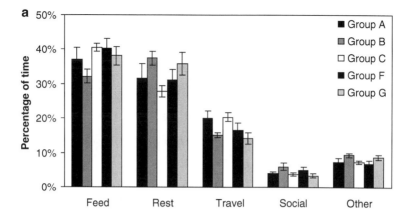

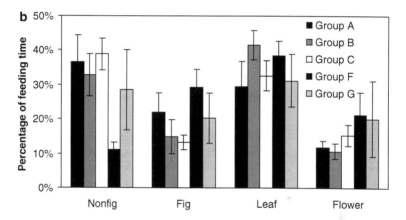

Fig. 6.2 (**a**) Mean ± SE percentage of time spent feeding, resting, traveling, engaging in social activities, and engaging in other activities by adults from May 2001 to April 2002. (**b**) Mean ± SE percentage of feeding time spent eating each type of plant food from May 2001 to April 2002. Insect feeding time (not shown) comprised <1% of feeding time for each group

The five most-frequently consumed food species for the set of five groups each comprised at least 1% of the annual diet of each study group (Table 6.1). However, the actual proportion of the diet consisting of plant parts from each species varied dramatically among the study groups. For example, all five groups ate substantial quantities of *Ficus* fruits, the most important food for the population, and *Dracontomelon dao* fruits (Table 6.1). However, *Clausena anisum-olens* and *Alangium javanicum* were among the most important foods for group G, but comprised <1% and <3% of the diets of all other study groups, respectively (Table 6.1).

Dietary overlap between groups ranged from 52% (groups A and G) to 74% (groups B and C). Mean overlap with all other groups was 60% for group G, 61% for group A, 63% for group F, and 65% for groups B and C.

Table 6.1 Top plant species (excluding nonfig lianas) for siamangs at Way Canguk, defined as plant species or genera comprising >1% of the diets of the five groups combined. Diets were calculated by first averaging the percentages of time spent eating each food for each 2-month block (e.g., Jan–Feb, Mar–Apr) in each study year for each group, and then averaging the means produced in the 2 years for each 2-month block, and then calculating an overall mean ± SE. %FT indicates percentage of feeding time ± SE. Only species comprising more than 1% of the diets of the combined average are shown, so species comprising <1% of the diet for the sample, but >1% of the diet for a specific group are not shown. Ranks were calculated for each group using all feeding data, including species not shown

Overall rank	Species	Part	Group A Rank	Group A % FT	Group B Rank	Group B % FT	Group C Rank	Group C % FT	Group F Rank	Group F % FT	Group G Rank	Group G % FT
1	Fi	Fruit	2	18.1±5.7	2	16.2±5.2	2	15.8±2.9	1	42.6±7.4	1	22.0±5.5
		Leaves		5.1±2.4		3.2±1.3		4.9±1.1		6.0±2.3		7.5±3.0
2	DD	Fruit	1	33.6±9.7	1	21.0±4.8	1	17.4±5.0	4	4.4±2.6	7	3.7±2.0
3	HG	Flowers	6	1.9±1.2	5	4.3±1.2	3	6.4±2.9	2	10.8±4.1	2	14.2±11.0
4	Po	Leaves	3	0.9±0.4	4	3.5±2.4	4	2.6±1.2	3	3.1±0.7	8	1.6±1.0
		Fruit		2.0±1.0		1.3±0.8		2.3±1.3		1.2±0.5		2.9±1.5
		Flowers		0.1±0.1		0		0.4±0.3		0.1±0.1		0.1±0.1
5	MP	Leaves	5	1.4±0.5	6	3.2±1.9	6	2.7±1.2	5	2.0±0.6	6	2.0±0.8
		Flowers		2.4±2.0		0		0.7±0.4		0.3±0.3		0.6±0.6
		Fruit		0.4±0.3		0.3±0.2		0.6±0.5		0.4±0.3		0.4±0.3
6	AJ	Fruit			11	1.7±1.7	7	2.6±2.6			3	10.9±10.9
7	CA	Fruit									4	10.4±10.4
8	SB	Leaves					11	1.5±1.5			5	4.3±4.0
9	AT	Fruit				0.5±0.5	5	3.3±1.7	9	0		
		Leaves				0.2±0.2		0.9±0.8		1.6±1.1		
		Flowers				0		0.7±0.6		0		
10	MC	Flowers	7	1.0±1.0		0		0		0.4±0.4	10	0
		Leaves		0.4±0.3		0.5±0.4		1.2±0.5		2.0±0.5		1.2±0.8
11	VQ	Leaves			7	3.4±0.9	10	1.8±0.6				
12	Kn	Fruit			3	6.3±6.3						
13	XN	Fruit					9	1.9±1.2			9	2.1±1.3
14	Pa	Fruit			8	2.7±2.1	8	2.0±1.4				

AJ, *Alangium javanicum* (Alangiaceae); AT, *Antiaris toxicaria* (Moraceae); CA, *Clausena anisum-olens* (Rutaceae); DD, *Dracontomelon dao* (Anacardiaceae); Fi, *Ficus* spp. (Moraceae); HG, *Hydnocarpus gracilis* (Flacourtiaceae); Kn, *Knema* sp. (Myristicaceae); MC, *Michelia champaca* (Magnoliaceae); MP, *Mitrephora polypirena* (Annonaceae); Pa, *Payena* sp. (Sapotaceae); Po, *Polyalthia* spp. (Annonaceae); SB, *Stelechocarpus burahol* (Annonaceae); VQ, *Vitex quinata* (Verbenaceae); XN, *Xerospermum noronhianum* (Sapindaceae).

Seasonal Variation

The percentages of time spent feeding and resting in a given 2-month period were significantly positively correlated for three out of ten pairs of groups, and the correlation coefficients were positive for nine out of ten pairs (Table 6.2). However, the mean correlation coefficient was low for both variables (0.401 for feeding, 0.299 for resting). Only two of ten pairs had significant correlations in their social time, and the correlation coefficients were only positive for six out of ten pairs (and zero for a seventh).

There were significant or near significant correlations between groups in the mean DPL in a given 2-month block for six out of ten pairs of groups (Table 6.2). However, for analyses involving group C, no correlations were significant, and the correlations coefficients were consistently low or negative.

There were strong positive correlations among groups in the proportion of feeding time spent eating nonfig fruits in a given 2-month period, with an average correlation coefficient of 0.738, and significant correlations for nine out of ten possible pairs of groups (Table 6.3). However, much of this effect appears to be due to heavy reliance on fruits of the species *Dracontomelon dao*, which is only available during some months of the year, by all five groups (mean correlation coefficient = 0.670; Table 6.3; Fig. 6.3a). When *D. dao* fruit-feeding time was excluded from the analysis, the mean correlation coefficient for nonfig fruit-feeding time decreased to 0.400, and the correlation was significant only for two out of ten pairs of groups, a pattern that does not differ from that expected by chance, although patterns of nonfig fruit-feeding for several additional species displayed fairly consistent patterns of seasonal variation among the study groups (Fig. 6.3b, c).

There was a significant positive correlation between two pairs of groups in the proportion of time spent eating fig fruits in a given study month, but five out of ten correlation coefficients were negative, suggesting that fig fruit-feeding time does not vary in a consistent manner for the population as a whole (Table 6.3). Similarly, the proportion of time spent eating leaves in a given month was only significantly positively correlated for one pair of groups (Table 6.3). This suggests that temporal variation in patterns of leaf consumption is not strongly correlated among siamang groups at Way Canguk.

The proportion of time spent eating flowers in a given month was positively correlated for each pair of groups, but the correlations were only significant for two out of ten groups and the correlation coefficients were relatively low (mean = 0.481), suggesting that there is no strong seasonal pattern of variation in flower-feeding in this population. However, if only the most important flower species, *Hydnocarpus gracilis*, was considered, the correlations in flower-feeding time between groups were significant and positive for five out of ten pairs of groups (Table 6.3). Group A spent substantially less time eating *H. gracilis* flowers than the other four study groups, and was not observed during the month when *H. gracilis* feeding time was highest in the four other groups. If group A is excluded from the analysis, then the average correlation coefficient was 0.777.

Table 6.2 Correlations between groups in the proportion of time in each 2-month period spent feeding, resting, and engaging in social activities. *Asterisks* (*) indicate significant correlations (whether positive or negative). *Shading* indicates significant positive correlations

Below diagonal	Group	A	B	C	F	G	Above diagonal
Resting time	A		0.550 (9)	0.709 (10)	0.600 (9)	0.036 (7)	Feeding time
r_s (n)							
p			0.062	0.011*	0.044*	0.470	
r_s (n)	B	0.683 (9)		0.721(10)	0.024 (8)	0.643 (7)	
p		0.021*		0.009*	0.478	0.060	
r_s (n)	C	0.467 (10)	0.467 (10)		0.433 (9)	0.476 (8)	
p		0.087	0.087		0.122	0.116	
r_s (n)	F	0.583 (9)	0.119 (8)	0.083 (9)		-0.179 (7)	
p		0.050*	0.389	0.416		0.351	
r_s (n)	G	-0.250 (7)	0.000 (7)	0.810 (8)	0.036 (7)		
p		0.294	0.500	0.007*	0.470		
Social time	A		0.821 (7)	0.143 (8)	0.771(6)	0.929(7)	DPLv
r_s (n)							
p			0.012*	0.368	0.036*	0.001*	
r_s (n)	B	0.667 (9)		0.300 (9)	0.900 (5)	0.750 (7)	
p		0.025*		0.216	0.018*	0.026*	
r_s (n)	C	0.479 (10)	0.685 (10)		-0.086 (6)	0.048 (8)	
p		0.081	0.014*		0.436	0.456	
r_s (n)	F	-0.717 (9)	-0.738 (8)	-0.367 (9)		0.943 (6)	
p		0.015*	0.018*	0.166		0.002*	
r_s (n)	G	0.607 (7)	0.286 (7)	0.500 (8)	0.000 (7)		
p		0.074	0.267	0.104	0.500		

Table 6.3 Correlations between groups in the proportion of time in each 2-month period spent feeding on specific plant parts or species. *Asterisks* (*) indicate significant correlations (whether positive or negative). *Shading* indicates significant positive correlations

Below diagonal		Group	A	B	C	F	G	Above diagonal
Fig fruits	r_s (n)	A		0.767 (9)	0.745 (10)	0.950 (9)	0.750 (7)	Nonfig
	p			0.008*	0.007*	<0.001*	0.026*	fruits
	r_s (n)	B	−0.033 (9)		0.612 (10)	0.738 (8)	0.571 (7)	
	p		0.466		0.030*	0.018*	0.090	
	r_s (n)	C	0.321 (10)	0.261 (10)		0.733 (9)	0.690 (8)	
	p		0.183	0.234		0.012*	0.029*	
	r_s (n)	F	−0.333 (9)	0.738 (8)	0.717 (9)		0.821 (7)	
	p		0.190	0.018*	0.015*		0.012*	
	r_s (n)	G	0.214 (7)	−0.179 (7)	−0.238 (8)	−0.107 (7)		
	p		0.322	0.351	0.285	0.410		
Flowers	r_s (n)	A		0.083 (9)	−0.261 (10)	−0.683 (9)	0.179 (7)	Leaves
	p			0.416	0.234	0.021*	0.351	
	r_s (n)	B	0.356 (9)		0.818 (10)	0.452 (8)	0.500 (7)	
	p		0.174		0.002*	0.130	0.127	
	r_s (n)	C	0.321 (10)	0.767 (10)		0.367 (9)	0.429 (8)	
	p		0.183	0.005*		0.166	0.145	
	r_s (n)	F	0.417 (9)	0.635 (8)	0.517 (9)		0.000 (7)	
	p		0.132	0.045*	0.077		0.500	
	r_s (n)	G	0.393 (7)	0.500 (7)	0.619 (8)	0.286 (7)		
	p		0.192	0.127	0.051	0.267		
D. dao	r_s (n)	A		0.252 (9)	−0.023 (10)	0.098 (9)	−0.304 (7)	H. gracilis
	p			0.256	0.475	0.401	0.253	
	r_s (n)	B	0.950 (9)		0.332 (10)	0.919 (8)	0.788 (7)	
	p		<0.001*		0.174	0.001*	0.018*	
	r_s (n)	C	0.830 (10)	0.745 (10)		0.848 (9)	0.873 (8)	
	p		0.001*	0.007*		0.002*	0.002*	
	r_s (n)	F	0.577 (9)	0.714 (8)	0.410 (9)		0.900 (7)	
	p		0.052	0.023*	0.137		0.003*	
	r_s (n)	G	0.668 (7)	0.611 (7)	0.791 (8)	0.401 (7)		
	p		0.050*	0.073	0.010*	0.186		

Inter-annual Differences

Time budgets did not differ significantly between study years for group A ($F_{4,8}$=2.474, p=0.128) and group B ($F_{4,10}$=1.089, p=0.413), but group C allocated its time significantly differently in the second year of the study than in the first year ($F_{4,11}$=4.660, Wilks's Lambda=0.371, p=0.019; Fig. 6.4). Analysis of each individual dependent variable, using a Bonferroni-adjusted significance level of $p<0.0125$,

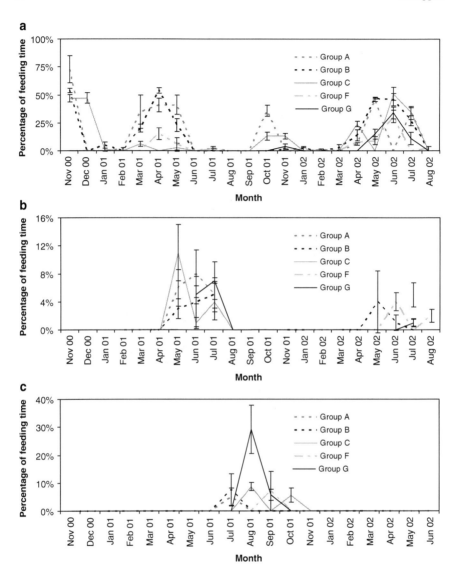

Fig. 6.3 Mean ± SE percentage of feeding time that each group spent feeding on fruits of three seasonally-important plant species in each month: (**a**) *Dracontomelon dao*; (**b**) *Diospiros macrophylla*; (**c**) *Xerospermum noronhianum*. Data from group F were only available from February 2001-August 2002, and data from group G were available from May 2001-August 2002

showed a significant effect of resting ($F=9.876$, df$=1$, $p=0.007$), but not of feeding ($F=4.660$, df$=1$, $p=0.049$), traveling ($F=5.229$, df$=1$, $p=0.038$), or social activities ($F=0.198$, df$=1$, $p=0.663$) on overall time allocation between the 2 years, although the results for feeding and travel approached significance. Mean DPL did not differ significantly between study years for any group (group A: $F_{1,10}=0.094$, $p=0.766$; group B: $F_{1,12}=0.634$, $p=0.443$; group C: $F_{1,13}=0.353$, $p=0.564$).

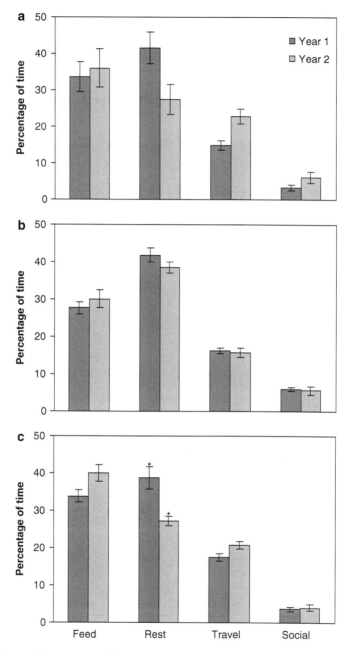

Fig. 6.4 Mean ± SE percentage of time that groups **a** (top), **b** (center), and **c** (bottom) spent feeding, resting, and engaging in social activities, in the first and second years of the study

Dietary overlap for the same groups in the two study years (October 2000–August 2001 vs. September 2001–August 2002) was 66% for group A, 60% for group B, and 68% for group C. The most pronounced differences in diets between years involved some of the most commonly consumed food items. For example, group B ate 7.5% less *Dracontomelon dao* and 8.9% more *Ficus* fruit in the second year of the study than in the first year, and group C ate 6.1% less *Ficus* fruit, 4.3% less *Polyalthia* fruit, and 5.3% more *Alangium javanicum* in the second year of the study than in the first year. However, rare items also had an impact: 6.3% of the diet of group B in the first year of the study comprised fruits from a single *Knema* sp. individual that did not produce fruit in the second year of the study.

Discussion

The behavior of siamangs at Way Canguk varied along several dimensions. Inter-group differences in activity patterns from May 2001 to April 2002 were significant, but modest in scale (Fig. 6.2), and despite variation in group size and home range size, no group was an extreme outlier in terms of their time budgets, daily path lengths, general patterns of use of plant parts, or patterns of plant species exploitation. Mean proportions of time spent feeding, resting, and socializing calculated over the course of an annual cycle for any group differed from average values of the five groups taken together by <3%. The mean proportions of time spent eating nonfig fruits, fig fruits, leaves, and flowers also varied among groups, but again, the differences were relatively small. The mean differences between the averages for all five groups and the mean values for individual groups ranged from 3 to 7%; subsamples including two groups did not show a markedly improved performance relative to single groups, but subsamples of three groups had mean values for these variables that differed from the grand mean by <3% on average for each variable. If these results are typical, they should reassure researchers that are forced to extrapolate from studies of very small numbers of groups or individuals to draw conclusions about the time budgets and ranging patterns of gibbons living in a specific habitat type. However, caution is warranted. A large, careful study of siamang demography at Way Canguk showed that siamangs in fire-damaged habitat suffer substantially higher infant and juvenile mortality than siamangs in healthy habitat (O'Brien et al. 2003). In this study, only the home range of group C included fire-damaged habitat. However, the activity patterns of group C and the plant part composition of group C's diet were similar to those of other groups, and the dietary overlap of group C with other groups was on the high end of the range of variation observed in this study. O'Brien et al.'s (2003) also noted that the home range of group C is of substantially higher quality than those of most other groups with home ranges including or adjacent to fire-damaged habitat. In the absence of information from O'Brien et al.'s study, my results might be interpreted as evidence that siamangs in fire-damaged habitat had similar diets and activity patterns to those in healthy habitat, which is clearly incorrect. This example clearly illustrates the desirability of sampling multiple groups in a specific habitat type whenever possible.

While the plant part compositions of siamang diets were broadly similar for all five study groups, the results suggested substantial spatial heterogeneity in the species composition of siamang diets. The five most important plant taxa together comprised about half the diet of each group, but the order of importance of some of these plants differed among the study groups, and plant taxa that were rarely or never consumed by some groups made substantial contributions to the diet of others (Table 6.1). These results suggest that a study of only one or a few groups is likely to substantially misrepresent the diets of a lowland rainforest population. Indeed, it is likely that the results from my sample of five groups only roughly approximate the actual diets of the whole population of siamangs in the study area.

While the effects of spatial heterogeneity within the Way Canguk research area on general activity patterns of siamangs appear to be relatively small, the results of this study highlight the potentially profound effects of temporal variation in food availability on siamang behavior, as well as the potential for interactions between spatial and temporal heterogeneity. The diets of groups A, B, and C differed as much in the 2 years of the study as they differed from those of other groups in the same year. There was also significant variation between 2-month periods in most behavioral parameters measured, including feeding time, DPL, the nonfig fruit and flower components of diets, and the species composition of diets, and temporal variation in siamang activity patterns and diets was not strongly correlated between groups for most variables. Only variation in time spent eating nonfig fruits and DPL (which are themselves correlated: Lappan 2009a) were strongly correlated between most pairs of study groups. These results suggest that seasonal variation in overall food availability in the study area associated with annual climatic cycles is often a less important determinant of temporal variation in diets and activity patterns of siamangs at Way Canguk than local or individual factors such as the local densities of specific plant species or competitors or variation in home range size, group size or composition, or female reproductive status among siamang groups.

So why were annual means for the activity patterns of siamangs and the plant part components of siamang diets relatively consistent across groups and years, while the species composition of siamang diets differed substantially over time and space within the study area? Some primates shift their diets to lower-quality foods during periods of low food availability (Knott 1998; Marshall and Wrangham 2007), but gibbons in tropical forests appear to seek out ripe fruits during both high-food and low-food periods (McConkey et al. 2002, 2003; Dominy et al. 2008). This may be accomplished by specializing on fruits found in small patches to reduce competition with other primate species (McConkey 2009), focusing on larger fruits or those found in larger patches to maximize fruit intake (McConkey et al. 2002), ranging further in search of fruits (Dominy et al. 2008), or being less selective about fruit characteristics during periods of relatively low fruit availability (McConkey et al. 2002). Therefore, the plant part components of gibbon diets may remain relatively constant even as the species composition varies.

In addition, I did not examine all aspects of siamang time allocation. For example, all types of social activities were considered together which may have the effect of obscuring important differences in rates of inter- and intragroup aggression related to feeding competition, and within-food-patch (foraging) movement and between-tree (travel) movement were also grouped. Overall patterns of time allocation are probably driven by energetic considerations that are fairly consistent across groups and time periods, but the allocation of time to different types of foraging or social activities may vary. Therefore, the observed similarity in overall patterns of time allocation across groups and years should not be interpreted as indicating an absence of variation in foraging strategies.

While overall fruit availability shows strong seasonal variation in some gibbon habitats (Bartlett 1999, 2007; McConkey et al. 2002, 2003; McConkey 2009), at Way Canguk, rainfall is not a significant predictor of fruit availability (Kinnaird and O'Brien 2005), and some highly-preferred plant species are present at low density or fruit at long or irregular intervals. Therefore, while fruit availability varies, it does not display pronounced and predictable seasonal peaks and troughs affecting all groups simultaneously. Furthermore, *Ficus* fruits comprise almost half of the overall fruit crop at Way Canguk, and in most months at least one large fig individual in each siamang home range produces fruit (Kinnaird and O'Brien 2005). As individual figs produce very large fruit crops that ripen over a period of several weeks, siamangs at Way Canguk generally have access to a steady supply of ripe fruits, which may buffer them to some extent from the effects of variation in preferred nonfig fruit, and permit them to maintain a relatively consistent fruit intake. These ecological differences between Way Canguk and other study sites caution against extrapolation of these results to more seasonal habitats.

While the plant part compositions of siamang diets were fairly consistent between groups and across years in this study, this does not mean that siamang diets are inflexible. In fact, siamang diets display substantial variation between study sites. Siamangs at Kuala Lompat in peninsular Malaysia ($N=1$ group; Raemaekers 1979, 1984) spend substantially more time eating leaves and less time eating fig fruits than siamangs at Ketambe in northern Sumatra ($N=2$ groups; Palombit 1997), whereas siamangs in this study spent more time eating nonfig fruits and much less time eating insects than siamangs at Ketambe and Kuala Lompat, along with intermediate levels of fig fruit and leaf consumption. The differences in siamang diets between sites reported to date far exceed the differences among groups observed at Way Canguk, which suggests that the effects of environmental variation between sites on siamang diets are more pronounced than the effects of local ecological variation or individual factors such as group size within sites.

The principle weakness of this study is that I did not measure actual variation in food availability in the home ranges of specific siamang groups. Therefore, my assumption that the differences in diets between groups and time periods were primarily related to differences in local food availability remains speculative.

A number of other factors, including differences in group composition, some reproductive parameters, and ranging patterns, may also affect activity patterns of gibbons, and the relative importance of these factors has yet to be rigorously quantified. However, the observation that time budgets and ranging patterns of siamangs differed relatively little among groups at Way Canguk despite pronounced variation in group composition and home range size suggests that the effects of these variables on overall activity patterns of siamangs may be relatively small.

The groups included in this study were not chosen at random; I chose groups that had a young infant in the group at the time when behavioral sampling began. Therefore, if the presence of an infant affects the time budgets, ranging patterns, or diets of siamangs, this study may have overestimated similarities between groups relative to those expected in a random sample of five groups. Adult female energetic needs and travel costs vary over time during a reproductive attempt, which can cause shifts in the activity patterns of females (Altmann 1980; Altmann and Samuels 1992; Barrett et al. 2006), and infant care by non-mothers may also affect both maternal time budgets (O'Brien and Robinson 1991; Ross and MacLarnon 2000) and the time budgets of the individuals providing care (Goldizen 1987; Price 1992). A detailed examination of the effects of infant care on adult behavior in the study groups suggested that female feeding and resting time, but not diet or DPL, were affected by lactation and infant care: females with young infants fed less and rested more than females with older infants (Lappan 2008). Accordingly, it is likely that the significant difference in resting time between the first and second years of this study observed in group C were related to changes in female reproductive status and infant development, rather than ecological variables. The diets, activity budgets, and ranging behavior of most males were not strongly affected by the presence of an infant, but the two males providing the most infant care fed somewhat less during the period of most intense male care of infants, which occurred in the second year of infant life (Lappan 2008). As infant care had opposite effects on the time budgets of males and females (females fed less in the first year of infant life, whereas males fed less in the second year), and this study examined average values of all adults in a group, however, the stronger effects of infants on female behavior should be more than compensate for the effects of infants on male behavior on the averaged values for each group.

Although this study spanned a period of almost 2 years, it did not sample the full range of conditions that siamangs at Way Canguk experience. Way Canguk experiences occasional long droughts associated with ENSO events separated by periods of several years during which rainfall is weakly seasonal. This study fell entirely within a relatively wet period – rainfall exceeded 100 mm every study month. A longer study or a study during a period with greater variation in rainfall would be likely to detect greater temporal variation in siamang behavior.

Researchers studying gibbon behavior face formidable logistical challenges due to gibbons' arboreal habits, rapid locomotion (often over difficult terrain), small group sizes, and generally fearful nature, which make habituation time-consuming in most cases. Nonetheless, the results of this study suggest that observations of several groups in a given habitat type may be required to accurately identify the most important food species for the population, to evaluate the importance of rare but preferred food species, to characterize temporal variation in diets for the population as a whole, or to examine the effects of local habitat variation within a landscape. Furthermore, studies spanning less than a year will underestimate the number of plant species that gibbons consume and overestimate the importance of specific plant foods that happen to be available during the study period. In this study, indices of dietary overlap were similar for the same group sampled in two consecutive years and for neighboring groups sampled in the same year, indicating that the effects of spatial variation on siamang diets are as strong as the effects of temporal variation. This suggests that a 1-year study of three to five groups is likely to reveal as much or more about gibbon diets than a 2-year study of only one or two groups. Therefore, researchers should consider their study questions carefully when deciding how much time to allocate for habituation vs. data collection: samples including a greater number of groups will often be preferable, even if they require a greater initial time investment.

All gibbon species are Endangered or Critically Endangered (IUCN 2008), and many populations are small, fragmented, and decreasing. Some local populations are likely to require habitat restoration to remain viable, and accurate information about gibbon diets and ranging behavior may be critical in allowing conservationists to develop effective restoration plans. Rehabilitation and reintroduction programs also require the establishment of behavioral baselines to which the behavior of reintroduced animals can be compared during rehabilitation and the post-release monitoring period, and detailed knowledge of what constitutes an appropriate habitat for reintroduction (Cheyne 2009). Conservation managers are sometimes forced to prioritize among primate populations or habitats for conservation activities, and they often do so in an atmosphere of considerable uncertainty. A better understanding of what constitutes "typical" behavior, diets, or habitats for primate taxa based on studies sampling larger numbers of groups from a variety of different habitat types, and a better understanding of gibbon ecological and behavioral flexibility and the consequences of variation in habitat quality may help to guide these decisions.

Acknowledgments I thank Sharon Gursky and Jatna Supriatna for their invitation to contribute to this volume, and for their patience throughout the writing process. Funding for this research was provided by the Leakey Foundation, Sigma Xi, the Fulbright Student Program, New York University, the New York Consortium for Evolutionary Primatology, the Margaret and Herman Sokol Foundation, and in the writing stages by Ewha University and the Amore Pacific Foundation. Permission to conduct research in Indonesia was granted by the Indonesian Institute of Sciences (LIPI), and permission to conduct research in the Bukit Barisan Selatan National Park was granted by the Indonesian Ministry of Forestry's Department for the Protection and Conservation of Nature (PHKA). I thank the American–Indonesian Exchange Foundation (AMINEF), *Universitas Indonesia* and the Wildlife Conservation Society–Indonesia Program for considerable logistical assistance in Indonesia, and Mohammad Iqbal, Anton Nurcayho, Maya Dewi Prasetyaningrum, Teguh Priyanto, Janjiyanto, Sutarmin, Tedy Presetya Utama, Abdul Roshyd, and Martin Trisunu Wibowo for their assistance in the field. Thanks to Tim O'Brien and Margaret Kinnaird for many helpful discussions and for sharing unpublished data, and thanks to Noviar Andayani for her consistent support for me and my research activities in Indonesia.

References

Altmann J (1980) Baboon mothers and infants. Harvard University Press, Cambridge, MA

Altmann J, Samuels A (1992) Costs of maternal care: infant-carrying in baboons. Behav Ecol Sociobiol 29:391–398

Barrett L, Halliday J, Henzi SP (2006) The ecology of motherhood: the structuring of lactation costs by chacma baboons. J Anim Ecol 75:875–886

Bartlett TQ (1999) Feeding and ranging behavior of the white-handed gibbon (*Hylobates lar*) in Khao Yai National Park, Thailand. Ph.D. dissertation, Washington University-St. Louis, St. Louis, MO

Bartlett TQ (2007) The Hylobatidae: small apes of Asia. In: Campbell CJ, Fuentes A, MacKinnon KC, Panger M, Bearder SK (eds) Primates in perspective. Oxford University Press, New York, pp 274–289

Bawa KS (1983) Patterns of flowering in tropical plants. In: Jones CE, Little RJ (eds) Handbook of experimental pollination biology. Scientific and Academic Editions, New York, pp 394–410

Cannon CH, Curran LM, Marshall AJ, Leighton M (2007) Beyond mast-fruiting events: community asynchrony and individual dormancy dominate woody plant reproductive behavior across seven Bornean forest types. Curr Sci 93:1558–1566

Chapman CA, Chapman LJ, Gillespie TR (2002) Scale issues in the study of primate foraging: red colobus of Kibale National Park. Am J Phys Anthropol 117:849–868

Cheyne SM (2009) The role of reintroduction in gibbon conservation: opportunities and challenges. In: Lappan S, Whittaker DJ (eds) The Gibbons: new perspectives on small ape socioecology and population biology. Springer, New York

Chivers DJ (1984) Feeding and ranging in gibbons: a summary. In: Preuschoft H, Chivers D, Brockelman W, Creel N (eds) The Lesser Apes: evolutionary and behavioural biology. Edinburgh University Press, Edinburgh, pp 267–284

Condit R, Ashton PS, Baker P, Bunyavejchewin S, Gunatilleke S, Gunatilleke N, Hubbell SP, Foster RB, Itoh A, LaFrankie JV, Lee HS, Losos E, Manokaran N, Sukumar R, Yamakura T (2000) Spatial patterns in the distribution of tropical tree species. Science 288:1414–1418

Cords M (1986) Interspecific and intraspecific variation in diet of two forest guenons, *Cercopithecus ascanius* and *C. mitis*. J Anim Ecol 55:811–827

Dominy NJ, Vogel ER, Haag L, van Schaik CP, Parker GG (2008) Fallback or fall forward: food dispersion, canopy complexity, and the foraging adaptations of apes in Southeast Asia. Am J Phys Anthropol S46:91–92

Elder AA (2009) Hylobatid diets revisited: the importance of body mass, fruit availability, and interspecific competition. In: Lappan S, Whittaker DJ (eds) The Gibbons: new perspectives on small ape socioecology and population biology. Springer, New York

Fan P, Jiang X (2008) Effects of food and topography on ranging behavior of black crested gibbon (*Nomascus concolor jingdongensis*) in Wuliang Mountain, Yunnan, China. Am J Primatol 70:871–878

Fan PF, Ni QY, Sun GZ, Huang B, Jiang XL (2008) Seasonal variations in the activity budget of *Nomascus concolor jingdongensis* at Mt. Wuliang, Central Yunnan, China: effects of diet and temperature. Int J Primatol 29:1047–1057

Fan P, Xiao W, Huo S, Jiang XL (2009) Singing behavior and singing functions of black-crested gibbons (*Nomascus concolor jingdongensis*) at Mt. Wuliang, Central Yunnan, China. Am J Primatol 71:539–547

Fuentes A (1999) Re-evaluating primate monogamy. Am Anthropol 100:890–907

Fuentes A (2000) Hylobatid communities: changing views on pair bonding and social organization in hominoids. Yearb Phys Anthropol 43:33–60

Fuentes A (2002) Patterns and trends in primate pair bonds. Int J Primatol 23:953–978

Gittins SP (1980) Territorial behavior in the agile gibbon. Int J Primatol 1:381–399

Goldizen AW (1987) Facultative polyandry and the role of infant-carrying in wild saddle-back tamarins (*Saguinus fuscicollis*). Behav Ecol Sociobiol 20:99–109

Holmes RT, Pitelka FA (1968) Food overlap among coexisting sandpipers on northern Alaskan tundra. Syst Zool 17:305–318

IUCN (2008) 2008 IUCN Red List of Threatened Species

Jiang X, Wang Y, Wang Q (1999) Coexistence of monogamy and polygyny in black-crested gibbons (*Hylobates concolor*). Primates 40:607–611

Kappeler M (1984) Diet and feeding behaviour of the moloch gibbon. In: Preuschoft H, Chivers DJ, Brockelman WY, Creel N (eds) The Lesser Apes: evolutionary and behavioural biology. Edinburgh University Press, Edinburgh, pp 228–241

Kinnaird MF, O'Brien TG (1998) Ecological effects of wildfire on lowland rainforest in Sumatra. Conserv Biol 12:954–956

Kinnaird MF, O'Brien TG (2005) Fast foods of the forest: the influence of figs on primates and hornbills across Wallace's Line. In: Dew JL, Boubli JP (eds) Tropical fruits and frugivores: the search for strong interactors. Springer, New York, pp 155–184

Knott CD (1998) Changes in orangutan diet, caloric intake and ketones in response to fluctuating fruit availability. Int J Primatol 19:1061–1079

Lappan S (2007a) Patterns of dispersal in Sumatran siamangs (*Symphalangus syndactylus*): preliminary mtDNA evidence suggests more frequent male than female dispersal to adjacent groups. Am J Primatol 69:692–698

Lappan S (2007b) Social relationships among males in multi-male siamang groups. Int J Primatol 28:369–387

Lappan S (2008) Male care of infants in a siamang (*Symphalangus syndactylus*) population including socially monogamous and polyandrous groups. Behav Ecol Sociobiol 62:1307–1317

Lappan S (2009a) Flowers are an important plant food for small apes in southern Sumatra. Am J Primatol 71 (early online), doi 10.1002/ajp.20691

Lappan S (2009b) Patterns of infant care in wild siamangs (*Symphalangus syndactylus*) in southern Sumatra. In: Lappan S, Whittaker DJ (eds) The Gibbons: new perspectives on small ape socioecology and population biology. Springer, New York

Leighton DR (1987) Gibbons: territoriality and monogamy. In: Smuts BB, Cheney DL, Seyfarth RM, Wrangham RW, Struhsaker TT (eds) Primate societies. University of Chicago Press, Chicago, pp 135–145

MacKinnon JR, MacKinnon KS (1980) Niche differentiation in a primate community. In: Chivers DJ (ed) Malayan forest primates: ten years' study in tropical rain forest. Plenum, New York

Malone NM (2007) The socioecology of the critically endangered Javan gibbon (Hylobates moloch): assessing the impact of anthropogenic disturbance on primate social systems. Ph.D. dissertation, University of Oregon, Eugene

Malone N, Fuentes A (2009) The ecology and evolution of hylobatid communities: causal and contextual factors underlying inter- and intraspecific variation. In: Lappan S, Whittaker DJ (eds) The Gibbons: new perspectives on small ape socioecology and population biology. Springer, New York

Marshall AG (2009) Are montane forests demographic sinks for Bornean white-bearded gibbons? Biotropica 41:257–267

Marshall AJ, Wrangham RW (2007) The ecological significance of fallback foods. Int J Primatol 28:1219–1235

McConkey KR (2009) The seed dispersal niche of gibbons in Bornean dipterocarp forests. In: Lappan S, Whittaker DJ (eds) The Gibbons: new perspectives on small ape socioecology and population biology. Springer, New York

McConkey KR, Aldy F, Ario A, Chivers DJ (2002) Selection of fruit by gibbons (*Hylobates muelleri x agilis*) in the rain forests of central Borneo. Int J Primatol 23:123–145

McConkey KR, Ario A, Aldy F, Chivers DJ (2003) Influence of forest seasonality on gibbon food choice in the rain forests of Barito Ulu, central Kalimantan. Int J Primatol 24:19–32

Nurcahyo A (1999) Studi perilaku harian siamang (*Hylobates syndactylus*) di taman nasional Bukit Barisan Selatan, Lampung. Unpubl. B.S. thesis, Universitas Gadjah Mada, Yogyakarta

O'Brien TG, Robinson JG (1991) Allomaternal care by female wedge-capped capuchin monkeys: effects of age, rank and relatedness. Behaviour 119:30–50

O'Brien TG, Kinnaird MF, Sunarto, Dwiyahreni AA, Rombang WM, Anggraini K (1998) Effects of the 1997 fires on the forest and wildlife of the Bukit Barisan Selatan National Park, Sumatra. Wildlife Conservation Society Working Paper No. 13. Wildlife Conservation Society, New York, NY

O'Brien TG, Kinnaird MF, Anton N, Prasetyaningrum MDP, Iqbal M (2003) Fire, demography and the persistence of siamang (*Symphalangus syndactylus:* Hylobatidae) in a Sumatran rainforest. Anim Conserv 6:115–121

O'Brien TG, Kinnaird MF, Anton N, Iqbal M, Rusmanto M (2004) Abundance and distribution of sympatric gibbons in a threatened Sumatran rain forest. Int J Primatol 25:267–284

O'Brien TG, Kinnaird MF, Nurcahya A, Nusalawo M (2008) Response of siamang and agile gibbons to climate fluctuations in Indonesia. XXII Congress of the International Primatological Society. August 3-8, 2008. Edinburgh, Scotland

Oates JF (1987) Food distribution and foraging behavior. In: Smuts BB, Cheney DL, Seyfarth RM, Wrangham RW, Struhsaker TT (eds) Primate societies. University of Chicago Press, Chicago, pp 197–209

Palombit RA (1997) Inter- and intra-specific variation in the diets of sympatric siamang (*Hylobates syndactylus*) and lar gibbons (*Hylobates lar*). Folia Primatol 68:321–337

Price EC (1992) The costs of infant carrying in captive cotton-top tamarins. Am J Primatol 26:23–33

Raemaekers JJ (1977) Gibbons and trees: comparative ecology of the siamang and *lar gibbons*. Unpubl. Ph.D. thesis, University of Cambridge, Cambridge

Raemaekers JJ (1979) Ecology of sympatric gibbons. Folia Primatol 31:227–245

Raemaekers JJ (1984) Large versus small gibbons: relative roles of bioenergetics and competition in their ecological segregation in sympatry. In: Preuschoft H, Chivers DJ, Brockelman WY, Creel N (eds) The Lesser Apes: evolutionary and behavioral biology. Edinburgh University Press, Edinburgh, pp 209–218

Reichard U (1995) Extra-pair copulations in a monogamous gibbon (*Hylobates lar*). Ethology 100:99–112

Reichard U (2009) Social organization and mating system of Khao Yai gibbons, 1992–2006. In: Lappan S, Whittaker DJ (eds) The Gibbons: new perspectives on small ape socioecology and population biology. Springer, New York

Riley EP (2008) Ranging patterns and habitat use of Sulawesi Tonkean macaques (Macaca tonkeana) in a human-modified habitat. Am J Primatol 70:670–679

Ross C, MacLarnon A (2000) The evolution of non-maternal care in anthropoid primates: a test of the hypotheses. Folia Primatol 71:93–113

Sakai S (2001) Phenological diversity in tropical forests. Popul Ecol 43:77–86

Sakai S, Momose K, Yumoto T, Nagamitsu T, Nagamasu H, Hamid AA, Nakashizuka T (1999) Plant reproductive phenology over four years including an episode of general flowering in a lowland dipterocarp forest, Sarawak, Malaysia. Am J Bot 96:1414–1436

Sangchantr S (2004) Social organization and ecology of Mentawai leaf monkeys. Ph.D. dissertation, Columbia University, New York

Savini T, Boesch C, Reichard U (2008) Home-range characteristics and the influence of seasonality on female reproduction in white-handed gibbons (*Hylobates lar*) at Khao Yai National Park, Thailand. Am J Phys Anthropol 135:1–12

Silva SSB, Ferrari SF (2008) Behavior patterns of southern bearded sakis (*Chiropotes satanas*) in the fragmented landscape of eastern Brazilian Amazonia. Am J Primatol 71:1–7

Silva MG, Tabarelli M (2001) Seed dispersal, plant recruitment and spatial distribution of Bactris acanthocarpa Martius (Arecaceae) in a remnant of Atlantic forest in northeast Brazil. Acta Oecol 22:259–268

Ungar P (1995) Fruit preferences of four sympatric primate species at Ketambe, Northern Sumatra, Indonesia. Int J Primatol 16:221–245

Whitten AJ (1982a) Diet and feeding behaviour of Kloss gibbons in Siberut Island, Indonesia. Folia Primatol 37:177–208

Whitten AJ (1982b) A numerical analysis of tropical rain forest, using floristic and structural data, and its application to an analysis of gibbon ranging behaviour. J Ecol 70:249–271

Whitten AJ, Damanik SJ, Anwar J, Hisyam N (2000) The ecology of Sumatra. Periplus, Hong Kong

Wich SA, van Schaik CP (2000) The impact of El Niño on mast fruiting in Sumatra and elsewhere in Malesia. J Trop Ecol 16:563–577

Yanuar A (2009) The population distribution and abundance of siamangs *(Symphalangus syndactylus)* and agile gibbons *(Hylobates agilis)* in west central Sumatra, Indonesia. In: Lappan S, Whittaker DJ (eds) The Gibbons: new perspectives on small ape socioecology and population biology. Springer, New York

Zar JH (1996) Biostatistical analysis. Prentice Hall, New Jersey

Chapter 7
Impact of Forest Fragmentation on Ranging and Home Range of Siamang (*Symphalangus syndactylus*) and Agile Gibbons (*Hylobates agilis*)

Achmad Yanuar and David J. Chivers

Introduction

Fragmentation always results in the reduction of forest area and isolation of forest remnants (Bierregaard et al. 1997). Primates are flexible animals in the usage of area and diet (Chapman 1988). The gibbon is one of the arboreal primates that persists in small forest fragments. Fortunately, gibbons that have small home ranges (Leighton 1987) may survive better in large, medium and small forest fragments, due to their ability to exploit young leaves, a food resource widely distributed in the forest (Kakati 2004). Many small-group animals with small home range are extremely tolerant to habitat changes, such as habitat fragmentation, because they are able to exploit leaves and have flexible home range size (Rylands and Keuroghlian 1988). Home range was initially defined as the area in which an animal spends most of its adult life (Burt 1943; Jewell 1966; Bates 1970). Thus, home range size and ranging patterns among primates may rely on social aspects and feeding behaviour strategies (Spironello 2001). The term home range is modified to specify a given period or duration of observation and, thus employed, to demonstrate changing patterns of range use over time (Harrison 1983). For a gibbon group, it can be defined as the total area traversed by the group within a given period (Gittins 1979).

Gibbons are among primate social groups that restrict their regular movements to a limited area of their habitat (DeVore 1965) and a group usually not only reveals a home range of some kind, but also vigorously defends part or all the area against neighbouring groups (Chivers 1969; Gittins 1983). In a fragment, the disparity in presence or absence of neighbouring groups of territorial animal could be among the reasons that affect home range sizes (Kakati 2004). The home range of a gibbon group is a distinct entity, with clear borders that are defended for the exclusive use of the resident group – a territory (Chivers et al. 1975; Ahsan 1994).

A. Yanuar (✉)
Conservation International Indonesia, Jl. Pejaten Barat 16 A, Jakarta, 12559, Indonesia
e-mail: ayanuar@conservation.org

S. Gursky-Doyen and J. Supriatna (eds.), *Indonesian Primates*,
Developments in Primatology: Progress and Prospects,
DOI 10.1007/978-1-4419-1560-3_7, © Springer Science+Business Media, LLC 2010

The overall distance an individual or group of animals or primates travels in a day is defined as day range. Travelling is energetically costly and relies on distribution and abundance of food (Altmann and Altmann 1970; Chivers 1974; Clutton-Brock 1975; Goodall 1977; Nellemann and Newton 2003), predators, mates and potential competitors for these mates (Struhsaker 1975; Rasmussen 1979), favourite rest trees (Mason 1968), night-sleeping trees (Kummer 1968), routes used (Fossey and Harcourt, 1977), weather conditions (Iwamoto and Dunbar 1983) and the availability of standing water (Wrangham 1977). Day range length, ranging pattern and home range size of the siamang and the agile gibbons are analysed, compared and discussed for each site and group, as well as compared to other studies of frugivores.

The results from this study seem similar to other studies of frugivorous primates, wherein the home range is smaller in fragments than in intact forest: (1) the guenon (*Cercopithecus cephus*) in Gabon (Tutin 1999), (2) Tana River red colobus (*Procolobus rufomitratus*) and crested mangabay (*Cercocebus galeritus galeritus*) in Tana River forest patches, Kenya (Decker 1994), and (3) hoolock gibbon in Bangladesh and Assam (Ahsan 2001; Kakati 2004).

The mean home range size or territory of an animal group is often relatively constant and this size is related to the abundance and distribution of the diet and habitat preference of that species (Eisenberg et al. 1972; Clutton-Brock and Harvey 1977; Gittins 1979). Generally, frugivorous animals need a larger area than folivores, due to their potential foods being limited throughout the year (Altman 1974). The home range size decrease as the amount of food in an area increases (e.g., *Cercopithecus aethiops*, Struhsaker 1974; *Cercocebus albigena*, Freeland 1979).

Hence, in this study, the aim is show how ranging and diet vary between the more frugivorous agile gibbon and the more folivorous siamang according to fragment size in relation to continuous forest.

Study Area

Four principal forest fragments were chosen as main study sites, on the basis of different history, but at similar altitude (100–400 m asl), as well as one principal study site in continuous forest (within the National Park of Kerinci-Seblat), where populations of siamang and agile gibbon are present (Table 7.1). These locations were also chosen on the basis of easy access and feasible logistic. All study fragments are located outside the east wing of the National Park at varying distance (Fig. 7.1).

Sungai (Sg.) Misang

There are several small and medium-size fragments situated near the city of Bangko, the administrative town of Kabupaten (district) of Merangin, Jambi provinces The forest fragments are separated by local gardens, primarily old and young rubber stands, village communities and asphalt, dirt and gravel roads. We selected

Table 7.1 Site, altitude, status and forest covers for five main study sites

Site name	Area (ha)	Altitude (m)	Status of forest	Surrounded by	Forest cover
Sg. Misang	100	100–150	Isolated	Rubber gardens, main roads and settlements	Primary and secondary
Lr. Gambir	45	100–150	Isolated	Rubber, cinnamon, oil palm gardens, dirt and main road and settlements	Primary and secondary
Sg. Mangun	55	250–300	Isolated	Oil-palm plantation	Primary and secondary
Sg. Tembalun	90	300–350	Isolated	Rubber, cinnamon, oil palm gardens, paddy field, roads and settlements	Primary, re-growth and old secondary
Kulai Tanang	200	350–400	Continuous	Main river and buffer areas	Primary

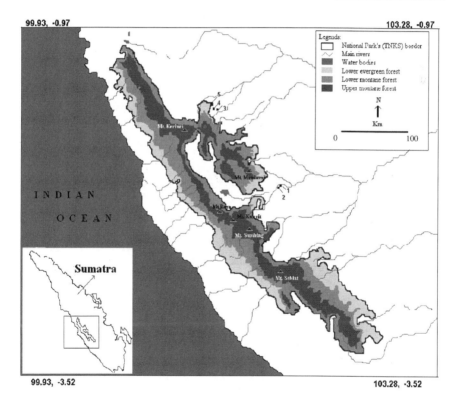

Fig. 7.1 Map of National Park (KSNP) and study areas (1) Sg. Misang; (2) Lr. Gambir; (3) Sg. Mangun; (4) Sg. Teambalun; and (5) Kulai Tanang

Sg. Misang as an intensive study site for siamang, based on advice by forestry authorities of Bangko. This area is situated away from Kerinci-Seblat National Park (ca 60 km) or ca 20 km from continuous forest.

Sg. Misang's forest fragment has mostly been logged from the 1960s until the 1970s, resulting in a vegetation type of predominantly old secondary and only a little original lowland forest remains, covering an area of less than 100 ha. This patch adjoins others, separated by shrubs, grasslands, ponds, gravel and dirt roads and local gardens, dominated by rubber plants as canopy gaps. Topographically, the area was lightly undulating, lowland, with steep inclines to small streams, although there were predominantly flatter areas, and elevation less than 200 m asl. Luckily, this forest fragment has been officially protected under the management of the local government, and it was declared as a city garden to attract local visitors to observe the native fauna and flora. There was no evidence of illegal felling activities in this area, except inhabitants normally made incisions in a tree (*Hevea brasiliensis*) to obtain rubber latex from this plant.

Four diurnal primate species were found in Sg. Misang: siamang, banded langur (*Presbytis melalophos*), long-tail macaque (*Macaca fascicularis*), pig-tail macaque (*Macaca nemestrina*), and the agile gibbon was absent from this forest fragment. It was also a home range for wild pig (*Sus scrofa*), tapir (*Tapirus indicus*), sun-bear (*Helarctos malayanus*) and muntjac deer (*Muntiacus muntjak*). These fragments had been established for more than ten years.

Lorong (Lr.) Gambir

Similar to Sg. Misang fragment, the area was also flatter, lightly undulating, with steep banks to streams. The distance between Sg. Misang and Lr. Gambir is about 3.5 km, separated by the city of Bangko, main roads and the Merangin River. No information on fragment history was available, despite the existence of these fragments reported by Dutch colonial officers (local people, pers. comm.). Forest cover of this area was predominantly old secondary and mixed, with harvest of traditional rubber plantations. This fragment is located at the side of main road, about 1km from Bangko city centre. Several small patches (less than 40 ha in size) with canopy gaps were inhabited by a small population of gibbon and other diurnal primate species, separated and surrounded by rubber, cinnamon and young oil-palm gardens, asphalt, dirt and gravel roads, as well as human settlers.

Each family living around fragment forests has one to two hectares of rubber gardens within these forests, so they will harvest rubber latex daily or weekly, mainly in the dry season. Today, the habitat of siamang is found only outside this area, because it is cut off by the large Merangin river. Agile gibbon, long-tail and pig-tail macaques, as well as banded langur, were commonly encountered within this patch and the Sumatran silvered langur was occasionally seen.

Sg. Mangun

The original forest remnant of this area is located on both sides of Mangun river. We selected a fragment forest on this river of 5,000 m (5 km) long and 100–200 m wide to study wild siamang. In fact, this forest adjoins with other fragments along the river and extends to continuous forest within the buffer zones of the National Park, cut-off by several bridges and roads and worker camps of the oil-palm plantation. The forest fragments were formed due to the original forests being converted for commercial oil-palm, namely Tidar Kerinci Agung Company estate (PT TKA) since 1984.

Previously, the original forest of Sg. Mangun had been logged intensively, before conversion to oil-palm estate since the 1970s. Unfortunately, due to poor security of this forest, small-scale illegal logging activities still occur, to fell remnant commercial woods. Several giant strangling figs (Moraceae), lianas, vines, epiphytes were still observed occasionally, on the side of Mangun River. Meanwhile, palm, rattan and manau trees were observed with some strangling figs on slight slopes.

The populations of siamang and agile gibbon have been isolated within this patch since 1985. Sadly, there were only two adult female agile gibbons, who lived separately and one group of siamang, who shared their territory with other primates in this area of about 50 ha. No silvered langur was reported. Banded langur, long-tail and pig-tail macaques were commonly observed, as well as three species of large frugivorous birds (rhinoceros, helmeted and Asiatic hornbills). They occasionally raided into the edge of the oil-palm plantation to forage on ripe fruit.

Sg. Tembalun

We predicted that there were 3 small groups of agile gibbon living in this area. Other groups of gibbon were observed within other small fragments separated by rural gardens, dirt (plantation) roads and rivers (Jujuhan river). Of about 90 ha, only 25% were dominated by original forest and 40% were re-growth vegetation (old secondary), 20% were young secondary forest and the rest was settled for agriculture, mostly rubber gardens. Presumably this area was lowland dipterocarp forest, as dipterocarp trees, as well as those of Fagaceae, observed along the ridge-top.

The existence of this area is recorded since the 1930s, surrounded by local agriculture, mainly rubber plantations and paddy fields. The size of this area has now decreased drastically due to expansion of oil-palm plantations, rural agriculture and human pressures. Nowadays, in and around the forest fragment there is harvest of traditional rubber, cinnamon gardens and modern oil-palm plantations of PT TSS and PT TKA.

There is an old rural settlement, namely Talao, with a population of 200–300 families. They migrated here from another village more than 100 years ago (local people, pers. comm.). Due to this recent population increase, many fragments are likely to vanish by felling and land clearing to make new agricultural areas.

Malayan sun-bear occasionally visited this fragment to forage and wild pigs were commonly observed to invade local gardens and oil-palm plantations. Banded langur and long-tail macaque were diurnal primates commonly encountered. Silvered langur was now absent, despite historic reports (local people, pers. comm.).

Kulai Tanang

Within the continuous forest, Kulai Tanang was selected as the main study site for siamang and agile gibbon, because both species are very common in these areas. This principal study area covers about 150 ha, and is situated in the foothills of the eastern slopes of the Gunung Tujuh complex within the forest edge of the National Park in the north-east (101°26′00 and 1°30′30), lying between two main rivers, Kulai Tanang and Batang Ganeh, just separated by a couple of ridges and streams. These forests are managed by TNKS, of which the HQ is in Sungai Penuh, about 100 km away. Unfortunately, there was no guard post of the National Park near this area, so there were few patrols by park rangers. Indeed, this area has been encroached by people recently.

It is surrounded by the oil-palm plantation of PT Sumatra Jaya Agro Lestari (PT SJ), buffered by secondary and primary forests as border areas, ranging from 50 to 300 m wide. This location can be reached by vehicle as far as the buffer area of the Park, since dirt and gravel roads have been built within the oil-palm complex. Topographically, the area was hilly, undulating, dipterocarp lowland rain-forest, with steep inclines to many streams running in deep valleys, which drain finally into Kulai Tanang and Ganeh rivers. Elevation ranges between 400 and 450 m asl.

Forest cover was occasionally damaged by wind, contiguous with a large area of undisturbed hill/lowland forest, dominated by dipterocarp trees, as well as areas disturbed by illegal felling, due to easy access from oil-palm plantation roads. The forests were also occasionally used by local people for hunting or trapping of terrestrial animals, collecting rattan manau, and gaharu tree (to obtain fragrant resin from the heartwood of this plant for sale), and fishing, and so forth. The study area had been extensively logged by Pasar Besar timber concession 15 years ago, and the original structure of the forest canopy had been destroyed. The prominence of the *Macaranga* trees is probably a mark of this disturbance, as these trees commonly grow along old logging roads (Whitmore 1975).

Methods

Day Range

Cycles of daily observation were followed at each location during 12-months (from October 2003 to September 2004) after the animals were habituated. In every month of the study, the locations of each of the three siamang and agile gibbon groups were recorded continuously through the day for 4–6 days, by an estimated distance and compass bearing from a known trail location (within trail system) or some other landmark during full-day follows, in addition to the 10-min scans of main activities. In order to monitor the range use of the study animals, a full animal-follow day involved observing the siamang and agile gibbon from the time they first left their sleeping trees in the morning until they entered sleeping positions in a tree in the afternoon.

The majority of the ranging data are drawn from the results of the 38 full-day follows for the siamang groups, consisting of 11 days in January, March, June and July 2004 in Sg. Misang, 15 in Sg. Mangun in April, May, June, July and August 2004 and 12 in Kulai Tanang in February, April, May and September 2004, as well as for agile gibbon groups in Lr. Gambir (10 days on January, April, May and June 2004), Sg. Tembalun (12 days in February, March, April, June and July 2004) and Kulai Tanang (11 days in April, May, July and September 2004).

This allowed a day range to be plotted, which represented the route covered from one sleeping site to the next. Day ranges were drawn by connecting the group movements from one location to another location on the map with lines. Normally, the routes of female siamang*s* and agile gibbon*s* were checked as the major route, because the adult female can be easily recognised and was used to represent the group. When the female was out of sight, the routes of the male and juvenile were used. Thus, day range was the sum of the distances travelled between points in a single day. Only the full animal-follow days were measured, using a mapping wheel to measure the home range area.

To provide a comparison to other studies, the route travelled by the groups of siamang and agile gibbon throughout the day were drawn on a grid map of the study area, from the reference of quadrat and numbers of trees recorded in each scan, after completing the dawn-to-dusk observations. Hence, we established grids of 50 m unit length throughout the study areas where the group was located during observation, so that each square represented an area of 2,500 m^2, equivalent to 0.25 ha. The number of hectares can be an areal expression of day range. The total number of 0.25ha quadrats entered by each group during the entire study period was also calculated and summed from scan samples. One scan was scored as one entry, unless a group was spread over more than one quadrat, but each quadrat entered was considered once each day, even if it was re-entered on the same day.

Because the day ranges of both species are influenced by cycles of production of plant parts of individual plant, we also investigated each phenophase (i.e. flowers, unripe and ripe fruit, young and mature leaves), which were noted as present or absent on each of plant in the study areas. This gave the proportion of plant parts available in a given month. Abundance of a particular plant part was presented as a percentage of trees producing a particular item.

Home Range

A home range area for siamang or agile gibbon group can be defined as the total area traversed by the group within a given period. Generally, the estimation of home-range area was measured by two methods used in most previous studies on arboreal primates. The first, is the taut string or perimeter-line method around the home range (Altman 1974), which measures the area enclosed by a line placed around the outer limit of all areas entered by animal group. Second, a grid of quarter-hectare (0.25 ha) quadrats entered was laid over the map, but it tends to overestimate home range areas, because large sections of some quadrats are often not used by the focal group. For this study, the number of quadrats entered and the number of 10-min scans during which each quadrat was occupied were counted on full-day follows.

Consequently, from the records on the map of the siamang and agile gibbon groups' ranging routes, the number of 0.25 ha quadrat entered of the map was used to count the hectares where the group ranged. The cumulative increase in home range for each siamang and agile group was plotted against the number of days sampled to verify if the sampling effort had been adequate. This follows the methods of other researchers to analyze home range area (Struhsaker 1975).

Results

Day Range

To represent the arboreal pathways used by each group of siamang and agile gibbon, each day range from full-day follows was drawn onto the home range area (Figs. 7.2 and 7.3). The total mean ($\pm$S.E) distance travelled (day range, DR) by two groups of siamang in fragmented forests, namely Sg. Misang and Sg. Mangun was shorter, compared within continuous forest, Kulai Tanang. Moreover, the total mean distance travelled by two groups of agile gibbon in Lr. Gambir and Sg. Tembalun both within fragmented forests was almost similar with a group in continuous forest. Yet, the mean day ranges of both species groups were different (Table 7.2).

The siamang groups' DRs were between 800 and 1200 m in all locations, where group in Sg. Mangun highest followed in Sg. Misang and Kulai Tanang, respectively. Otherwise, the highest of agile gibbons DRs were between 1,600 and 2,000 m in Sg. Tembalun, while in Kulai Tanang and Lr. Gambir this gibbon normally had DRs between 1,200 and 1,600 m (Fig. 7.4).

There were no significant variation in the total mean DR among locations, both for siamang and agile gibbon, as well as between group, such as between siamang in Sg. Misang and Sg. Mangun; and between Sg. Misang and Kulai Tanang. Meanwhile, a total mean DR of this species showed significant differences between Sg. Mangun and Kulai Tanang. There was highly significant difference only in DR for agile gibbon group, between Lr. Gambir and Sg. Tembalun (Table 7.3).

Siamang groups in Sg. Misang and Kulai Tanang, over monthly observations of range use, showed positive correlation between DR and production of plant parts, such as ripe and unripe fruits, and flowers, but not young leaves in Sg. Misang, and unripe fruit in Kulai Tanang. Monthly ripe fruit production was highest for all study sites in dry season. Moreover, the highest production of young leaf was recorded in wet season at Lr. Gambir, Sg. Mangun, Sg. Tembalun, and Kulai Tanang. Meanwhile, flowering for each month in Lr. Gambir, Sg. Mangun, and Kulai

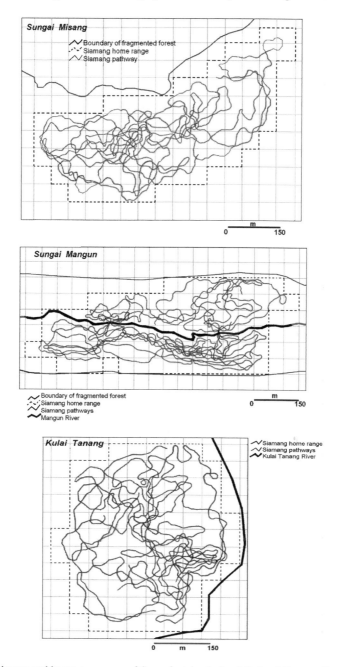

Fig. 7.2 Pathways and home range map of *S. syndactylus* during full-day follows at Sungai (Sg) Misang, Sg. Mangun and Kulai Tanang

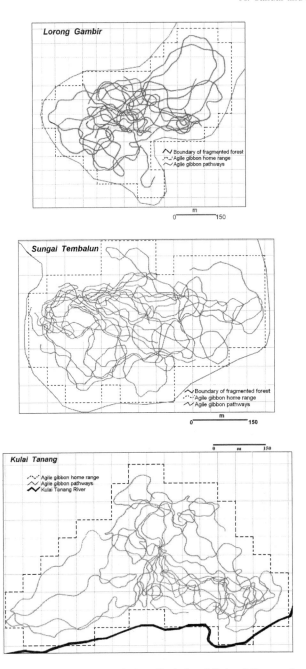

Fig. 7.3 Patways and home range map of *H. agilis* during full-day follows at Lorong Gambir, Sg. Tembalun and Kulai Tanang

Table 7.2 Mean home range sizes (ha) and day ranges (m) of the *S. syndactylus* and *H. agilis* group

Species/fragment size (ha)	Site	Home range (ha)	Day range (m [±S.E])	No. of days
S. syndactylus				
100	Sg. Misang	19.75	923 (±123.46)	11
55	Sg. Mangun	20.25	770 (±81.62)	15
Continuous (200)	Kulai Tanang	21.50	1158 (±120.89)	12
H. agilis				
50	Lr. Gambir	23.75	1280 (±82.39)	10
90	Sg. Tembalun	21.25	1542 (±93.10)	12
Continuous (200)	Kulai Tanang	24.75	1500 (±100.45)	11

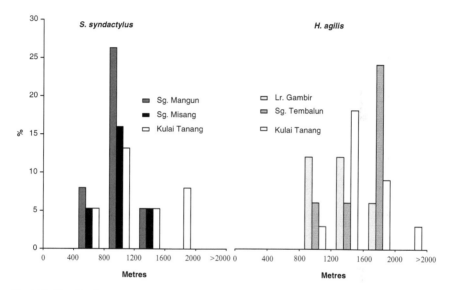

Fig. 7.4 The distribution of day ranges of *S. syndactylus* and *H. agilis*

Tanang was commonly produced in the dry season. There was also a positive correlation between DR and average monthly rainfall in both locations. By contrast, there was no correlation between DR and production of different plant parts and average monthly rainfall in Sg. Mangun (Table 7.4). Furthermore, the agile gibbon's day range, showed no correlation between DR and production of plant parts, and rainfall in all locations (Table 7.4).

Home Range Size

The plots of cumulative number of 0.25-ha quadrats entered during full-day follows showed increase in home range against the number of days sampled, approaching asymptotes for the three groups of siamangs and the three group of agile gibbons

Table 7.3 Results of the Kruskal–Wallis (H) and Mann–Whitney (U) tests for comparison of day ranges of *S. syndactylus* and *H. agilis* among and between sites. Significant correlations are highlighted in *bold*

Species/between site	Mean ± S.E (m)	Species/between site	Mean ± S.E (m)
S. syndactylus		*H. agilis*	
Kruskal–Wallis		Kruskal–Wallis	
H	4.469	H	4.567
df	2	df	2
p	ns	*p*	ns
Mann–Whitney		Mann–Whitney	
Sg. Misang – Sg. Mangun		Lr. Gambir – Sg. Tembalun	
U	80.0	U	27.50
p	ns	*p*	**0.030***
Sg. Misang – Kulai Tanang		Lr. Gambir – Kulai Tanang	
U	43.0	U	34.00
p	ns	*p*	ns
Sg. Mangun – Kulai Tanang		Sg. Tembalun – Kulai Tanang	
U	46.5	U	61.50
p	**0.032***	*p*	ns

Results significant at *$p<0.05$, *ns* not significant

by 12–15 days (Figs. 7.5 and 7.6). Home range sizes of the groups of siamangs and agile gibbons were not correlated with fragment size. Siamang home range size in the smaller fragment (Sg. Mangun) was larger than in Sg. Misang fragment. Similarly, the agile gibbon in smaller fragment in Lr. Gambir had a home range size larger than in Sg. Tembalun fragment. Both siamangs and agile gibbons in continuous forest had larger home ranges than in the fragments.

The overall mean home range size from full-day follows of siamang in fragment forests was 20.0 ha, compared with 21.5 ha in continuous forest, which was highly significant ($\chi^2 = 39.4$, df = 2, $p < 0.001$). Meanwhile, the agile gibbon groups had a mean home range size of 22.5 ha in fragments, and 24.75 ha in continuous forest, which also were highly significant ($\chi^2 = 22.3$, df = 2, $p < 0.002$).

Number of 0.25-ha Quadrats Entered

The total mean (±S.E) number of 0.25-ha quadrats entered daily by two groups of siamang in fragmented forests was lower, while a group in continuous forest was higher. Siamang and agile gibbon group varied for each location (Table 7.5).

The distribution of number of 0.25-ha quadrats for both siamang and agile gibbon was shown in Fig. 7.7. Daily scores of 20–30 quadrats accounted common of observations for siamang in Sg. Mangun and Sg. Misang, lower for Kulai Tanang; while in Sg. Misang, 10–20 quadrats accounted highest of observations. For agile gibbons 20–30 quadrats accounted for high observations in Lr. Gambir and Kulai Tanang, and low in Sg. Tembalun.

Table 7.4 Spearman-Rank correlation (rs) between day range and unripe and ripe fruit, young leaves, flower, and rainfall for each locations. Significant relationships are highlighted in *bold*

| | Fruit | | | | Leaf | | Flower | | Rainfall | |
| | Unripe | | Ripe | | Young | | | | | |
	r_s	P	r_s	P	r_s	P	r_s	P	r_s	P
S. syndactylus										
Sg. Misang (n=6)	0.986	**0.001***	−0.912	**0.011****	−0.029	n.s	0.899	**0.015***	−0.943	**0.005****
Sg. Mangun (n=7)	0.224	n.s	0.0.38	n.s	0.468	n.s	0.09	n.s	0.045	n.s
Kulai Tanang (n=6)	−0.676	n.s	0.971	**0.001***	0.928	**0.008***	−0.841	**0.036***	0.928	**0.008****
H. agilis										
Lr. Gambir (n=5)	0.3	n.s	−0.462	n.s	0.2	n.s	−0.3	n.s	0.3	n.s
Sg. Tembalun (n=5)	0.0371	n.s	0.377	n.s	−0.143	n.s	0.086	n.s	−0.143	n.s
Kulai Tanang (n=5)	−0.051	n.s	−0.025	n.s	0.205	n.s	−0.359	n.s	0.359	n.s

Results significant **p<0.01 and *p<0.05, n.s not significant

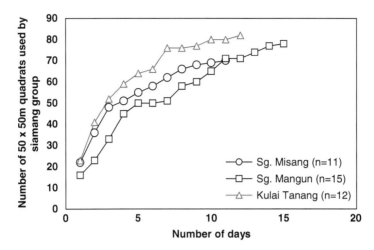

Fig. 7.5 The cumulative number of 0.25 ha quadrats entered by *S. syndactylus* group during full-day follows

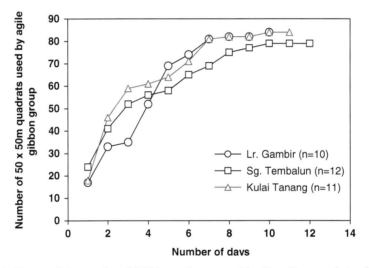

Fig. 7.6 The cumulative number of 0.25 ha quadrats entered by *H. agilis* group during full-day follows

The total mean number of 0.25-ha quadrat varied significantly between study sites for siamang, but not between study sites for agile gibbon groups. There was highly significant variation in the number of 0.25-ha quadrat used daily for siamang groups, such as between Sg. Misang and Sg. Mangun, as well as between Sg. Mangun and Kulai Tanang, but there was no significant difference between groups in Sg. Misang and Kulai Tanang.

Meanwhile, the number of 0.25-ha quadrats varied significantly only between group of agile gibbon in Lr. Gambir and Sg. Tembalun. Conversely, the frequency

Table 7.5 Number of 0.25 ha quadrat entered daily by *S. syndactylus* and *H. agilis* during full-day follows, and results of the Kruskal–Wallis (H) test and Mann–Whitney (U) tests for comparison of number of 0.25 ha quadrat entered among and between sites. Significant correlations are highlighted in *bold*

Species/between site	Mean (m [±S.E])	Range (ha)	No. of days
S. syndactylus			
Sg. Misang	29.1 ± 1.88	19–40	11
Sg. Mangun	18.4 ± 0.99	12–24	15
Kulai Tanang	28.0 ± 2.05	18–41	12
Kruskal–Wallis			
H	18.525		
df	2		
p	**0.0001****		
Mann–Whitney			
Sg. Misang – Sg. Mangun			
U	11.0		
p	**0.0001****		
Sg. Misang – Kulai Tanang			
U	60.5		
p	ns		
Sg. Mangun – Kulai Tanang			
U	18.0		
p	**0.0001****		
H. agilis			
Lr. Gambir	24.5 ± 2.03	17–36	10
Sg. Tembalun	29.7 ± 1.38	23–41	12
Kulai Tanang	28.3 ± 2.35	17–41	11
Kruskal–Wallis			
H	3.668		
df	2		
p	ns		
Mann–Whitney			
Lr. Gambir – Sg. Tembalun			
U	29.5		
p	**0.043***		
Lr. Gambir – Kulai Tanang			
U	41.5		
p	ns		
Sg. Tembalun – Kulai Tanang			
U	54.5		
p	ns		

Results significant at **$p < 0.01$ and *$p < 0.05$, *ns* not significant

distribution of number of 0.25-ha quadrats tends not to be significant between agile groups in Lr. Gambir and Kulai Tanang, and between groups in Sg. Tembalun and Kulai Tanang (see also Table 7.5). The siamangs and agile gibbons use of their home range was observed by super-imposing a hectare quadrat grid over each study

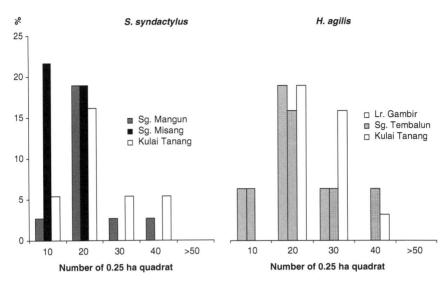

Fig. 7.7 The distribution of 0.25 ha quadrats entered of *S. syndactylus* and *H. agilis*

area (Figs. 7.8 and 7.9). All of the 79, 81 and 86 0.25-ha quadrats were visited during the 11, 15 and 12 full-day follows in siamang study sites, namely Sg. Misang, Sg. Mangun and Kulai Tanang, respectively (Fig. 7.8).

On the other hand, within study sites of agile gibbon in Lr. Gambir, Sg. Tembalun and Kulai Tanang, there were 95, 85 and 100 0.25-ha quadrats visited during the 10, 12 and 11 full-day follows. During 10-min scan sampling of the full-day follows, a total of 578 visits, one quadrat scored 37 times and 12 quadrats were used only once in Sg. Misang, while the quadrat used most frequently scored for 48 and 24, and at least 10 and 11 quadrats were visited only once of the 783 and 612 entries in Sg. Mangun and Kulai Tanang. Similarly, the quadrat entered most frequently accounted for 32, 29 and 35 of the 499, 574 and 557 quadrat entries and 29,18 and 12 quadrats were entered only once in the study sites of agile gibbon in Lr. Gambir, Sg. Tembalun and Kulai Tanang (Fig. 7.9).

The peak number of scans noted in a single quadrat was in Sg. Mangun due to two favourite food trees used by siamang group, namely a strangler fig (*Ficus sp.* [aro kapas]) and argus pheasant tree (*Dracontomelum dao*). The peak use of a single quadrat by agile gibbon group was observed in Kulai Tanang, because of the presence of the same two tree species, and it is an overlapping area with another gibbon group.

Overlap

The only case of conspecific overlap was for agile gibbon (U3) in Kulai Tanang (continuous forest), which shared about 20–25% of home range with two neighbouring groups of agile gibbon (and one group of siamang). A group of siamang

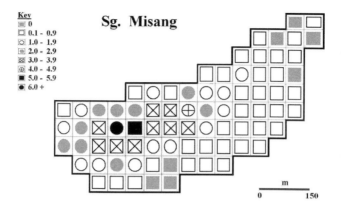

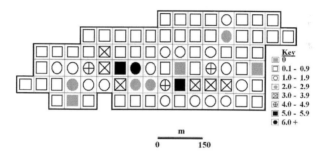

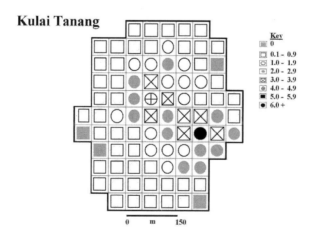

Fig. 7.8 The percentage of the total observation of *S. syndactylus* in each 0.25 ha

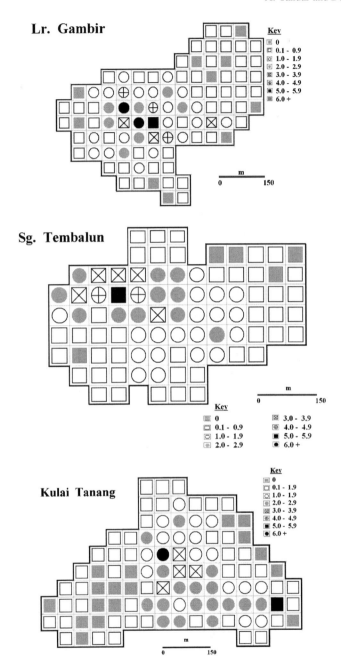

Fig. 7.9 Percentage of the total observation of *H. agilis* in each 0.25 ha quadrat

(S3) shared 10 ha (40%) of its home range with three groups of agile gibbon in the same forest. It was presumably due to habitat conversion by oil-palm plantation in the edge of the forest and as a result, it had been forced groups to share or to overlap their territory. The study area of Kulai Tanang had a high agile gibbon density.

Moreover, in Sg. Mangun, a group of siamang (S2) shared 1.5 ha (5%) of its home range with a single agile gibbon along one edge for 50–100 m. Siamang group did not share their home range with neighbours in Sg. Misang. Similarly, there were no overlaps encountered for agile gibbons with neighbours since they were separated by far more than 200 m in Lr. Gambir. Although three groups of agile gibbon were recorded in the fragment of Sg. Tembalun, home ranges did not overlap.

Discussion

In primates, the home range is a home in reality (Chivers 1974). To accurately determine accurately home range size in primates is difficult (Bartlett 1999), but in this study, superimposed grids of 0.25 ha method were used to count the hectares where the siamang and agile gibbon groups ranged, and this is the method most often used (e.g., Marsh and Wilson 1981; Robinson and Redford 1986; Nunes 1995). Smaller grids give smaller, and presumably more accurate estimates of the home range (Bartlett 1999). Home range size for siamangs and agile gibbons in this study were fairly similar, such that siamangs in forest fragments were 19.75 and 20.25 ha and in continuous forest was 21.50 ha, while for agile gibbon were 23.75 and 21.25 ha in forest fragments and 24.75 ha in continuous forest. Apparently, the size of agile gibbon home range in continuous forest, Kerinci-Seblat National Park, is similar to the mean lar gibbon (2 groups) home range (23.4 ha) at Khao Yai study site (Bartlett 1999).

Indeed, siamang and agile gibbon average home range size in smaller fragments is larger than in large fragments; home range size for siamang and agile gibbon was not correlated with fragment size. Similarly, in Assam study sites, hoolock gibbon home range did not correlate with the size of fragment, or with the disturbance level, but was strongly influenced by the presence or absence of neighbouring groups and amount of food (Kakati 2004).

Home range size for hoolock gibbons in forest fragments varied widely in the smallest fragments (16–17 ha) and the medium-sized fragments (13–24 ha), but were similar in the large forest (24–28 ha) (Kakati 2004). Again, the gibbons that have the smallest home range, the number of sources providing fruit in their home range was also lower (Kakati 2004). Home range size may increase as the number of primates competitors in an area increase (e.g., *Cercocebus albigena*, Chalmers 1968).

In general, the size of home range is related to the size of the group, as well as the group weight or biomass (Milton and May 1976). Pair will defend territory to provide for up three offsprings. Gittins (1979) reported that there was no significant difference between the home range size for lar and agile gibbons.

The size of home range will normally vary within and between species, within certain limits, according to the richness of habitat and population densities (Gittins 1979). Siamang and agile gibbon average home range size in both forest fragments and continuous forest were seemingly smaller (19–25 ha) than previously reported from other study sites of siamang and agile gibbon.

Siamangs and gibbons are more likely to confine their day range to within their territory rather than move out of their home range, as a functional response to human disturbance (Marsh and Wilson 1981). The total mean distance travelled by siamang and agile gibbon groups within their home range in forest fragments was shorter than their counterpart in continuous forest, while the total mean travelled for agile gibbon was longer than siamang. Similarly, Kakati (2004) reported that day ranges of hoolock gibbons at two of the three disturbed habitats (Borajan and Buridehing) were also shorter than that of the other groups. The short day ranges of the hoolock gibbon in Borajan were positively linked to the proportion of fruit and negatively linked to that of leaf in their diet (Kakati 2004). Day ranges of the hoolock gibbon groups were longer when they ate more fruit (Kakati, 2004). Travel distance decreases as the amount of food on trees decreases (Schoener 1971; Raemaekers 1977).

The total mean day range patterns of animals are also related to body weight of animal, energy cost use and season and weather (Chivers 1974; Raemaekers, 1977; Johns 1983; Whitten 1984; Ahsan 1994). As the body weight decreases, the day range increases. The energy cost of travel/unit body weight is lower in larger animals, but the metabolic needs are higher in small animals (Schmidt-Nielsen 1972). Day range is normally smaller during rainy and cloudy weather and higher on dry and sunny days (Chivers 1974), with no movement during heavy rain (Whitten 1984; Ahsan 1994).

Moreover, the agile gibbon's day range in this study showed no correlation with the production of plant parts and mean rainfall in forest fragments and continuous forest. These results are apparently similar to data from lar gibbon and siamang in Peninsular Malaysia, with regard to correlation between day range and rainfall (not significant), but positively correlated with fruit abundance (Raemaekers 1977).

In this study, siamang groups in Sg. Misang (forest fragment) and Kulai Tanang, showed a positive correlation between day range and fruit abundance. There was also a positive correlation between day range and average monthly rainfall in both locations. According to Raemaekers (1979), when siamang became folivorous, they use this strategy, travelling shorter distance, half or less, than lar gibbon, which eat relatively more fruit. Interestingly, *Dracontomelum dao* (Anacardiaceae) is the favourite food of siamang and agile gibbon, as has been observed in Bukit Barisan Selatan National Park, Sumatra (O'Brien et al. 2003), but this species is apparently not present in Ketambe, Gunung Leuser National Park (de Wilde and Duyfjes 1996; Palombit 1992), as well as in Peninsular Malaysia study sites (Chivers 1974).

Acknowledgments We would like to express our gratitude to Dr Ir Willie Smits and the Board Members of the Gibbon Foundation for their encouragement and financial support of this study. Thank you to Dr Kim McConkey for her comments on this paper. We would like to thank the

Director-General of Nature Conservation, for sponsorship and support, who gave us permits to carry out this study in Kerinci-Seblat National Park and surrounds. Our thanks also to go to Dinas Kehutanan Solok District at Solok and Merangin District at Bangko, for helping us with permits to work in local protected forests within oil-palm plantations in South Solok and local gardens in Bangko. In the field, we also thank the Director and staff of oil-palm plantations, PT Tidar Kerinci Agung, PT Tidar Sungkai Sawit and PT Sumatra Jaya Agro Lestari for permission to work in their operational area and for use of their facilities. We would like to thank University of Andalas, Padang and Dr Wilson Noavarino for their help and co-operation and allowing us to use their students, and Pak Sudirman, Sahar, Jon, Hen, Pak Gadimel, Pak Muas and Rauh, who were our field assistants.

References

Ahsan MF (1994) Behavioural ecology of the hoolock gibbon (*Hylobates hoolock*) in Bangladesh. Ph.D. dissertation, University of Cambridge, Cambridge

Ahsan MF (2001) Socio-ecology of the hoolock gibbon (*Hylobates hoolock*) in two forests in Bangladesh. In: Chicago Zoological Society (eds) The Apes: challenges for the 21st century. Chicago Zoological Society, Brookfield, Illinois, pp 286–299

Altmann SA, Altmann J (1970) Baboon ecology: African field research. University of Chicago, Chicago

Altman J (1974) Observational study of behaviour: sampling methods. Behaviour 49:227–267

Bartlett TQ (1999) Feeding and ranging behavior of the white-handed gibbon (*Hylobates lar*) in Khao Yai National Park, Thailand. Ph.D. dissertation, University of Washington

Bates BC (1970) Territorial behavior in primates: a review of recent field studies. Primates 11:271–284

Bierregaard RO Jr, Laurance WF, Sites JW Jr, Lynam AJ, Didham RK, Anderson M, Gascon C, Tocher MD, Smith AP, Viana VM, Lovejoy TE, Sieving KE, Kramer EA, Restrep C, Moritz C (1997) Key priorities for the study of fragmented tropical ecosystems. In: Laurance WF, Bierregaard RO Jr (eds) Tropical forest remnants ecology, management and conservation of fragmented community. The University of Chicago, Chicago, pp 515–525

Burt WH (1943) Territoriality and home range concepts as applied to mammals. J Mammals 24:352–364

Chalmers NR (1968) Group composition, ecology, and daily activities of free living mangabeys in Uganda. Folia Primatol 8:247–262

Chapman CA (1988) Patterns of foraging and range use by three species of neotropical primates. Primates 29:177–194

Chivers DJ (1969) On the daily behaviour and spacing of howler monkey groups. Folia Primatol 10:48–102

Chivers DJ (1974) The siamang in Malaya. Contrb Primatol 4:1–335

Chivers DJ, Raemaekers JJ, Aldrich-Blake FPG (1975) Long-term observations of siamang behaviour. Folia Primatol 23:1–49

Clutton-Brock TH (1975) Ranging behaviour of red colobus (*Colobus badius tephrasceles*) in the Gombe National Park. Anim Behav 23:706–722

Clutton-Brock TH, Harvey PH (1977) Species difference in feeding and ranging behaviour in primates. In: Clutton-Brock TH (ed) Primate ecology: studies in feeding and ranging behaviour in lemurs, monkeys and apes. Academic, London, pp 557–584

Decker BS (1994) Effects of habitat disturbance on the behavioural ecology and demographics of the Tana River red colobus (*Colobus badius rufomitratus*). Int J Primatol 15(5):703–734

DeVore I (1965) Primate behavior: field studies of monkeys and apes. Holf, Rinehart and Winston, New York deWilde WJJO, Duyfjes BEE (1996) Vegetation, floristic and plant bio-geography in Gunung Leuser National Park. In: van Schaik CP, Supriatna J (eds) Leuser: a Sumatran sanctuary. Yayasan Bina Sains Hayati Indonesia, Jakarta Indonesia

Eisenberg JF, Muckenhirn NA, Rudran R (1972) The relation between ecology and social structure in primates. Science 176:863–874

Fossey D, Harcourt AH (1977) Feeding ecology of free-living mountain gorillas (*Gorilla gorilla beringei*). In: Clutton-Brock TH (ed) Primate ecology: studies in feeding and ranging behaviour in lemurs, monkeys and apes. Academic, London, pp 415–447

Freeland WY (1979) Mangabay (*Cercocebus albigena*) social organization and population density in relation to food use and availability. Folia Primatol 32:108–124

Gittins SP (1979) The behaviour and ecology of the agile gibbon (*Hylobates agilis*). Ph.D. dissertation, University of Cambridge, UK

Gittins SP (1983) The use of the forest canopy by the agile gibbon. Folia Primatol 40:134–144

Goodall J (1977) Infant killings and cannibalism in free-living chimpanzees. Folia Primatol 28:259–282

Harrison MJS (1983) Patterns of range use by the green monkey, *Cercopithecus sabaeus*, at Mt Assirik, Senegal. Folia Primatol 41:682–694

Iwamoto T, Dunbar RIM (1983) Thermoregulation, habitat quality, and behavioural ecology of gelada baboons. J Anim Ecol 52:357–366

Jewell PA (1966) The concept of home range in mammals. In: Symposia Zoological Society of London, vol 18, pp 85–109

Johns AD (1983) Ecological effects of selective logging in a west Malaysian rain forest. Ph.D. dissertation, University of Cambridge, Cambridge, UK

Kakati K (2004) Impact of forest fragmentation on the hoolock gibbon in Assam, India. Ph.D. dissertation, University of Cambridge, UK

Kummer H (1968) Social organization of hamadryas baboons: a field study. Karger, Basel

Leighton D (1987) Gibbons: territoriality and monogamy. In: Smuts BB, Cheney D, Seyfarth R, Wrangham R, Struhsaker T (eds) Primate societies. University of Chicago Press, Chicago, pp 135–145

Marsh CW, Wilson WL (1981) A survey of primates in peninsular Malaysia forests final report for the Malaysian Primates Research Programme. Universiti Kebangsaan Malaysia and University of Cambridge, UK

Milton K, May ML (1976) Body weight, diet, and home range area in primates. Nature 259:459–462

Nellemann C, Newton A (2003) The great apes – the road ahead. WCMC-UNEP, Cambridge

Nunes A (1995) Foraging and ranging patterns in white-bellied spider monkeys. Folia Primatol 65:85–99

O'Brien TG, Kinnaird MF, Nurcahyo A, Prasetianingrum M, Iqbal M (2003) Fire, demography and the persistence of siamang (*Symphalangus syndactylus*: Hylobatidae) in a Sumatran rainforest. Anim Conserv 6:115–121

Palombit AR (1992) Pair bond and monogamy in wild siamang (*Hylobates syndactylus*) and white-handed gibbon (*Hylobates lar*) in northern Sumatra. Ph.D. dissertation, University of California, Davis

Raemaekers JJ (1977) Gibbons and trees: comparative ecology of the siamang and lar gibbons. Ph.D. dissertation, University of Cambridge

Raemaekers JJ (1979) Ecology of sympatric gibbons. Folia Primatol 31:227–245

Rasmussen DR (1979) Correlates of patterns of range use of a sites, impregnable females, births, and male emigrations and immigrations. Anim Behav 27:1098–1112

Robinson JG, Redford KH (1986) Body size, diet and population density of neotropical forest mammals. Am Nat 128:665–680

Rylands AB, Keuroghlian A (1988) Primate populations in continuous forest and forest fragments in central Amazonia. Acta Amazonica 18:291–307

Schmidt-Nielsen K (1972) Locomotion: energy cost of swimming, flying, and running. Science 177:222–228

Schoener TW (1971) Theory of feeding strategies. Ann Rev Ecol Syst 2:369–404

Spironello WR (2001) The brown capuchin monkey (*Cebus paella*): ecology and home range requirements in central Amazonia. In: Bierregaard RO, Gascon C, Lovejoy TE, Mesquita RCG (eds) Lessons from Amazonia. The ecology and conservation of a fragmented forest. Yale University Press, Conneticut, pp 271–289

Struhsaker TT (1974) Correlates of ranging behaviour in a group of red colobus monkey (*Colobus badius tephrosceles*). Am Zool 14:177–184

Struhsaker TT (1975) The red colobus monkey. Chicago University Press, Chicago

Tutin CEG (1999) Fragmented living: behavioural ecology of primates in forest fragment in the Lopé Reserve, Gabon. Primates 40:240–265

Whitmore TC (1975) Tropical rain forests of the far east. Clarendon, Oxford

Whitten AJ (1984) Ecological comparisons between Kloss gibbons and other small gibbons. In: Preuschoft H, Chivers DJ, Brockelman WY, Creel N (eds) The lesser apes: evolutionary and behavioural biology. Edinburgh University Press, Edinburgh, pp 219–227

Wrangham WR (1977) Feeding behaviour of chimpanzees in Gombe National Park, Tanzania. In: Clutton Brock TH (ed) Primate ecology. Academic, London, pp 504–539

Chapter 8
Behavioural Ecology of Gibbons (*Hylobates albibarbis*) in a Degraded Peat-Swamp Forest

Susan M. Cheyne

Introduction

Gibbons are small arboreal apes inhabiting the rainforests of South-East Asia, Northwest India and Bangladesh (Carpenter 1940; Chivers 1977). The taxonomy of gibbons is under dispute, as the status of several taxa as species or subspecies is uncertain. Within the family Hylobatideae, there are four genera of gibbons: *Bunopithecus* (hoolock gibbon), *Hylobates*, *Nomascus* (crested gibbons) and *Symphalangus* (siamangs), and at least 12 species (Brandon-Jones et al. 2004). Apart from the sympatric *Hylobates agilis*/*Hylobates lar* and siamangs in Sumatra and peninsular Malaysia, gibbons are allopatric (Leighton 1987; Reichard and Sommer 1997). Some hybrids have been found within the genus *Hylobates*, including populations in Borneo (*Hylobates albibarbis* and *Hylobates muelleri*: Mather 1992), in Thailand (*H. lar* and *Hylobates pileatus*: Brockelman and Gittins 1984) and in peninsular Malaysia (*H. lar* and *H. agilis*: Brockelman and Gittins 1984). The Bornean agile or southern gibbon (*H. albibarbis*) occurs in southern Borneo, between the Kapuas and Barito rivers (Brandon-Jones et al. 2004). Its taxonomic status is unclear, but recent molecular evidence identifies it as a separate species, rather than a subspecies of *H. agilis* (Brandon-Jones et al. 2004; Geissmann 2007; Groves 2001).

Although their diet also includes young leaves and flower buds, gibbons are mostly frugivorous (Cheyne 2007a; Gittins 1982, 1983; Gittins and Raemakers 1980; McConkey et al. 2002). Thus they appear to play a primary role in forest regeneration as high quality seed dispersers (Gittins 1982; McConkey 2000; O'Brien et al. 2003).

S.M. Cheyne (✉)
Wildlife Conservation Research Unit, Department of Zoology,
Oxford University, Abingdon Road, Tubney OX13 5QL, UK
e-mail: susan.cheyne@zoo.ox.ac.uk

Orang-utan Tropical Peatland Project, Centre for the International Cooperation in Management of Tropical Peatlands (CIMTROP), University of Palangka Raya, Central Kalimantan, Indonesia

S. Gursky-Doyen and J. Supriatna (eds.), *Indonesian Primates*,
Developments in Primatology: Progress and Prospects,
DOI 10.1007/978-1-4419-1560-3_8, © Springer Science+Business Media, LLC 2010

Gibbons live in small family groups of two to six individuals (Gittins and Raemakers 1980; Leighton 1987) with an average group size of four (Gittins and Raemakers 1980). They are socially monogamous, with males and females forming stable pairs (Gittins and Raemakers 1980; Leighton 1987; Mitani 1987b, 1990), but several long-term studies have reported extra-pair copulations and reproductive patterns may differ from the social system (Palombit 1994; Reichard 1995; Reichard and Sommer 1997). All gibbon species exhibit territorial behaviour, with mated pairs defending exclusive territories (Gittins and Raemakers 1980; Reichard and Sommer 1997). Home ranges may, however, overlap, especially in areas where gibbon densities are high (Gittins and Raemakers 1980; Mitani 1990; Reichard and Sommer 1997). Most gibbon pairs use a sequence of calls called a duet to defend their territory and strengthen their pair bond (Mitani 1987a; Cowlishaw 1992; Geissmann and Orgeldinger 2000; Cheyne et al. 2007a): in the morning the resident pair of each territory utters a series of calls, the more characteristic and most easily recognisable one being the female's great call (Brockelman and Srikosamatara 1993). It must be noted that the Kloss gibbon (*Hylobates klossii*) and the Moloch or Javan gibbon (*Hylobates moloch*) do not duet, though both sexes do still sing (Bartlett 2007). This great call, which can be heard up to 1 km away in flat, dense rainforest (Brockelman and Ali 1987), can be used for auditory sampling method in surveys (Brockelman and Ali 1987; Brockelman and Srikosamatara 1993; Nijman and Menken 2005; Cheyne 2008a).

All Indonesian gibbons are faced with threats to their survival, both because of habitat loss through logging, encroachment and forest fires, and because of hunting for food or capture for the pet trade, and are placed on the URL CITES Appendix 1. Following new genetic research and on the recommendation of the Asian Primate Red List Workshop, held in 2006 in Phnom Penh, Cambodia, the species' status has been recognised as *H. albibarbis* and classified as Endangered, partly because of the rapid rate of destruction of peat-swamp forests (PSF), which constitute a large part of its range (Geissmann 2007; Campbell et al. 2008). *Hylobates albibarbis* was previously classified as a sub-species of *H. agilis* and listed as low risk. The status of the species Hylobates albibarbis has been changed to Endangered following the most recent IUCN Red list assessment (IUCN Red List 2008). Although much is known about the behaviour and ecology of gibbons, limited data are available on the population status of *H. albibarbis*, and the Asian Primate Red List Workshop concluded that more recent population estimates and rigorous monitoring of those populations are needed for the species' conservation (Geissmann 2007).

Although gibbons have been the focus of many behavioural and ecological studies (e.g. Gittins 1980; Brockelman and Srikosamatara 1984; Chivers 1984; Mitani 1990; Bartlett 2007; Cheyne 2007a, 2008a; Fan et al. 2009), there is little information on their habitat requirements, in particular in PSF (Buckley et al. 2006; Geissmann 2007; Campbell et al. 2008; Cheyne et al. 2008b). No clear explanation for the important variation in gibbon densities between field sites has been established (Leighton 1987); since vegetation correlates of primate densities have been found for other species (Ross and Srivastava 1994; Muoria et al. 2003; Wieczkowski 2004; Rovero and Struhsaker 2007), it is possible that they exist for gibbon populations as well. As acknowledged in recent workshops on the conservation status of gibbons in Indonesia

(Geissmann 2007; Campbell et al. 2008), one of the major habitats of the Bornean agile gibbon is PSF. Peat-swamp forests occur in the Indo-Malayan region, principally in Kalimantan and Sumatra (Page et al. 1999a, 2002). Kalimantan has extensive peatlands, covering about 6 million hectares of its lowlands (Rieley 2002; Rieley et al. 1993, 1996, 2004), a very small proportion of which is protected within national parks (Morrogh-Bernard 2003). Peat-swamp forests are seasonally flooded, waterlogged lowland forests. Because they were thought to harbour little biodiversity, and because they contain tree species of commercial interest (Rieley et al. 1996; Page et al. 2002; Morrogh-Bernard et al. 2003), PSF have received little conservation attention and have been extensively cleared and/or converted to cultivated land. Logging activities, in addition to the removal of large trees, are detrimental to the ecosystem as the canals dug within the forest to carry felled trunks to adjacent rivers drain the peat, making the soil dry and prone to wildfires (Morrogh-Bernard 2003; Cheyne 2007a). However, more recent studies have highlighted the importance of PSF for conservation. Despite their waterlogged, highly acidic, nutrient-poor soil, PSF have tree species diversity comparable to other forests on mineral soils, and feature a number of commercially valuable trees (Rieley et al. 1993; Felton et al. 2003). Reports on the fauna in PSF are scarce but surveys have recorded 57 species of mammals, 237 species of birds, 55 species of fish, as well as reptiles and amphibians (Page et al. 2002). In Central Kalimantan, between the Kapuas and Barito rivers, lies the protected PSF of the Sabangau catchment. This area covers 5,300 km² and was gazetted as a national park in November 2004, after having been allocated to logging companies for timber extraction for 30 years. Previous work in the area has focused on ecological and hydrological studies as well as forest regeneration monitoring (Page et al. 1999b) and has shown that the Sabangau catchment harbours the largest remaining wild population of Bornean orang-utans (*Pongo pygmaeus*) (Morrogh-Bernard et al. 2003) and an extensive wild population of Bornean agile gibbons (*Hylobates albibarbis*) (Buckley *et al.* 2006; Cheyne 2008a). Ongoing work is being carried out on behavioural aspects and feeding ecology of gibbons and orang-utans in PSF, aiming to fill gaps in scientific knowledge about apes in this unique ecosystem (Page et al. 2002). Survey work is also ongoing on populations of nocturnal primates (Bornean slow loris *Nycticebus coucang menagensis* and western tarsier *Tarsius bancanus borneanus*: Blackham 2005; Nekaris et al. 2008), flying foxes (*Pteropus vampyrus natunae:* Struebig et al. 2007) and wild felids (Cheyne 2008a).

Historical Note

Gibbons have been described as far back as ca. 200 AD by Aelian (Claudius Aelianus) a Roman who wrote about animal classification. Darwin in his Descent of Man (1871) describes siamangs and agile gibbons from Sumatra (Indonesia) and discusses their singing in detail (Chaps. 18 and 19). The first (western) study of gibbons was conducted by Clarence Ray Carpenter in 1937 as part of the Asiatic Primate Expedition – everything we know about gibbons in general came from this work. In China, the gibbon has been known since at least the

Zhou Dynasty (1027–221 BC; van Gulik 1967), where gibbons are described as "the aristocrat among apes and monkeys." Despite the prevalence of records and stories about gibbons throughout history, both in the east and west, we are still learning about these fascinating small apes. Most importantly, we need to learn how to conserve them, as gibbons are threatened throughout their range (IUCN Red List 2008).

IUCN Status and Threats

Following the Asian Primate Red List Workshop (Geissmann 2007) and the Indonesian Gibbon Conservation and Management Workshop (Campbell et al. 2008), all Indonesian gibbons are now listed as "endangered" on the IUCN Red List 2008 recognising the ongoing threats and population decline of all gibbon species.

Population estimates of the Bornean agile or southern gibbon (*Hylobates albibarbis*) are rarely reported. Probably the largest extant and contiguous population is in the Sabangau Catchment, Central Kalimantan, with an estimated population of at least 30,000 individuals (Cheyne et al. 2007b). Information on population status and trends are lacking from other areas, particularly non-protected areas.

A wide variety of threats to *H. albibarbis* were identified by the Kalimantan Working Group of the Indonesian Gibbon Conservation and Management Workshop (Table 8.1) (Campbell et al. 2008). I will return to these threats at the end of the chapter and offer some solutions based on the behaviour and ecology presented below.

Hylobates albibarbis

Following Groves (2001) and Geissmann (2007) the Bornean agile or southern gibbon has been recognised as a separate species designated *H. albibarbis*. This species occurs within the Kalimantan regions of Indonesian Borneo east of the Kapuas River (west Kalimantan), west of the Barito River (Central Kalimantan), south of the Busang River (Central Kalimantan) and to the north and west of the Schwaner Mountains (see Fig. 8.1). Some of the main areas with substantial populations are Central Kalimantan: Sabangau Catchment, Tanjung Puting NP, Bukit Baka Bukit Raya NP, Hampapak Nature Reserve (NR), Tahura, Lamandau, Tuanan, Kendawangan NR, Arut Blantikan. West Kalimantan: Rongga Perai LH, Gunung Palung NP, Sungai Putri (Campbell et al. 2008).

Males weigh 6.1–6.9 kg and females weigh 5.5–6.4 kg (Cheyne unpublished data). Pelage can vary but generally *H. albibarbis* has black hands and feet with brown arms and legs. White eyebrows and/or cheeks are common but not universal, and the main body colour is usually dark brown with dark chest and head cap. Newborns are very pale brown with no hair on face, palms or soles of the feet (all of which are black, Cheyne pers. obs).

Hylobates albibarbis are found in primary forest and disturbed (logged) secondary forest and in lowland and montane habitat including PSF (Cheyne 2007b), thus as

Table 8.1 Threats to Indonesia's gibbons

Threat	Reason
Oil palm plantations	Forested land is cleared for plantations instead of using already cleared land
Acacia plantations	Forested land is cleared for plantations instead of using already cleared land
Legal logging	Companies do not follow the rules of the logging concession
Illegal logging	Uncontrolled logging in protected (and un protected areas)
Fire	Fires destroy gibbon habitat and create palls of smoke that can last for several months and are detrimental to gibbon health (and to humans)
Habitat fragmentation	Gibbons cannot disperse from small fragments to create new groups; thus small fragments reach carrying capacity very quickly
Pet trade	Devastates gibbon groups, is unsustainable, 5 gibbons die for each infant which makes it to market (pers. obs.)
Mining	Forest is cleared to expose large areas of land for open-cast mining and oil drilling
Global warming and climate change	Indirect effects through increased intensity of fires and direct effects through unpredictable food availability for gibbons
Clearing forest for urban expansion	Gibbon habitat is being encroached upon to allow expansion of villages, towns and cities
	Status of protected gibbon habitat is changed to allow for urban expansion
Hunting gibbons (not for pet trade)	Adult gibbons are hunted by local communities for bush meat, thus contributing to population decline. While this practice is not presently very intensive, the practice is gaining in popularity as a status symbol
Harvest of non-timber forest products	Gibbon habitat is encroached upon by people gathering orchids, hunting flying foxes, gemur tree (*Alseodaphne coriacea* (Lauraceae) anti-malarial properties), agar wood.
Dam development for electricity	Gibbon habitat is flooded when dams are built (more of a threat for gibbons in mountainous areas).

a result of the habitat, the average heights of gibbons in the forest are very dependent on the type of forest they inhabit.

Females reach sexual maturity in about 48months (SD 3.67) (Cheyne and Chivers 2006; Cheyne 2008b) and the menstrual cycle lasts 26 days (SD 0.65) with the peri-ovulatory swelling phase lasting about 6.3 days (Cheyne and Chivers 2006). Gibbons do not have reproductive suppression and mature sub-adults are ejected from the natal group by the same-sex parent (Chivers 1972).

Study Area and Gibbons

The Sabangau Catchment (SC) encompasses one of the largest PSF in the world and is recognised as a low-productivity habitat (Page et al. 1999b; Morrogh-Bernard et al. 2003). The area is home to probably the largest population of

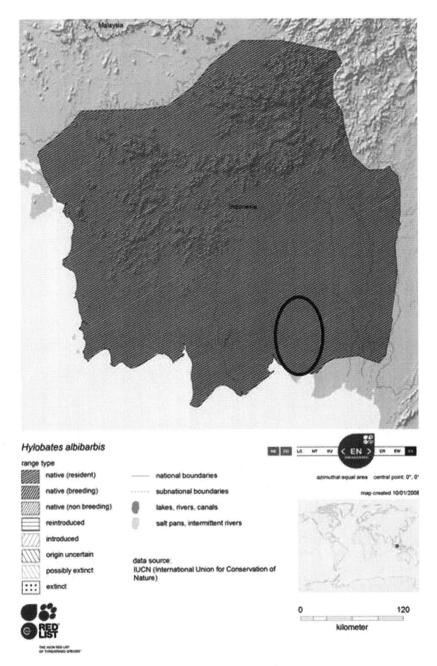

Fig. 8.1 Distribution of *H. albibarbis* from IUCN Red List 2008

Bornean agile gibbon (*Hylobates albibarbis*), with an estimated population of about 30,000 individuals (Cheyne et al. 2007b). The research was carried out in the Natural Laboratory of Peat-swamp Forest, Central Kalimantan, Indonesia (2° 31″ S and 113° 90″ E), operated by the Centre for International Cooperation in Management of Tropical Peatland (CIMTROP). The research site is located in the north-eastern edge of the Sabangau Catchment between the rivers Sabangau and Katingan. The SNP is also home to the worlds' largest orang-utan population, *Pongo pygmaeus wurmbii*, (Morrogh-Bernard et al. 2003).

The area is deep-PSF consisting of three distinct habitat sub-types (Shepherd et al. 1997; Page et al. 1999a). The forest is flooded annually from October–June and has a varied canopy structure having been subjected to 30 years as a logging concession followed by several years of illegal logging before being declared as a national park in 2004. The altitude is about 10 m a.s.l. and the whole park is about 5,300 km² with the core study area consisting of 9 km². Temperature averages 26°C (range 18–38°C) and rainfall averages 232 mm (4–532 mm) (Husson et al. 2008).

Eleven groups of gibbons and two unmated males have been habituated and at least two groups per day have been followed since June 2005. Data collection is ongoing, but results are presented here from July 2005 to July 2008. Group composition during the study period is presented in Table 8.2. Only data from paired adults are presented here. In this chapter, I will describe the findings from this study, which highlights the importance of long-term datasets, and I will compare these results to those of other gibbon species. I will also discuss the conservation implications for gibbons.

Table 8.2 Total study animals by age/sex class

Group name (habituated)	Adult F	Adult M	Sub-adult F	Sub-adult M	Juvenile M	Juvenile F	Infant	Group total
A	1	1	1	0	0	0	1	4
C	1	1	1	0	1	0	1	5
E	1	1	1	0	0	1	1	5
K	1	1	1	0	1	0	1	5
N	1	1	1	0	1	0	1	5
2	1	1	1	0	0	1	0	4
H	1	1	0	1	0	1	0	4
M	1	1	0	1	1	0	0	4
J	1	1	0	1	1	0	0	4
T	1	1	1	0	0	1	0	4
S	1	1	1	1	0	0	1	5
				Study site total number of gibbons				48
				Average gibbons/group (not inc. single male)				4.45

Aims

- To monitor gibbon population size, distribution, behavioural ecology, feeding ecology and health.
- To monitor food availability and overall productivity in the forest
- To monitor gibbon energy intake, and to assess how this is governed by food availability and how it relates to the behaviour and health of the population.
- To monitor the affects of anthropogenic disturbance and conservation measures on the gibbon population and food availability/forest productivity.
- To investigate the changes in gibbon singing behaviour in response to changing climactic conditions.
- Collect measures of forest productivity, including fruiting/flowering patterns, litter fall, tree growth rates and mortality rates.
- Train local staff and CIMTROP personnel in field methods and data analysis; and produce a training DVD.
- Disseminate information locally through CIMTROP's patrol team, regionally to forest management authorities and conservation organisations and internationally through scientific papers and conferences.

Methods

Gibbon density was estimated by three groups of two observers stationed from early in the morning at three listening posts. Gibbon pairs sing duets most mornings for the purposes of maintaining pair-bonds and defining each pair's territory. Observers recorded the direction of each calling group heard and the time and estimated distance of the call to enable triangulation of group locations. This was repeated for four consecutive days to ensure that all groups within hearing distance are heard singing at least once. Gibbon density was estimated from these data using a standard formula (Cheyne 2008a; Hamard 2008). This method uses indicators of presence as opposed to actual counts of animals, and consequently additional parameters (e.g. calling frequency) are required to estimate animal density.

Measurements of habitat quality are made along each transect, including canopy cover at 20 m; density of medium (>10-cm dbh); large (>20-cm dbh) and very large (>30-cm dbh) trees (NB tree size in PSF is generally smaller than in other habitats); and an assessment of overall tree biomass by measuring the total basal area of all trees >7-cm dbh in small plots. Trees are identified (using the CIMTROP Herbarium as required) and indices of species diversity obtained.

A focal adult is followed during its active period (dawn to dusk) and behavioural data collected using standardised 5-min instantaneous sampling techniques (Altmann 1974). This yields a range of behavioural variables that are monitored from year-to year to assess behavioural ecology, nutritional and physical health. These include population age–sex structure, length of daily active period, activity budgets, dietary composition and daily travel distance. Nutritional health was assessed by calculating energy intake using known caloric values of foods and energy expenditure, thus calculating the relative energy balance.

Six permanent forest plots of total area 2.4 ha and containing >2,500 tagged stems have been established in the core research area. These are monitored monthly for flower and fruit production, and also provide long-term data on tree growth and mortality rates, and changes in forest biomass. Sixteen 1-m litterfall traps are spaced apart in the forest and fallen leaves, small branches and reproductive parts will be collected, separated, dried and weighed on a monthly basis. All of these measures are useful indicators of forest productivity and biomass, and respond to changes in season, rainfall, soil nutrient content and smoke pollution from forest fires. Readings of temperature, rainfall and light intensity are made daily and air quality readings are obtained from the monitoring station in nearby Palangka Raya.

Habituation of Gibbons

The purpose of habituation is to acclimatise the animals to human presence so that they engage in their normal activity. In animal behaviour studies in the field, investigators rely on the study animals becoming habituated to the presence of the investigator. Additionally, researchers have a responsibility to the animals to ensure that the habituation process does not result in the animals becoming more susceptible to poachers.

Habituation techniques are frequently overlooked in the literature, though getting the habituation right is vital to effective behavioural studies on primates. Though most PhD (and some masters) theses present detailed information on habituation of primates, much of this information is not as accessible as peer-reviewed publications. A search on Web of Science resulted in one published paper describing how to habituate wild primates (Sykes monkeys, *Cercopithecus mitis albotorquatus*) in Kenya, (Moinde et al. 2004) and one general book chapter (Williamson and Feistner 2003).

Habituation Techniques

- Green/camouflaged or forest coloured clothing is worn by all researchers and field staff. (No bright colours are worn.)
- Minimum distance between researcher and gibbons is 10 m.
- Researchers act submissively – not looking directly at the gibbons as they can feel threatened by direct staring, hide and sit down, pretend to eat leaves.
- No loud talking.
- Once the estimated locations of the gibbon territories were known (through triangulation mapping), systematic searching within these territories began.

To locate gibbons, two researchers walk the trails in a systematic manner, no more than 1 trail apart and in radio contact. Once gibbons are sighted, the other person is radioed and comes to meet the first researcher.

Signs that the Gibbons are Becoming More Habituated

- Fleeing and hiding behaviour becomes uncommon
- Gibbons can be sighted and followed from sleeping tree – sleeping tree or for most of their active period (about 9 h from 04:30 to 12:30 h).

Searching for and Locating Gibbons

- The best way to find gibbons is to use their morning songs to locate them.
- Once gibbons begin to sing in the morning it is essential that researchers do not immediately approach the gibbons' position. Researchers should approach to a close distance without disturbing the gibbons. Researchers then wait until singing has completely stopped before approaching. Gibbons use morning duetting to defend their territory. If gibbons are constantly being disturbed by humans while singing, they will frequently interrupt the singing and flee or hide quietly. This is disruptive to both the pair-bond of the gibbons and to their territorial defence. During the habituation process I strongly recommend that gibbons are not constantly interrupted in their singing.
- Never chase the gibbons. Always mirror their behaviour i.e. stop when they stop and move when they move. You can move around on the forest floor to obtain a clear view of your focal gibbon without having to be directly under the tree.
- When tagging trees, wait until the gibbons have moved out of the tree before tagging. If the tree is a sleeping tree, wait until the following morning before tagging it, to ensure that the gibbons did not move in the night.
- To find the sleeping tree the next morning, tie cotton thread from the tree to the nearest point on the transect.

This method does mean that often by the time you have approached the gibbons after they stop singing, they have already travelled away. This makes the habituation slow and frustrating but I believe this is the best method as you avoid disturbing the gibbons while they are singing. After several months of this method we were able to sit directly under the gibbons as they were singing.

Once gibbons have been located, the following process begins. Two researchers try to follow gibbons, by choosing a focal animal and trying to follow this individual. Researchers must remain quiet and unobtrusive and should not use machetes to cut through vegetation, even if off-trail. Researchers should not approach the gibbons too closely. If the gibbon approaches, a minimum distance of 10 m should be observed at all times. If gibbons are lost then researchers should continue to try and locate them, using the territory maps (from triangulation study) as a guide for searching. If gibbons are lost between 13:00 and 14:00 h researchers can assume that they have entered the sleeping tree. Thus, next morning return to this area to wait for the gibbons to start singing again. Researchers should be at the sleeping tree/sleeping area by 04:30 h.

When this project was initiated, the gibbons fled at the sight of humans. Now 11 groups and one lone male (another lone male died as a result of an inter-group encounter) can be followed from morning to night sleeping trees. Before habituation can take place it is important to know the distribution of the primates (to ensure accurate identification of animals). Additionally, researchers need to understand the biology and behaviour of the primates and to tailor their responses to them (e.g. no bared teeth or eye contact) and the need to ensure that your habitation of the primates does not put them at risk from poachers.

Home Range Size, Group Encounter Rates and Territorial Overlap

Since July 2005, both GPS waypoints taken when following each gibbon group from daily night-tree to night-tree and digitised hand-drawn maps produced during the follows. Trails were about 250 m apart (see Fig. 8.7 for map of the study area). Trails were marked at 25- and 50-m intervals and at crossroads, which allowed the teams to precisely map individuals' travel paths. Observed travel routes were digitised in ARC/INFO 3.4; the lengths of the routes were measured using ArcView 3.0a software. Data presented here are from July 2005 to July 2008. Estimates of home-range sizes were based on all observed travel routes using the minimum convex polygon method (White et al. 1996; Linnell et al. 2001; Savini et al. 2008). Minimum polygon outlines were then digitised in ARC/INFO 3.4, and the areas of the polygons were calculated using ArcView 3.0a software to obtain the actual home-range sizes.

Sabangau home ranges average 47 ha (range 39–52; Table 8.3) (Chivers 1974; Raemaekers 1979; Kappeler 1981; Leighton 1984; Feeroz and Islam 1992; Ahsan 2001; Chivers 2001; Cheyne 2008a; Savini et al. 2008). Based upon detailed maps of the territories and digitised GIS maps, the overlap of the territories is about 15%, considerably less than the 64% found by Reichard and Sommer (1997) in Khao Yai National Park in Thailand. Few other data are available on territorial overlap in gibbons. The reasons behind this large difference in overlap could be due to the comparative low densities of gibbons in Sabangau (average 2.6 gibbons/km^2 Cheyne et al. 2007b; Hamard 2008), compared to other sites (Table 8.4) (Chivers 1974; Tenaza 1975; Rodman 1976; MacKinnon 1977; Brockelman et al. 1998; Cheyne et al. 2007a).

A low density of gibbons is not an ultimate cause for the low observed overlap, it is possible that the food abundance and availability (and available energy) are ultimately responsible for the low overlap. Khao Yai National Park gibbons have

Table 8.3 Density of gibbons in this and other studies

Species	Location	Site	Source	Density (km^2)
syndactylus	Malaysia	Ulu Gombak	Chivers 1974	4.7
		Kuala Lompat	Mackinnon 1977	4.5
	Sumatra	Ranun	Mackinnon 1977	6.6
		Ketambe	Rijksen 1978	16
Klossi	Mentawi Islands	Siberut	Tenaza 1975	30
muelleri	Borneo	Ulu segama	Mackinnon 1977	10.5
Agilis	Malaysia	Sungai Dal	Chivers 1974	4.6–6.3
Lar	Malaysia	Ulu Gombak	Chivers 1974	6.8
		Ulu Sempan	Chivers 1974	1.4
		Kuala Lompat	Chivers 1974	4.9
		Kuala Lompat	Mackinnon 1977	4.1
	Thailand	Khao Yai	Brockelman et al. 1998	5
albibarbis	Borneo	Sabangau	Cheyne 2007a; Hamard 2008	2.6

Table 8.4 Home range sizes of gibbon in this and other study sites

Species	Location	Site	Source	Home range size (ha)
Lar	Thailand	Khao Yai	Savini et al. 2008	28
	Malaysia	Kuala Lompat	Raemaekers 1979	14
albibarbis x muelleri	Borneo	Barito Ulu	Chivers 2001	18
muelleri	Borneo	Kutai	Leighton 1984	44
Moloch	Java	Ujung Kulon	Kappeler 1981	17
Kloss	Mentawi Islands	Siberut	Whitten 1980	32
hoolock	Assam	Lawachara	Feeroz and Islam 1992	35
			Farid Ahsan 1993	63
		Chunati	Farid Ahsan 1993	26
			Feeroz and Islam 1992	33
siamang	Malaysia	Ulu Sempan	Chivers 1974	15
albibarbis	Borneo	Sabangau	Cheyne 2008a	47

small home ranges and the area appears to have high food abundance (see section on Diet and Feeding Ecology below).

Groups average 4.45 individuals (Cheyne et al. 2007a) with group size ranging from 3 to 5. Two lone males have been identified and followed (though one subsequently died) and an additional third lone male is known to have dispersed from his natal group (behavioural data are being collected on this male but are not presented here). Due to the larger territories and lower densities of gibbons, it is to be expected that group encounters and fights are low. Gibbons were involved in inter-group interactions on 25 of 550 follow days (4.55%). Thus, the encounter rate between groups is about once every 22 days. For now, the question of monogamy in this population cannot be answered and no extra-pair copulations have been witnessed. A paternity project to test DNA from known, habituated gibbons is planned for 2009.

Activity Patterns and Time Budgets

Of great importance in a species ecology is the proportion of time that animals spend in different activities and the distribution of these activities throughout the active period. Activity profiles can help understand how a species uses resources and adapts to its environment.

It was not always possible to determine at what time the gibbons left sleeping trees, in the majority of cases. Initially the gibbons were not habituated enough to allow full follows from morning to evening sleeping tree. It was not possible to determine the time the gibbons awoke, but time of leaving and entering sleeping trees was recorded. Once one of the adults was awake, all members of the group would awake within several minutes of each other. Defecation and urination took place, often while the gibbon was hanging in the sleeping tree and the group would move off together. Gibbons would leave the sleeping tree −77 to +35 min (mean 0.2 min, median 14.5 min, $n = 189$) before/after sunrise. *Hylobates albibarbis* has an activity start time similar to other small-bodied gibbons, though quite different from its closest relative, *H. agilis* from Malaysia, where there are no orang-utans (Fig. 8.2) (Gittins 1982). These differences could be attributed to the proximity of the SNP to the equator and the relatively constant day/night lengths (Cheyne 2007a).

The gibbons would enter the sleeping tree 219–297 min (mean 257 min, median 255 mins, $n = 142$) before sunset, regardless of time of year. Once in the sleeping trees there would often be grooming between the adults before one adult moved to another sleeping tree. The gibbons could be in the sleeping tree for 12–54 min (mean 26, median 29) before activity ceased and the gibbons were assumed to be asleep. *Hylobates albibarbis* enter the sleeping tree significantly earlier than other species (Fig. 8.3). The main difference between all these sites is that orang-utans are only present in the SNP.

Gibbons avoid prolonged resting in the middle of the day, characteristic of monkeys. Thus, the early entry to sleeping trees could be to avoid food competition with monkeys and orang-utans, which can be active up to and beyond sunset (pers. obs.).

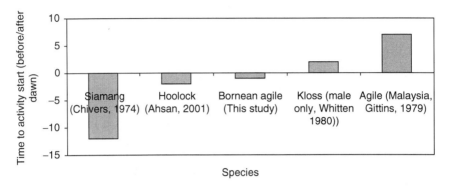

Fig. 8.2 Onset time of activity for different gibbon species (times are before/after dawn – data from Ahsan 2001; Chivers 1974; Gittins 1979; Whitten 1980)

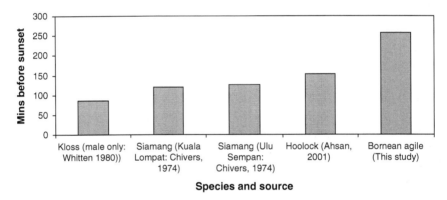

Fig. 8.3 Times of entering sleeping trees. Sources Ahsan (2001); Chivers (1974); Gittins (1979); Whitten (1980)

Food species overlap with orang-utans is about 69% thus the early entry to sleeping tree could be a behavioural adaptation to solve the food supply problem and avoid niche overlap with orang-utans. Thirty-two percent (31 of 96) of all aggressive encounters are with orang-utans and, of these 31 incidences, 29 involved feeding trees. In all 29 cases the gibbons vacated the feeding tree, allowing the orang-utan to take possession of the food tree.

SNP gibbons average 29% resting (range 28–31, SD 0.94), 29% feeding (range 27–31, SD 1.78), 29% travelling (range 26–32, SD 1.92), 9% singing (range 7–10, SD 1.07) and 4% in social activities (range 2–8, SD 0.57, Fig. 8.4). These values varied significantly between groups ($\chi^2 = 18.75$, df $= 3$, $p = 0.01$). Groups without ventral infants fed more and groups with ventral infants rested more (Kruskal–Wallis $H = 9.76$, df $= 2$, $p = 0.01$).

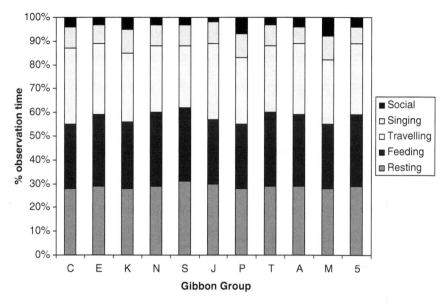

Fig. 8.4   Activity budgets of SNP gibbons (averaged across years)

There is no significant variation between activities of the adult males and females within and between groups (ANOVA, $p > 0.05$) and no significant differences within and between groups when monthly activity budgets were compared (ANOVA, $p > 0.05$).

Secondary Activities (Fig. 8.5)

All are weighted for follow time on males and females to account for the fact that some individuals were followed more than others. Feeding will be covered in more detail in the diet section. As with all gibbon species, brachiation is the dominant form of locomotion, followed by leaping, where the main propulsion comes from the legs rather than the arms. Gibbons possess hard pads on their rear called *ischial callosites*, providing more comfort when sitting. Lying in the tree was predominantly seen when gibbons had entered their afternoon sleeping tree. Duetting was deemed to have begun once the first great call was heard from the female. Hooting refers to the warm-up phase in the mornings before the adults of the group had fully coordinated their singing. Alarm calling was classified as calling (including duetting) which took place after 10:00 h i.e. after the end of the normal singing period. Of 96 encounters with other species resulting in alarm calling 2% were sun bears (*Helarctos malayanus*), 13% pig-tailed macaques (*Macaca nemestrina*), 26% another gibbon group, 27% red langurs (*Presbytis rubicunda*) and 32% orang-utans. Grooming was the dominant

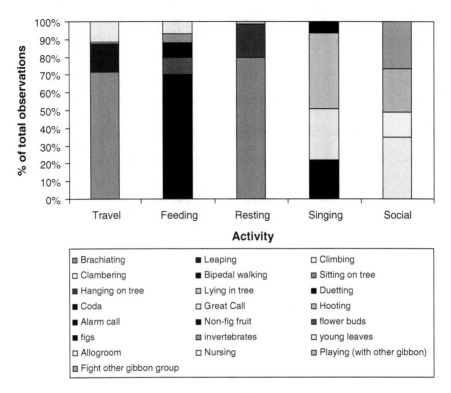

Fig. 8.5 Breakdown of average activity budgets of SNP gibbons (averaged across groups and years)

social activity between adults followed by playing generally between an adult and sub-adult, juvenile or infant, rarely between the adults.

Social Interactions Within the Group

The cohesion between the adult pair was assessed every 5 min using the following categories based on the position of the focal adult: (1) physical contact in view; (2) close enough for physical contact but without contact in view; (3) 1–5 m apart in view; (4) 6–10 m apart in view; (5) 11–25 m apart in view; (6) cannot see other adult. Data presented are from 16,095 observations (Fig. 8.6).

In 51% of all observations, the adults were within visual range of each other, though not necessarily in the same tree. Overall, the average distance between males and females without infants was 8.54 m and between males and females with infants 4.86 m ($n=779$) was these distances were significantly different between the two groups (Mann–Whitney U-test $W=10.5$, $p=0.003$).

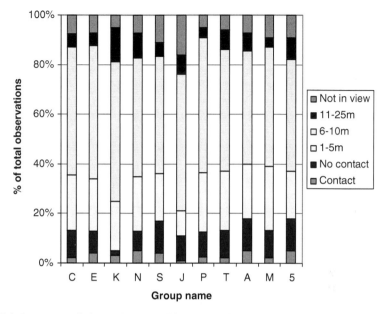

Fig. 8.6 Percentage of observation time gibbons spend in each PA category across all 11 groups

All members of the group (excluding ventral infants) were found together in 86.6% ($n=4{,}573$), and with one member missing (usually the sub-adult or the adult male) 7.4% ($n=2{,}367$). The minimum % of time all members of the group were together during feeding was 90.6% ($n=8{,}753$), during resting was 81.6% ($n=5{,}673$) and for social activities (usually playing) and singing, all members of the group were together 100% of the time (social $n=1{,}759$; singing $n=4{,}563$). No significant difference was found in the sightings of all group members when scans were compared between groups (χ^2 test$=5.46$, df$=3$, $p=0.08$).

These values are similar to those reported by Ahsan (2001) for hoolock gibbons (*Hylobates hoolock*). There is little doubt that, despite the large home range sizes in *H. albibarbis*, cohesiveness between the group members is important and an integral part of group movement during the active period.

Of all social activities 35% were spent allogrooming. The majority of allogrooming was between the adults with male-to-female grooming accounting for 47% ($n=86$), female-to-male 33% ($n=62$) and female-to-sub-adult or -juvenile 20% ($n=37$). Incidences of allogrooming from females to ventral infants were not recorded. Bartlett (2003) reports up to 73% of grooming by adults with the recipients being mainly the sub-adults and juveniles.

Grooming periods normally last for 7.65 min (range 9 s–16.5 min, number of calls included in analysis$=276$). Grooming (and indeed all social activity) peaked from 10:00 to 11:00 h when the gibbons were resting. The other peak was from

13:30 to 14:00 h before the gibbons moved towards the sleeping tree. Chivers (1974) reports that male siamangs initiated grooming and bouts lasted for 12.9 min at Kuala Lompat and 11.5 min Ulu Sempam. Ahsan (2001) reports a mean of 5.9 min for *H. hoolock*. Male-to-female grooming is more frequent than female-to-male grooming.

Play is an important part of the social life of a gibbon, encompassing 24% of all social activity time. The majority of all play bouts involved the infant or juvenile swinging from, grappling with and biting the adults or sub-adults (pers. obs.).

As noted by Bartlett (2003), having long-term data on well-habituated gibbons is highlighting previously unknown or unseen behaviours and necessitates a re-evaluation of the importance of social behaviours in gibbon groups.

From five births which have occurred since the study began, births take place from November to May, the gestation period for these five females ranged from 7 to 9 months (210–270 days), which is towards the high end of gibbon gestations (Geissmann 1991). Inter-birth interval data from this study are still limited but from two groups where we have data on consecutive infants, the inter-birth interval is 28 months (2.4 years). Palombit (1995) reports inter-birth intervals of 22 and 31 months for siamangs (one female), and 26.1 months for lar gibbons. Chivers and Raemaekers (1980) report 48 months for a siamang and 120 months for a lar gibbon. Reichard and Barelli (2008) report a mean inter-birth interval for lar gibbons in Khao Yai of 41 months and Zhou et al. (2008) report a mean of 24 months for Hainan gibbons.

Copulation (between adults) occurred in all months and mean duration of copulation is 34.6 s (range 17.5–39.6, $n=43$). Copulation was classified as when mounting was achieved and pelvic thrusts observed from (Chivers 1974).

Other studies have shown that adult females are most likely to lead a group. Chivers (1974) found that the adult female of a siamang (*Symphylangus syndactylus*) group led 65% of travel bouts. Ahsan (2001) found that for hoolock gibbons (*Hylobates hoolock*) the adult female led 61% of travel bouts with the adult male leading 33% of bouts and the juvenile leading 6% of bouts. Gittins (1980) found that for agile gibbons in Malaysia (*Hylobates agilis*) the female led in 36% of travel bouts and the male in 52% of bouts. In Sabangau the situation is more complex than these other studies would suggest (Table 8.5).

The adult female does take the lead in the majority of cases (46% of all observations); however, in every case where the sub-adult led, the adult female was carrying a ventral infant. In the groups where the adults led there was no ventral infant present. It is possible that when a vulnerable infant is present, the female will not lead and the male will remain close to her for protection.

Conflict Between Groups

A total of 15% of all interactions observed involved inter-group conflict between gibbons (681 of 4,561 observations). All encounters involved duetting, chasing, branch-throwing and branch-shaking as a display. On only one occasion was the

Table 8.5 Travel order of group members

Travel order	% Total observations	Ventral infant present?
AF-AM-Sub-Juv	46	N
AM-AF-Sub-Juv	30	N
Sub-AF-AM-Juv	13	Y
Sub-AM-AF-Juv	7	Y
Sub-Juv-AM-AF	2	Y
Sub-Juv-AF-AM	2	Y

conflict between a group and a lone individual, suggesting that the gibbons must be well habituated in order to observe this behaviour and that lethal aggression is a rare event. I describe the attack and subsequent events here as this event is rare enough to warrant further description. On March 24, 2006 an aggressive territorial encounter was witnessed between a resident gibbon group (C) and a lone male (Yoga). The lone male entered the eastern edge of the group's territory, about 500 m from where he was normally seen. The aggressive encounter lasted 2 h, and the death of the lone male occurred 7 h after the conflict initially started (5 h after the cessation of conflict). The adult male of the group was the only individual to attack physically the other male, but all members of the group, including the adult female with a ventrally-carried infant, displayed, shook branches, alarm called and harassed the lone male, indicating that territorial defence is a group activity. The increased levels of aggression in males may indicate a greater responsibility for resource defence (territory and, by association, food). The most likely explanation for the level of aggression directed to a lone male is reproductive competition, suggesting that territorial males are primarily defending reproductive access to the females, and resource defence is a by-product of this. In this case, reproductive competition is most likely to have been the reason behind the attack, as Yoga and Group C's ranges did not normally overlap.

It is clear that we have a very limited understanding of lethal aggression among gibbons including frequency of occurrence and why it happens. More studies are needed in areas where there are several habituated groups and known lone individuals. It is difficult to ascribe intentions to a non-human primate without sounding anthropomorphic. In this instance, Group C could have continued the attack to ensure Yoga was dead. Instead they disabled the intruder and left. More research is needed to tease out differences between lethal aggression towards a group male and that directed towards a lone male (Cheyne et al. 2010).

Ranging Behaviour and Travel Distances

All distances are weighted by follow effort to allow comparison of data on groups which were followed for more time than others. Data are based on digitised maps and GPS data (Fig. 8.7). Average daily travel distance (day range length) is 2,433 m

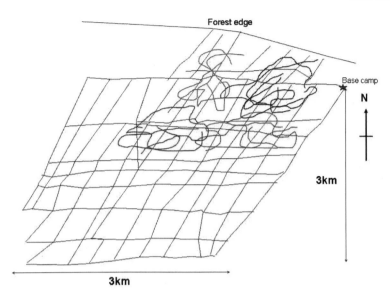

Fig. 8.7 Example of day travel routes of four gibbon groups from June 2008 based on GPS positions taken every 30 min. Number of day follow routes shown in brackets. Group C in *blue* (4), Group E in *green* (3), Group K in *brown* (3) and Group N in *purple* (3)

(range 1,030–5,310, $n = 550$ follow days). There is a significant difference in day range between wet and dry season (Fig. 8.8 Kruskal–Wallis $H = 18.74$, $p = 0.001$).

Ranging distance was generally longer from 05:00 to 12:00 h than after 12:00 h. Day range varied between groups (Kruskal–Wallis $H = 19.65$, df $= 55$, $p = 0.001$) – groups without infants travelled farther than groups with ventral infants (Fig 8.9). It was assumed that non-lactating females were either pregnant or attempting to become pregnant following the weaning of the infant. Thus the differences in day length could be explained by the possibility that the (pregnant) adult female is trying to build up fat reserves to last during lactation (see section on Diet and Energy Intake below). Knott (1999) has already demonstrated that orang-utans can go into negative energy balance if their energy expenditure exceeds that of their intake from the diet, thus leading to knock-on effects on pregnancy and successful parturition. This may also be applicable to gibbons given the extensive overlap in diet (67%) in Sabangau, though more data are needed to establish this. Alternatively it is the infants themselves which are limiting the travel distance, especially once weaned and travelling alone, they are less able to maintain the pace of the older individuals. Newly independent infants would often stop and call for the female who would have to retrace her route to retrieve the infant (pers. obs). Gibbons are frequently moving around their range, with feeding bout length varying from 1 to 226 min ($n = 8,967$). The availability and distribution of foods within their range will play an important role in the distance travelled and the frequency of travel bouts for each group, as noted by Raemaekers (1980).

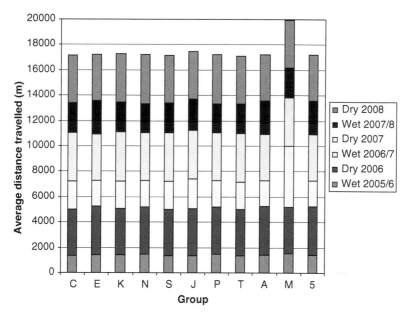

Fig. 8.8 Average daily travel distance for all groups for both wet and dry seasons since the study began

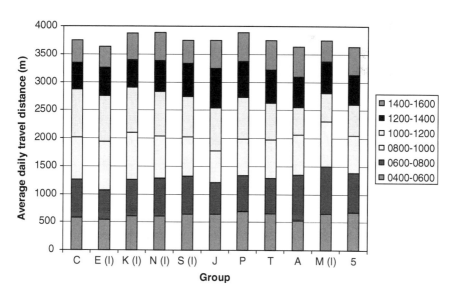

Fig. 8.9 Travel distances by time of day for May–September (dry season) 2008 showing groups with and without dependant infants (indicated by (I)) next to group name

Vertical Use of Forest Canopy

There is no significant difference in canopy height use between males and females (Kruskal–Wallis H-test: Fig. 8.10). When all 11 groups were considered separately for canopy-height use there was a significant difference ($\chi^2 = 37.25$, df = 7, $n = 5842$, $p < 0.001$). This suggests that each group has to adapt to varying canopy conditions within their territories. From habitat work done by Thompson (2007), there is a significant difference between the availability of canopy heights in all group territories ($\chi^2 = 25.889$, df = 3, $n = 360$, $p < 0.001$). These differences are a likely result of human disturbance. Railways were constructed while the area was a logging concession stretching initially 27 and 13 km into the forest, respectively, and the vegetation on either side of it has been severely damaged as a result.

Jacob's D Value is an index used to test for preference between different strata and has previously been used to test between food selection and abundance (Jacobs 1974).

$$D = (r - p)/(r + p - 2rp)$$

r = relative frequency of use and p = relative frequency of availability.

The Jacob's D values for canopy use were calculated for all 11 groups. Groups favour higher canopy heights ($\chi^2 = 18.37$, df = 2, $n = 4{,}679$, $p < 0.001$) and are actively avoiding heights below 10 m ($\chi^2 = 19.91$, df = 2, $n = 3{,}412$, $p < 0.001$, Fig. 8.11). Cannon and Leighton (1994) also found that gibbons avoided lower height-classes, and strongly preferred the emergent layer. The SNP gibbons do not seem to prefer the emergent canopy (>26 m). The quality of forest and tree heights will differ between each study, but the fundamental result remains the same, gibbons are selecting higher height-classes. Average available canopy height in the SNP is 16–20 m (range 0–45 m). From Fig. 8.11 it is clear that gibbons are selectively using the 11–15 m height category more than it is available in the forest, suggesting that gibbons are preferentially selecting this canopy layer.

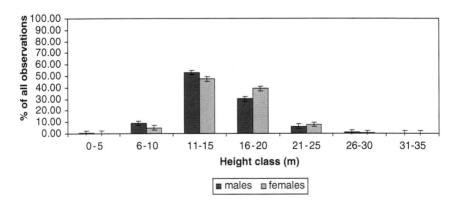

Fig. 8.10 Use of the canopy by males and females showing standard deviation

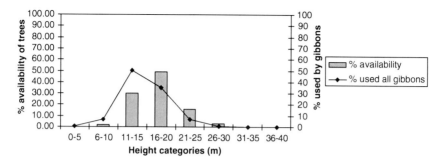

Fig. 8.11 Percentage availability of trees of different height and the percentage of each height category used by gibbons

Diet and Feeding Ecology

Forest productivity was measured by surveying 6×0.4 ha productivity plots (2.4 ha total area) surveyed monthly for fruit and flower production expressed as percentage gibbon fruit/flower trees with food/month. Data were collected on food species eaten, feeding rates and part eaten from direct observations of the gibbons. Samples of the foods were then sent for nutritional analysis to calculate energy intake from each species. Forest productivity was measured monthly using established plots.

Of the 77 species identified as gibbon food, all were trees, figs or woody lianas (except for one species of unidentified epiphyte). Eighty-three percent of all feeding time is on just 20 species and seven of these have asynchronous fruiting cycles. These 20 species are from the plant families Anacardiaceae, Annonaceae, Apocynaceae, Ebenaceae, Euphorbiaceae, Gnetaceae, Icacinaceae, Moraceae, Myrtaceae, Sapindaceae, Sapotaceae and Tetrimeristaceae. Based on the species accumulation curve (e.g. Lwanga et al. 2001), the gibbon diet in Sabangau should plateau at about 118 species and a further 2 years of data are required to ensure we have assessed the complete variety of species in the gibbon diet. The number of species consumed within each year varies from 25 to 42 species (mean 34, SD 8.62).

Lappan (pers. comm.) reports 75 species consumed by siamangs at Way Canguk from 2000 to 2002 and Malone (pers. comm.) reports only 34 species consumed by Javan gibbons (time-frame unknown) and hybrid gibbons at Barito Ulu consume 114 species (10 years of data, PBU reports).

Some of these families, e.g. Annonaceae and Moraceae, contain species which have an asynchronous fruiting cycle, i.e. the trees do not all fruit at the same time, thus providing food for gibbons nearly all year round. They are extremely important, therefore, for gibbons during the low-fruiting months.

These 20 species are thus critical components of gibbon diet in the Sabangau. Additionally, there are species which do not constitute a large part of gibbon diet overall, but are important at certain times of year. The three most-eaten species in each month were identified for each of the 37 months of the study (Table 8.6).

Table 8.6 Seasonally important species

Family	Species	Local name	Overall ranking	# of months species is in top three foods
Ebenaceae	*Diospyros bantamensis*	Malam Malam	2	15
Sapotaceae	*Palaquium cochlearifolium*	Nyatoh Gagas	7	9
Myrtaceae	*Syzygium garcinfolia*	Jambu Buring	1	9
Gnetaceae	*Gnetum sp. 1*	Bajakah Luaa	8	6
Clusiaceae	*Mesua sp 1*	Tabaras akar tinggi	5	6
Moraceae	*Parartocarpus venenosus*	Lilin Lilin	3	6
Anacardiaceae	*Campnosperma coriaceum*	Terantang	6	6
Sapotaceae	*Palaquium sp. 3*	Nyatoh Burung	4	6
Moraceae	*Ficus sp.*	Lunuk Bunyer	10	3
Moraceae	*Ficus sp.*	Lunuk Buhis	9	3
Annonaceae	*Mezzettia umbellata*	Pisang Pisang (Kambalitan Hitam)	13	3
Euphorbiaceae	*Blumeodendron kurzii*	Kenari	18	3
Sapotaceae	*Madhuca mottleyana*	Katiau	12	3
Sapotaceae	*Palaquium pseudorostratum*	Nyatoh Babi	11	3
Tetrimeristaceae	*Tetramerista glabra*	Pornak	14	3
Sapindaceae	*Nephellium lappaceum*	Rambutan Hutan	16	3
*Myrtaceae	*Syzygium havilandii*	Tatumbu	25	3
*Clusiaceae	*Garcinia bancana*	Manggis	28	6
*Elaeocarpaceae	*Elaeocarpus mastersii*	Mangkinang	27	3
*Gnetaceae	*Gnetum sp. 2*	No local name	47	3
*Clusiaceae	*Callophyllum hosei*	Bintangor/jinjit/ mentangor	23	3
*Menispermaceae	*Fibraurea tinctoria*	Liana Kuning	38	3
*Meliaceae	*Aglaia rubiginosa*	Kajalaki	42	3
*Meliaceae	*Sandoricum beccanarium*	Papong	21	3
*Fabaceae	*Koompassia malaccensis*	Kempas	51	3
*Annonaceae	*Polyalthia hypoleuca*	Alulup (Rewoi)	36	3
*Clusiaceae	*Garcinia cf. parvifolia*	Gandis	53	3
*Polygalaceae	*Xanthophyllum cf. ellipticum*	Kemuning	30	3
Anacardiaceae	*Campnosperma squamatum*	Nyating	22	3

Species not in top 20 indicated with asterisk* (from Cheyne and Shinta 2007.)

These species include 16 of the 20 species and six additional species including *Garcinia bancana*. These species are either favoured when they come into season, or important fall-back foods when more preferred species are not in fruit. *Diospyros bantamensis* was in the top three species for 15 months of the study (fruit) and *Palaquium cochlearifolium* (flowers) and *Syzygium garcinfolia* (fruit) for 9 months of the study and are therefore extremely important foods for gibbons in Sabangau.

There is no significant difference in diet between groups based on individual seasons (MANOVA $F_{1,10} = 2.6$, $p > 0.05$), but there is a significant difference in all groups diets between seasons in different years (MANOVA $F_{1,8} = 1.7$, $p = 0.01$), as different foods were available in each subsequent wet and dry throughout the study period: some species in this area bear fruit/flower biannually. During the dry season, gibbons ate flowers and leaves in far greater quantities than previously noted for any other gibbon species (3–23%, Fig. 8.12). Fan et al. (2009) noted an even more pronounced shift in diet in the black-crested gibbons (*Nomascus concolor*) from 3.1 to 61.9%. Fan et al. (2009) propose that this is due to seasonal stress, similar to the variations in food availability seen in Sabangau. Gibbons feed on a wide variety of food types and identifying these is important in understanding food choice. Splitting the diet into the classic categories gives an average of fruit 63%, young leaves and leaf shoots 25%, figs 6%, flower buds 5% and invertebrates 1% but this does not account for seasonal differences, nor for the wide variety of food types eaten by gibbons (Fig. 8.13). The nine types of food presented here

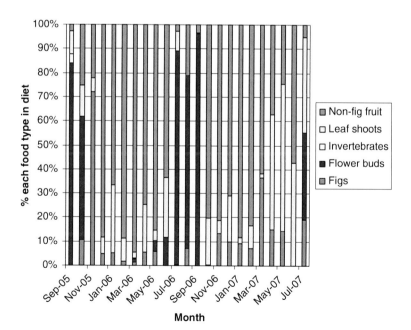

Fig. 8.12 Breakdown of diet averaged for all groups throughout the study period

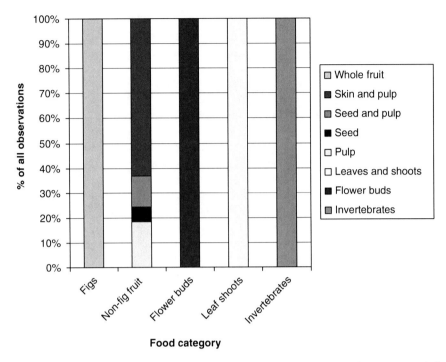

Fig. 8.13 Breakdown of overall diet into all components consumed (mean of all groups and all study time)

(averaged for all adult gibbons over the whole study) indicate that gibbons are selecting different parts of a food item (seed and pulp means that both parts were consumed). Data have been collected on the chemical composition and mechanical properties of all gibbon food species and preliminary analysis suggests that very hard foods are avoided, as are foods high in tannins e.g. seeds but that pH has no impact on food selection; however, the shape and toughness are important in determining how easy the food is to manipulate with hands and open with teeth.

Food availability for the gibbons in the forest was measured as the percentage of trees within productivity plots, which contained edible gibbon foods and were measured monthly (based on the parts eaten by gibbons from Fig. 8.13). Food availability is highly variable and changes between months and seasons (Fig. 8.14). McConkey et al. (2003) report that gibbon food choice was strongly influenced by the availability of flowers, despite non-fig fruit compromising most of the diet (52–64%). This variability will also play a large role in gibbon food selection, where fruit consumption dominates except in the dry seasons, where flower bud consumption exceeds fruit consumption. All groups (based on adults feeding) consumed the same available species each month; there was no variation in diet between groups. The variety of foods in the diet is also highly variable with the

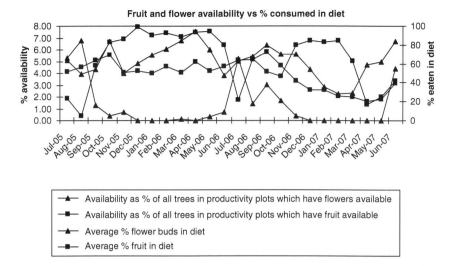

Fig. 8.14 Percentage of fruit and flower bud availability and consumption averaged for all groups throughout the study

number of species consumed ranging from 47 (wet season) to 6 (dry season: mean 18, median 15, SE 1.78).

Gibbons are very adaptable in both diet and behaviour and can exploit low-productivity forest well. It also sheds light on possible fallback food species exploited in times of food scarcity. Detailed feeding ecology data have provided an insight into how gibbons cope with food availability in the Sabangau. Additionally, data on species which dominate the diet (i.e. present for more than 3 months per year) and fallback foods for different forest habitats are needed to make effective management plans. More data are needed to understand the implications of diet and food availability on gibbon population sizes and carrying capacity.

Energy Intake

Standard laboratory analysis methods and physiological fuel values used (e.g. Conklin-Brittain et al. 1998; Knott 1998, 2001, 2005): energy intake was calculated in Kcal (Energy intake = Σ energy *content* × dry *weight* × feeding *rate* × bout *length*). Females not carrying infants could be building up their reserves while pregnant to offset the costs of lactation; hence, the significant difference between females carrying ventral infants and those without energy intakes (Fig. 8.15, Mann–Whitney $U = 9.47$, $p = 0.001$: with $n = 5$, without $n = 6$). Further work is underway to elucidate more about energy intake and availability and how this affects gibbons.

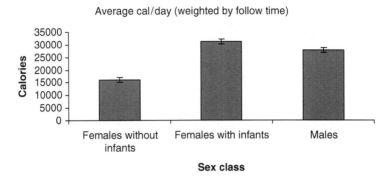

Fig. 8.15 Showing differences in Kcal intake for different sex classes (adults only). Bars show standard deviation

Table 8.7 Characteristics sleeping trees

Variable	Average range (m)	Range
dbh (cm)	21–30	11–50
Height of tree (m)	21–25	16–30
Height of gibbon sleeping position (m)	21–25	11–30
Height of first branching (m)	21–25	6–25
Number of other gibbons in tree	1	0–4

Sleeping Trees

The sleeping tree project is in the very early stages so I will present only summary data based on 26 trees: Dipterocarpaceae (2), Ebenaceae (1), Fabaceae (Leguminosae) (1), Moraceae (1), Anacardiaceae (1), Anisophyllaceae (1), Burseraceae (1), Clusiaceae (Guttiferae) (1), Lauraceae (1), Myrtaceae (1), Sapotaceae (1), Tetrameristaceae (1), Unknown (1).

All groups were selecting tall trees, with high first branching and large dbh. This suggests that stability of the tree is important, as is inaccessibility (security, Table 8.7). There were no significant differences in selection of tree height or dbh between groups (Kruskal–Wallis $H = 23.4$, df $= 4$, $p = 0.19$).

It is interesting that 7 of 19 trees had vines. Whitten (1982) found that almost none of his sleeping trees had vines, a fact he attributed to gibbon safety – vines provide easier access for predators, such as pythons and cats. The presence of vines in some of the Sabangau sleeping trees could be due to a lack of availability of non-vine trees. Trees with vines were slightly higher than non-vine trees, but the result was not significant (Mann–Whitney $W = 14.53$, $p = 0.062$). A larger sample should help elucidate if sleeping trees with vines are indeed taller than non-vine trees.

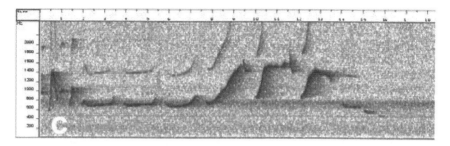

Fig. 8.16 Sonogram of female *H. albibarbis*

Singing

The most distinctive part of the singing is the female's great call consisting of long notes with a loud peak note in the middle of the sequence (Fig. 8.16). The great call is produced by most females over age 4 years though juvenile and sub-adult females practice with the adult female. The male song is called a coda and consists of short notes in rapid succession. Produced by most juvenile and sub-adult males though they require practice by imitating the adult male (Cheyne 2007a, 2008b; Cheyne et al. 2007a).

The level of flexibility for each female within the population was first assessed using the Kruskal–Wallis, ANOVA following Haimoff and Gittins (1985), Haimoff and Tilson (1985), Dallmann and Geissmann (2001) and Cheyne et al. (2007a). Following this, a sample great-call sequence from each female was analysed in the same way. A total of 11 females were included providing a total of 374 great call sequences for analysis. The following song variables were found to be significantly different between females (to at least 0.01): duration of great call(s), number of notes/great call, duration of climax note(s), highest frequency of peak note (Hz) and notes/duration (n/d).

No significant relationship was found between astronomical twilight cues (month, time of sunrise, time of moonrise, nocturnal illumination index, day and night length: (Cheyne 2007a)). This is likely due to the proximity of the study site to the equator.

Rain (light and heavy combined) was recorded on 140 of the 550 gibbon study days. Of these 140 rain days, no singing was recorded on 111 days, or 79% of the time. This effect was significant (111/140: MANOVA $F_{1,9} = 614.39$, $p < 0.05$). On mornings when there was rain and singing, the gibbons started to sing much later, between 09:00 and 10:00 h ($n = 23$), instead of between the usual 04:30 and 06:00 h ($n = 6$). Temperature did not vary significantly between singing ($n = 435$) and non-singing days ($n = 115$) (singing: median = 27.90, SD = 0.095; non-singing: median = 27.45, SD = 0.105; MANOVA $F_{1,12} = 2.76$, $p > 0.05$). Cloud cover had no significant effect on the onset time of singing ($n = 293$; singing days: MANOVA $F_{1,10} = 7.43$, $p > 0.05$).

The association between windiness (light and gale) and days when there was no singing was significant ($n=231$; singing days: MANOVA $F_{2,10}=13.5$, $p<0.05$).

Onset times of singing rarely varied and all singing occurred between 04:00 and 09:30 h. The average number of great calls/singing bout was 12 (SD 0.23) and the average duration of singing bouts was 48 min (SD 0.5).

Conclusions

The results reported here highlight the variation among the gibbon species, and how intricately this is tied to habitat availability and behavioural and feeding ecology requirements. The differences between gibbons across their range in South-east Asia is still being explored and extrapolating from a few study sites to all the species is risky as it gives a too-narrow picture of gibbon behavioural ecology. Peat-swamp forest, already highly important for carbon storage and maintaining the ecosystem (Rieley et al. 1993, 2004; Page et al. 1999b, 2002; Aldhous 2004), has been well documented, yet the importance of this habitat for the fauna has not received much attention. In order to maintain the tropical PSF, we must conserve the fauna. To conserve we must first understand, and this is the importance of long-term studies.

Conservation Recommendations

More local students and researchers need to be involved in the conservation of gibbons and other primates. To this end, more details on habituation techniques and survey methods should be published and made available. To this end, the Orang-utan Tropical Peatland and Conservation Project is producing a DVD to help train researchers in carrying out auditory sampling of gibbon populations. This method is far more accurate than line transects (Brockelman and Ali 1987; Brockelman and Srikosamatara 1993; O'Brien et al. 2004; Cheyne et al. 2007b; Hamard 2008).

Only with sufficient information on feeding behaviour (energy intake and food selection), ecology and the effects of habitat degradation on gibbons, can the habitat be managed to ensure their long-term survival. The following recommendations are suggested to ensure the continued survival of gibbons in the SNP and other parts of Indonesia (based on this study and the suggestions of the delegates of the Indonesian Gibbon Conservation and Management Workshop 2008) (Campbell et al. 2008).

1. Old logging canals must be dammed to prevent drainage of the peat, leading to fires.
2. Fires started in the forests must be tackled quickly; using bores to extract ground water (Limin et al. 2004).
3. Permits for plantations (oil palm and acacia in particular) should not be given for areas where there is standing forest, only for areas which are already deforested.

4. Optimise law enforcement in protected areas by BKSDA to stem the flow of gibbons to the market.
5. Education of people about the impacts of their activities on gibbon habitat and gibbon populations.
6. Consider forest outside of protected areas and logging concession and logged-over areas as potential gibbon habitat.
7. No more mining in gibbon habitat (including open cast or oil).
8. Address the problem of land-use planning. Determine clear boundaries between protected areas and districts agreed between local government and the forestry management.
9. Build corridors between fragments.
10. Improve law enforcement for protected areas.
11. Implement IUCN and ITTO guidelines on biodiversity conservation and sustainable use of tropical timber production forest.
12. Development and implementation of a control system by the Department of Forestry and local government (DINAS Kehutanan) for logging concession and plantation companies.

Acknowledgements I thank Jatna Supriatna and Sharon L. Gursky for the invitation to contribute to this book. This work was supported by grants from the Department of Anthropology, George Washington University; Primate Conservation Inc., Cambridge University Philosophical Society, Rufford Small Grants for Conservation, Columbus Zoo and IdeaWild. I gratefully acknowledge the contribution of all the researchers who assisted with the project: Adul, Ambut, Andri Thomas, Iwan, Ramadhan, Santiano, Twentinolosa, Yudhi Kuswanto, Zeri, Mark E. Harrison, Claire J.H Thompson, Grace V. Blackham, Bernat Ripoll, Dave A. Smith, Andrea Höing, Lindy Thompson and Marie Hamard. This work was carried out as part of the biodiversity monitoring research and conservation efforts of the Orang-utan Tropical Peatland Project. I thank the Indonesian Institute of Science and the Indonesian Department of Forestry for permission to carry out research in the Sabangau Catchment and I thank the Center for International Cooperation in Sustainable Management of Tropical Peatland (CIMTROP), University of Palangka Raya for sponsoring my research and providing invaluable logistical support. I thank David J. Chivers and an anonymous reviewer who commented on early drafts of this chapter.

References

Ahsan MF (2001) Socio-ecology of the hoolock gibbon (*Hylobates hoolock*) in two forests of Bangladesh. In: The apes: challenges for the 21st century. Brookfield Zoo, Chicago, pp 286–299

Aldhous P (2004) Borneo is burning. Nature 432:144–146

Altmann J (1974) Observational study of behavior: sampling methods. Behaviour 49:227–267

Bartlett TQ (2003) Intragroup and intergroup social interactions in white-handed gibbons. Int J Primatol 24:239–259

Bartlett TQ (2007) The Hylobatidae: small apes of Asia. In: Campbell CJ, Fuentes A, Panger M, Bearder SK (eds) Primates in perspective. Oxford University Press, Oxford, pp 274–289

Blackham GV (2005) Pilot survey of nocturnal primates, *Tarsius bancanus borneanus* (western tarsier) and *Nycticebus coucang menagensis* (slow loris) in peat swamp forest, Central Kalimantan, Indonesia. Oxford Brookes University, Oxford

Brandon-Jones D, Eudey AA, Geissmann T, Groves CP, Melnick DJ, Morales JC, Shekelle M, Stewart CB (2004) Asian primate classification. Int J Primatol 25:97–164

Brockelman WY, Ali R (1987) Methods of surveying and sampling forest primate populations. In: Mittermeier RA, Marsh RW (eds) Primate conservation in the tropical rainforest. Alan Liss, New York, pp 23–62

Brockelman WY, Gittins SP (1984) Natural hybridisation in the *Hylobates lar* species group: implications for speciation in gibbons. In: Preuschoft H, Chivers DJ, Brockelman WY, Creel N (eds) The lesser apes: evolutionary and behavioural biology. Edinburgh University Press, Edinburgh, pp 498–532

Brockelman WY, Srikosamatara S (1984) Maintenance and evolution of social structure in gibbons. In: Preuschoft H, Chivers DJ, Brockelman WY, Creel N (eds) The lesser apes: evolutionary and behavioural biology. Edinburgh University Press, Edinburgh, pp 298–323

Brockelman WY, Srikosamatara S (1993) Estimation of density of Gibbon groups by use of loud songs. Am J Primatol 29:93–108

Brockelman WY, Reichard U, Treesucon U, Raemaekers JJ (1998) Dispersal, pair-formation and social structure in gibbons (*H. lar*). Behav Ecol Sociobiol 42:329–339

Buckley C, Nekaris KAI, Husson SJ (2006) Survey of *Hylobates agilis albibarbis* in a logged peat swamp forest: Sabangau catchment, Central Kalimantan. Primates 47:327–335

Campbell C, Andayani N, Cheyne SM, Pamungkas J, Manullang B, Usman F, Wedana M, Traylor-Holzer K (2008) Indonesian Gibbon conservation and management workshop final report. IUCN/SSC Conservation Breeding Specialist Group, Apple Valley, MN, USA

Cannon CH, Leighton M (1994) Comparative locomotor ecology of gibbons and macaques: selection of canopy elements for crossing gaps. Am J Phys Anthropol 93:505–524

Carpenter CR (1940) A field study in Siam of the behaviour and social relations of the gibbon *Hylobates lar*. Comp Psychol Monogr 16:1–212

Cheyne SM (2007a) Effects of meteorology, astronomical variables, location and human disturbance on the singing apes: *Hylobates albibarbis*. Am J Primatol 40:1–7

Cheyne SM (2007) Indonesian Gibbons – *Hylobates albibarbis*. In: Rowe N, editor. All the World's Primates. Charlestown, RI: Pogonias Press Inc.

Cheyne SM (2008a) Feeding ecology, food choice and diet characteristics of gibbons in a disturbed peat-swamp forest, Indonesia. In: Lee PC, Honess P, Buchanan-Smith H, MacLarnon A, Sellers WI (eds) 22nd Congress of the international primatological society (IPS). Top Copy, Bristol, Edinburgh, UK, p 342

Cheyne SM (2008b) Gibbons – *Hylobates albibarbis*. In: Rowe N (ed) All The World's primates. Pogonias, Charlestown, RI, USA

Cheyne SM, Chivers DJ (2006) Sexual swellings of female gibbons. Folia Primatol 77:345–352

Cheyne SM, Monks EM, Kuswanto Y (2010) An observation of lethal aggression in bornean agile gibbons *Hylobates albibarbis*. Gibbon Journal 6.

Cheyne SM, Shinta E (2007) Important tree species for gibbons in the Sabangau peat swamp forest. Report to the Indonesian Department of Forestry (PHKA), Jakarta and Palangka Raya. CIMTROP, Palangka Raya, p 7

Cheyne SM, Chivers DJ, Sugardjito J (2007a) Covariation in the great calls of rehabilitant and wild gibbons *Hylobates agilis albibarbis*. Raffles Bull Zool 55:201–207

Cheyne SM, Thompson CJH, Phillips AC, Hill RMC, Limin SH (2007b) Density and population estimate of gibbons (*Hylobates albibarbis*) in the Sabangau catchment, Central Kalimantan, Indonesia. Primates 49:50–56

Chivers DJ (1972) The siamang and the gibbon in the Malay peninsula. Gibbon Siamang 1:103–135

Chivers DJ (1974) The Siamang in Malaya: a field study of a primate in tropical rainforest. Karger, Basel

Chivers DJ (1977) The ecology of gibbons: some preliminary considerations based on observations in the Malay Peninsula. In: Prasad MRN, Kumar TC (eds) The use of non-human primates in biomedical research. Indian Science Research Academy, New Delhi, pp 85–105

Chivers DJ (1984) Feeding and ranging in gibbons: a summary. In: Preuschoft H, Chivers DJ, Brockelman WY, Creel N (eds) The lesser apes: evolutionary and behavioural biology. Edinburgh University Press, Edinburgh, pp 267–281

Chivers DJ (2001) The swinging singing apes: fighting for food and family in Far-East forests. In: The apes: challenges for the 21st century. Brookfield Zoo, Chicago, pp 1–28

Chivers DJ, Raemaekers JJ (1980) Long term changes in behaviour. In: Chivers DJ (ed) Malayan forest primates: ten years study in tropical forest. Plenum, New York, pp 109–160

Conklin-Brittain NL, Wrangham RW, Hunt KD (1998) Dietary responses of chimpanzees and cercopithecenes to seasonal variations in fruit abundanceII. Macronutrients. Int J Primatol 19:971–998

Cowlishaw G (1992) Song function in gibbons. Behaviour 121:131–153

Dallmann R, Geissmann T (2001) Individuality in the female songs of wild silvery gibbons (*Hylobates moloch*) on Java, Indonesia. Folia Primatol 71:220

Fan P, Ni Q, Sun G, Huang B, Jiang X (2009) Gibbons under seasonal stress: the diet of the black crested gibbon (*Nomascus concolor*) on Mt. Wuliang, Central Yunnan, China. Primates 50:47–54

Feeroz MM, Islam MA (1992) Ecology and behaviour of Hoolock gibbons in Bangladesh. In Multidisciplinary Action Research Centre, Dhaka

Felton AM, Engstrom LM, Felton A, Knott CD (2003) Orangutan population density, forest structure and fruit availability in hand-logged and unlogged peat swamp forests in West Kalimantan, Indonesia. Biol Conserv 114:91–101

Geissmann T (1991) Reassessment of the age of sexual maturity in gibbons (*H. lar*). Am J Primatol 23:11–22

Geissmann T (2007) Status reassessment of the gibbons: results of the Asian Primate Red List Workshop 2006. Gibbon J 3:5–15

Geissmann T, Orgeldinger M (2000) The relationship between duet songs and pair bonds in siamangs, *Hylobates syndactylus*. Anim Behav 60:805–809

Gittins SP (1979) The behaviour and ecology of the agile gibbon (*Hylobates agilis*). [phd]. Cambridge: University of Cambridge

Gittins SP (1980) Territorial behaviour in the agile gibbon. Int J Primatol 1:381–399

Gittins SP (1982) Feeding and ranging in the Agile Gibbon. Folia Primatol 38:39–71

Gittins SP (1983) Use of the forest canopy by the Agile Gibbon. Folia Primatol 40:134–144

Gittins SP, Raemakers JJ (1980) Siamang, agile and lar gibbons. In: Chivers DJ (ed) Malayan forest primates: ten years study in tropical rainforest. Plenum, New York, pp 258–266

Groves C (2001) Primate taxonomy. Smithsonian Institute Press, Washington, DC

Haimoff EH, Gittins SP (1985) Individuality in the songs of wild agile gibbons (*H. agilis*) of Peninsular Malaya. Am J Primatol 8:239–247

Haimoff EH, Tilson RL (1985) Individuality in the female song of wild Kloss' gibbons (*Hylobates klossii*) on Siberut Island, Indonesia. Folia Primatol 44:129–137

Hamard MCL (2008) Vegetation correlates of gibbon density in the peat-swamp forest of Sabangau National Park, Central Kalimantan, Indonesia. MSc, Oxford Brookes University, Oxford

Husson SJ, Cheyne SM, Morrogh-Bernard H, D'Arcy L, Harrison ME (2008) Orang-utan tropical peatland and conservation project annual report. University of Cambridge and University of Oxford, Cambridge

IUCN (2008) Red List of threatened species. IUCN, New York

Jacobs J (1974) Quantative measurement of food selection: a modification of the forage ratio and Ivlev's electivity test. Oecologica 14:413–417

Kappeler M (1981) The Javan silvery gibbon: Hylobates moloch, habitat, distribution and numbers. Dissertation, University of Basel

Knott CD (1998) Changes in orangutan caloric intake, energy balance, and ketones in response to fluctuating fruit availability. Int J Primatol 19:1061–1079

Knott CD (1999) Reproductive, physiological and behavioural responses of orangutans in Borneo to fluctuations in food availability. PhD thesis, Harvard University

Knott CD (2001) Female reproductive ecology of the apes: implications for human evolution. In: Ellison PT (ed) Reproductive ecology and human evolution. Walter de Gruyter, New York, pp 429–463

Knott CD (2005) Energetic responses to food availability in the great apes: implications for human evolution. In: Brockman D, van Schaik C (eds) Primate Seasonality: implications for human evolution. Cambridge University Press, Cambridge

Leighton D (1984) Monogamy and territoriality in muellers gibbon PhD, University of California, Davis

Leighton DR (1987) Gibbons: territoriality and monogamy. In: Smuts BB, Cheney DL, Seyfarth RM, Wrangham RW, Struhsaker TT (eds) Primate societies. University of Chicago Press, Chicago, pp 135–145

Limin S, Jaya A, Siegert F, Rieley JO, Page SE, Boehm HDV (2004) Tropical peat and forest fire in 2002 in Central Kalimantan, its characteristics and the amount of carbon released. In: Päivänen J (ed) Wise use of peatlands (vol I) proceedings of the 12th international peat congress, Tampere, Finland. Saarijärven Offset Oy, Saarijärvi, Finland, pp 679–686

Linnell J, Andersen R, Kvam T, Andren H, Liberg O, Odden J, Moa P (2001) Home-range size and choice of management strategy for lynx in Scandinavia. Environ Manage 27:869–879

Lwanga JS, Butynski TM, Struhsaker TT (2001) Tree population dynamics in Kibale National Park, Uganda 1975–1998. Afr J Ecol 38:238–247

MacKinnon JR (1977) A comparative ecology of Asian apes. Primates 18:747–772

Mather R (1992) A field study of hybrid gibbons in Central Kalimantan, Indonesia. PhD, University of Cambridge

McConkey KR (2000) Primary seed shadow generated by gibbons in the rainforests of Barito Ulu, Central Borneo. Am J Primatol 52:13–29

McConkey KR, Aldy F, Ario A, Chivers DJ (2002) Selection of fruit by gibbons (*Hylobates muelleri* x *agilis*) in the rain forests of Central Borneo. Int J Primatol 23:123–145

McConkey KR, Ario A, Aldy F, Chivers DJ (2003) Influence of forest seasonality on gibbon food choice in the rain forests of Barito Ulo, Central Kalimantan. Int J Primatol 24:19–32

Mitani JC (1987a) Species discrimination of male song in gibbons. Am J Primatol 13:413–423

Mitani JC (1987b) Territoriality and monogamy among agile gibbons (*H. agilis*). Behav Ecol Sociobiol 20:265–269

Mitani JC (1990) Demography of agile gibbons (*H. agilis*). Int J Primatol 11:411–424

Moinde NN, Suleman MA, Higashi H, Hau J (2004) Habituation, capture and relocation of Sykes monkeys (*Cercopithecus mitis albotorquatus*) on the coast of Kenya. Anim Welf 13:343–353

Morrogh-Bernard H (2003) Behavioural ecology of orang-utan (*Pongo pygmaeus*) in a disturbed deep-peat swamp forest, Central Kalimantan, Indonesia. University of Cambridge, Cambridge

Morrogh-Bernard H, Husson S, Page SE, Rieley JO (2003) Population status of the Bornean orang-utan (*Pongo pygmaeus*) in the Sabangau peat swamp forest, Central Kalimantan, Indonesia. Biol Conserv 110:141–152

Muoria PK, Karere GM, Moinde NN, Suleman MA (2003) Primate census and habitat evaluation in the Tana delta region, Kenya. Afr J Ecol 41:157–163

Nekaris KAI, Blackham GV, Nijman V (2008) Conservation implications of low encounter rates of five nocturnal primate species (*Nycticebus* spp.) in Asia. Biodivers Conserv 17:733–747

Nijman V, Menken S (2005) Assessment of census techniques for estimating density and biomass of gibbons (Primates: Hylobatidae). Raffles Bull Zool 53:169–179

O'Brien TG, Kinnaird MF, Nurcahyo A, Prasetyaningrum M, Iqbal M (2003) Fire, demography and the persistence of siamang (*Sympgylangus syndactylus*: Hylobatidae) in a Sumatran rainforest. Anim Conserv 6:115–121

O'Brien TG, Kinnaird MF, Nurcahyo A, Iqbal M, Rusmanto M (2004) Abundance and distribution of sympatric gibbons in a threatened Sumateran rainforest. Int J Primatol 25:267–284

Page SE, Rieley JO, Shtyk OW, Weiss D (1999a) Interdependence of Peat and vegetation in tropical peat swamp forest. Philos Trans R Soc London 354:1885–1897

Page SE, Rieley JO, Shotyk ØW, Weiss D (1999b) Interdependence of peat and vegetation in a tropical peat swamp forest. Philos Trans R Soc Lond B 354:1885–1897

Page SE, Siegert F, Boehm HDV, Jaya A, Limin S (2002) The amount of carbon released from peat and forest fires in Indonesia during 1997. Nature 420:61–65

Palombit RA (1994) Extra pair copulations in a monogamous ape. Anim Behav 47:721–723

Palombit RA (1995) Longitudinal patterns of reproduction in wild siamang (*H. syndactylus*) and white-handed gibbons (*H. lar*). Int J Primatol 16:739–760

Raemaekers J (1979) Ecology of sympatric gibbons. Folia Primatol 31:227–245

Raemaekers J (1980) Causes of variation between months in the distance traveled daily by gibbons. Folia Primatol 34:46–60

Reichard U (1995) Extra Pair Copulation in a monogamous Gibbon (*H. lar*). Ethology 100:99–112

Reichard U, Barelli C (2008) Life history and reproductive strategies of Khao Yai *Hylobates lar*: implications for social evolution in apes. Int J Primatol 29:823–844

Reichard U, Sommer V (1997) Group encounters in wild gibbons (*H. lar*): agonism, affiliation and the concept of infanticide. Behaviour 134:1135–1174

Rieley JO (2002) STAPEAT state of knowledge report: ecology. Unpublished report

Rieley JO, Sieffermann RG, Page SE (1993) The origin, development, present status and importance of the lowland peat swamp forests of Borneo. Suo 43:241–244

Rieley JO, Ahmad-Shah A, Brady MA (1996) The extent and nature of tropical peat swamps. In: Maltby E, Immirzi CP, Safford RJ (eds) Tropical lowland peatlands of Southeast Asia. IUCN, Gland, Switzerland, pp 17–53

Rieley JO, Page S, Wuest R, Weiss D, Limin S (2004) Tropical peatlands and climate change: past, present and future perspectives. In: Päivänen J (ed) Wise use of peatlands (vol i) proceedings of the 12th international peat congress, Tampere, Finland. Saarijärven Offset Oy, Saarijärvi, Finland, pp 713–719

Rijksen HD (1978) A fieldstudy on Sumatran orang utans (*Pongo pygmaeus abelii*, lesson 1827): ecology, behaviour and conservation. H. Veenman, Netherlands

Rodman PS (1976) Synecology of Bornean primates. In: Montgomery GG (ed) Ecology of arboreal folivores. Smithsonian Institute Press, Washington DC

Ross C, Srivastava A (1994) Factors influencing the population density of the Hanuman langur (*Presbytis entellus*) in Sariska Tiger Reserve. Primates 35:361–367

Rovero F, Struhsaker TT (2007) Vegetative predictors of primate abundance: utility and limitations of a fine-scale analysis. Am J Primatol 69:1242–1256

Savini T, Boesch C, Reichard U (2008) Home-range characteristics and the influence of seasonality on female reproduction in white-handed gibbons (*Hylobates lar*) at Khao Yai National Park, Thailand. Am J Phys Anthropol 135:1–12

Shepherd PA, Rieley JO, Page SE (1997) The relationship between forest structure and peat characteristics in the upper catchment of the Sungai Sabangau, Central Kalimantan. In: Rieley JO, Page SE (eds) Biodiversity and sustainability of tropical peatlands. Samara Publishing, Cardigan, UK, pp 191–210

Struebig MJ, Harrison ME, Cheyne SM, Limin SH (2007) Intensive hunting of large flying-foxes (*Pteropus vampyrus natunae*) in the Sabangau Catchment, Central Kalimantan, Indonesian Borneo. Oryx 41:1–4

Tenaza RR (1975) Territory and monogamy among Kloss' gibbons (*Hylobates klossii*) in Siberut Island, Indonesia. Primates 24:60–80

Thompson CJH (2007) Gibbon locomotion in a disturbed peat-swamp forest: Sabangau, Central Kalimantan. University of Cambridge, Cambridge

White P, Saunders G, Harris S (1996) Spatio-temporal patterns of home-range use by foxes (*Vulpes vulpes*) in urban environments. J Anim Ecol 65:121–125

Whitten AJ (1980) The Kloss Gibbon in Siberut. PhD, Dissertarion, University of Cambridge

Whitten AJ (1982) Diet and feeding behaviour of Kloss gibbons on Siberut Island, Indonesia. Folia Primatol 37:177–208

Wieczkowski J (2004) Ecological correlates of abundance in the Tana mangabey (*Cercocebus galeritus*). Am J Primatol 63:125–138

Williamson EA, Feistner ATC (2003) Habituating primates: processes, techniques, variables and ethics. In: Setchell JM, Curtis DJ (eds) Field and laboratory methods in primatology. Cambridge University Press, Cambridge, pp 25–39

Zhou J, Wei F, Li M, Lok CBP, Wang D (2008) Reproductive characters and mating behaviour of wild *Nomascus hainanus*. Int J Primatol 29:1037–1046

Chapter 9
Effect of Habitat Quality on Primate Populations in Kalimantan: Gibbons and Leaf Monkeys as Case Studies

Andrew J. Marshall

Introduction

Primate ecologists seek to answer fundamental questions about how and why particular ecological factors influence primate individuals, groups, and populations (Isbell 1991; Sterck et al. 1997; van Schaik 1983; Wrangham 1980). While prima-tologists have used a variety of approaches to address these questions, the study of a particular taxon across a range of ecological conditions provides a particularly useful framework for investigating key questions about the interactions between primates and their habitats (Davies 1994; Doran et al. 2002; Morrogh-Bernard et al. 2009; Strum and Western 1982; van Schaik et al. 2009). Studies conducted on small spatial scales may be especially useful in this context because they permit investigation of the effects of variation in some ecological conditions while controlling for others (e.g., Caldecott 1980; Chapman and Chapman 1999; Dunbar 1992a; Iwamoto and Dunbar 1983). While such studies may address fundamental ecological questions in ways that other research designs cannot, they remain relatively underutilized. Understandably, given the difficulties of sampling long-lived, generally rare vertebrates, most primate studies have focused on a single or small number of groups. Equally reasonably, most studies are conducted in relatively high quality habitats, where behavioral data can be most efficiently collected. The focus on a small number of groups and disproportionate sampling of high quality habitats may limit our ability to observe variation in a species' ecology, hamper examination of the full range of behavioral plasticity that a primate species exhibits, and bias our understanding of how ecological factors affect primate populations.

In this chapter, I provide an overview of selected results from my study of gibbon and leaf monkey populations inhabiting seven distinct forest types at the Cabang Panti Research Station (CPRS) in Gunung Palung National Park (GPNP), West Kalimantan, Indonesia. These seven forest types comprise the full range of habitats

A.J. Marshall (✉)
Department of Anthropology, University of California, Davis, CA, USA
e-mail: ajmarshall@ucdavis.edu

S. Gursky-Doyen and J. Supriatna (eds.), *Indonesian Primates*,
Developments in Primatology: Progress and Prospects,
DOI 10.1007/978-1-4419-1560-3_9, © Springer Science+Business Media, LLC 2010

that these species occupy at GPNP, thereby providing the unusual opportunity to address two fundamental ecological questions: (1) what determines habitat quality? and (2) what are the consequences of variation in habitat quality for individuals, groups, and populations? Gibbons and leaf monkeys have fundamentally different diets (frugivores vs. gramnivores/folivores, respectively), social systems (socially monogamous vs. polygynous), and life histories ("slow" vs. "fast"), and simultaneous investigation of these two taxa can indicate how such characteristics mediate a species' response to environmental variation. CPRS is an exceptional study site that permits sampling of a large number of primate groups across a broad range of forest types, the quality of which markedly, providing a rare opportunity to examine how ecology and social systems interact under a wide range of environmental conditions.

Here, I provide a brief introduction to the study site, species, and field methods; consider what determines habitat quality for gibbons and leaf monkeys; describe some important effects that habitat quality has on these species; and discuss the theoretical and practical relevance of these results.

Study Site and Subjects

CPRS, located in southwestern Borneo (Fig. 9.1), is composed of seven distinct, contiguous forest types, determined by elevation, soils, and drainage:(1) *peat swamp forest* on nutrient-poor, bleached white soils overlain by variable amounts of organic matter (5–10 m asl); (2) *freshwater swamp forest* on nutrient-rich, seasonally flooded, poorly drained gleyic soils (5–10 m asl); (3) *alluvial forest* on rich sandstone-derived soils recently deposited from upstream sandstone and granite parent material (5–50 m asl); (4) *lowland sandstone forest* on well-drained sandstone-derived soils with a high clay content and sparse patches of shale (20–200 m asl); (5) *lowland granite forest* on well-drained, granite-derived soils (200–400 m asl); (6) *upland granite forest* on well-drained, granite-derived soils (350–800 m asl); and (7) *montane forest* on largely granite-derived soils (750–1100 m asl). These forest types differ substantially in their floristic composition, temporal patterns of food availability, structure, and temperature (Cannon et al. 2007a, b). As a result, these habitats support densities of gibbons and leaf monkeys that differ by more than an order of magnitude (Table 9.1). This substantial variation occurs over a very small spatial scale (~5–10 km), so that variation in predators, disease, biogeography, and climate that confound comparative studies of more distant sites is controlled (Chapman and Chapman 1999; Marshall and Leighton 2006). The site had been the location of a long-term research since the mid 1980's (e.g., Knott 1998; Curran and Leighton 2000; Cannon et al. 2007b), but was closed between 2002 and 2006 because of tensions with illegal loggers.

Bornean White-bearded Gibbons (*Hylobates albibarbis,* hereafter referred to as "gibbons") and Red Leaf Monkeys (*Presbytis rubicunda rubida,* here "leaf monkeys") are an excellent pair of species for comparative study because they differ

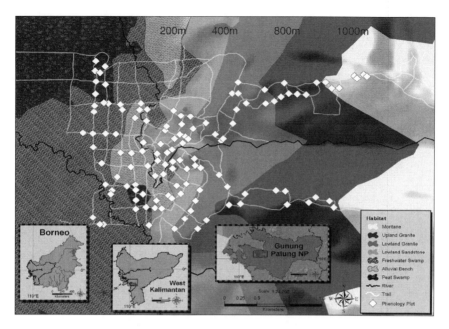

Fig. 9.1 Site map of the Cabang Panti Research Station (CPRS), located in GPNP. Shading depicts the seven forest types: shading top to bottom on legend = habitats right (montane) to left (peat and freshwater swamp) on figure. The 70 botanical plots (10 per forest type) are indicated by white diamonds; 2 vertebrate census routes (not shown) are also located in each forest type. Elevation of major contours is indicated along the top of the figure.

substantially in their diets, social systems, and life histories. Gibbons are frugivores; their diet at CPRS comprises mainly the pulp of ripe fruits (65% of the diet on average, range 0–95%, based on data collected between 1985 and 1992), augmented by ripe figs (23%, range 0–75%), flowers (6%, range 0–28%), leaves (3%, range 1–25%), and seeds (3%, range 0–8%; Marshall and Leighton 2006; Marshall et al. 2009a). On the other hand, leaf monkeys are seed and leaf specialists (seeds: 52%, range 25–95%; leaves: 25%, range 0–42% during the same period as the gibbon data), and also consume unripe fruit pulp (13%, range 2–72%; confined to plant taxa that are dispersed by bats and whose nutritional quality is similar to that of leaves), figs (5%, range 0–25%), and flowers (5%, range 0–20%; Marshall 2004; Marshall et al. 2009b). Gibbons at GPNP fall back on figs (Marshall and Leighton 2006), whereas leaf monkeys fall back on a combination of young and mature leaves (Marshall 2004). Gibbons are socially monogamous (each of the 33 groups observed at GPNP between 2000-2002 contained one adult male and one adult female), whereas leaf monkeys are polygynous (each of the 13 study groups during the same period contained a single adult male and 1–4 females). Primate life history data from CPRS are limited, but suggest that gibbon life histories are roughly half as fast as leaf monkey life histories (e.g., gibbon inter-birth intervals are approximately twice as long as leaf monkeys': Mitani 1990; Marshall 2009; Marshall et al. 2009b),

Table 9.1 Overview of key variation among forest types at GPNP. For each forest type the following data are listed from left to right: mean altitude (m asl); structural data (stems per ha of trees in two size classes, woody climbers, and figs); fruit productivity (for trees, woody climbers, and figs); and the population density of gibbons and leaf monkeys (based on census routes walked between 2000–2002). Data from Marshall (2004, 2009) and Cannon et al. (2007a, b)

| Forest type | Mean altitude | Stem densities (stems/ha) | | | | Fruit productivity per month | | | Gibbon density ± SE (indiv/km²) | Leaf monkey density ± SE (indiv/km²) |
		Small trees (15–25 cm dbh)	Big trees (≥25 cm dbh)	Woody climbers	Figs	All trees (m²/ha)	Woody climbers (m²/ha)	Figs (stems/ha)		
Peat swamp	5	238.0	160.4	112.1	6.7	0.97	59.19	0.39	7.28 ± 1.25	2.52 ± 1.06
Freshwater swamp	5	241.5	152.6	138.5	11.5	0.99	79.99	0.37	5.90 ± 1.34	7.79 ± 2.16
Alluvial bench	34	171.0	148.7	61.7	15.7	1.39	44.78	0.49	7.10 ± 1.70	10.53 ± 2.81
Lowland sandstone	120	157.0	162.1	79.5	9.0	1.66	82.74	0.30	10.27 ± 2.12	5.85 ± 1.76
Lowland granite	255	194.4	203.8	46.9	7.6	1.21	23.51	0.17	6.23 ± 1.33	7.26 ± 1.87
Upland granite	545	244.4	215.2	65.7	8.6	1.24	61.31	0.12	4.17 ± 1.17	6.89 ± 1.76
Montane	880	285.7	140.0	30.0	0.0	0.65	10.82	0.00	0.44 ± 0.25	1.24 ± 0.64

a result that is in accordance with more general comparisons between primate species showing that ape life histories are generally slower than monkey life histories (Schultz 1968; Smith 1989).

Field Methods

Between August 2000 and August 2002, three local assistants and I carried out direct observations of animals along transects to systematically measure the habitat-specific densities of gibbon and leaf monkey populations and to augment data on gibbon and leaf monkey group composition (see below). We established a pair of replicate census routes in each of the seven forest type at CPRS and walked a total of 409 censuses (1,374 km); with an average of 58 censuses (range 38 to 87) in each forest type (Marshall 2004; 2009). Census routes averaged 3.5 km in length and followed existing trails through the forest. We walked each route at least twice per month (starting at opposite ends) at the same speed and time of day (beginning at 05:30 h), and gathered standard line transect data for all vertebrates encountered (e.g., perpendicular sighting distance, group size, group spread). Whenever gibbon or leaf monkey groups were encountered, we ensured that full group counts and information on group composition were recorded by following the group until these data were collected.

We followed standard methods for the analysis of line transect data using Distance 5.2 (Thomas et al. 2006), calculating detection functions separately in each forest type and controlling for size bias in sampling (Buckland et al. 2001). For analyses in which territory- or home range-specific indices of habitat quality were required, I calculated an index of the population density that could be supported in a particular group's territory or home range (i.e., the carrying capacity of the territory or home range). For groups whose entire range was contained within one forest type, this number was the habitat-specific density for the forest type that the group occupied (determined from line transects). For groups whose territory spanned multiple forest types, I summed the habitat-specific density of all forest types occupied, scaled by the proportion of the territory in each. For example, if 80% of a gibbon group's territory was in peat swamp and 20% in freshwater swamp, then this index would be equal to 0.8 * habitat-specific gibbon density for peat swamp + 0.2 * habitat-specific gibbon density for freshwater swamp (i.e., 0.8 * 7.28 individuals/km^2 + 0.2 * 5.90 individuals/km^2 = 7.00 individuals/km^2).

Although data gathered during censuses generally provided complete and accurate data on group size and demographic structure, in 2002, I closely observed and followed all gibbon and leaf monkey groups detected on the census routes to estimate the extent of each territory and to ensure that my field assistants and I had accurately counted the total number of individuals in each group. In order to increase the sample size of groups, I thoroughly searched the study site to identify additional groups that might have been missed on the census routes. This resulted in reliable demographic data on 33 groups of gibbons and 13 groups of leaf monkeys. I observed each of the 46 groups for a minimum of 3 consecutive hours in each of 3 separate months, although in most cases, sample sizes were far greater: each gibbon group was observed a mean of 11.5 times (range 3–73) and for an average of 11.0 months (range 3–23 months); each leaf monkey group was observed a mean of 25.2 times (range 3–96) and for an average of 17.1 months (range 6–25 months).

In order to assess the size of gibbon territories and leaf monkey home ranges, I plotted all group sightings for each species on a map superimposed with a 50 m × 50 m grid. I counted the number of squares inside the smallest polygon that included all group observations, and multiplied this number by 0.25 ha to estimate the home range size of each group (in ha). In cases where sample sizes were sufficiently large (i.e., $n \geq 10$ observations over at least 6 months), these home range estimates were accurate because groups tended to deflect at territorial boundaries during the course of longer group follows. Observed inter-group encounters at territorial boundaries provided useful additional information while estimating a group's home range. Nevertheless, in many cases, sample sizes were inadequate to accurately determine the full home range size. In addition, as estimated home range size increased in a curvilinear fashion with the number of observations of a group, comparing home range sizes among groups with different numbers of observations would be inappropriate. Therefore, I used the home range residuals (HRR, the residuals from the polynomial regression of home range size on observation number) as an unbiased estimate of home range size.

To compile a list of food taxa utilized by gibbons and leaf monkeys, I used a large sample of independent feeding observations from M. Leighton's long-term census data gathered between April 1985 and December 1991, additional opportunistic feeding observations from the same period, and my own data from August 2000 – August 2002 ($n_{GIBBONS} = 536$; $n_{MONKEYS} = 895$). When coupled with phenological data from the same period, I was able to identify preferred foods for gibbons and leaf monkeys, and foods that were eaten during periods when such foods were scarce (i.e., fallback foods, see section "What Determines Habitat Quality for Primates?"). Details are provided in Marshall (2004) and Marshall and Leighton (2006).

We measured the habitat-specific availability of food resources by randomly placing ten 0.5 ha plots in each forest type ($n_{total plots} = 70$). In these plots, we conducted a full census of all fig roots and liana stems with diameters at breast height (dbh, 137 cm above the ground) greater than 4.5 cm, and all trees with boles greater than 14.5 cm dbh. The dbh and botanical identification (using scientific nomenclature) of each stem was recorded ($N = 9,282$ total stems). Details on sampling methodology and botanical nomenclature are provided in Marshall and Leighton (2006). On the basis of stem density data from these plots, we computed the habitat-specific density of each plant taxon. From these data, the density of particular preferred or fallback foods, or the total preferred or total fallback foods, in a habitat could easily be calculated.

Habitat-Specific Population Densities

Population density differs substantially among forest types. Point estimates of gibbon densities range from 0.44 individuals/km^2 in montane forest to 10.27 individuals/km^2 in lowland sandstone habitats; estimates of leaf monkey density range from 1.24 individuals/km^2 in montane forests to 10.53 individuals/km^2 in alluvial bench forest (Table 9.1). During the observation period, population density was stable within each habitat; no changes in population size were detected on surveys conducted bi-monthly between September 2000 and July 2002. In order to ascertain whether population densities were stable over longer periods, I examined the data collected using an identical protocol between May 1985 and January 1992. Populations were stable over this longer period, and there were no significant differences in habitat-specific encounter rates of either species between 1985–1992 and 2000–2002 (Leighton and Marshall, unpublished data). These results indicate that habitat-specific population densities of both species did not fluctuate over time, an interpretation that fits expectations for species exhibiting risk-averse life-history strategies and low intrinsic rates of population increase (Charnov and Berrigan 1993; Marshall 2004).

What Determines Habitat Quality for Primates?

The most widely used index of habitat quality is the population density that a habitat can support at carrying capacity; high quality habitats support high population densities and low quality habitats support low population densities (Begon et al. 1996; Krebs 2001). Thus, habitat quality (i.e., habitat-specific population density) should be a function of the net energy available to support primate biomass, or the balance between habitat-specific energy availability and habitat-specific costs (Caldecott 1980; Iwamoto and Dunbar 1983). Most previous work exploring the ecological determinants of primate population density has focused on variation in habitat-specific benefits– essentially, different levels of energy input (i.e., food availability; e.g., Davies 1994; Chapman and Chapman 1999). Until recently, measures of food availability have been relatively simplistic, largely failing to consider potentially important variation in what types of food are available. There is both growing empirical evidence and widening conceptual realization that distinct classes of foods can have quite different effects on primate populations on ecological and evolutionary time scales (Laden and Wrangham 2005; Lambert 2007; Lambert et al. 2004; Marshall et al. 2009b; Rosenberger 1992; Vogel et al. 2008; Wrangham et al. 1998).

Particular attention has been paid to the relative importance of preferred and fallback foods. Preferred foods are positively selected (i.e., disproportionately used relative to their abundance; cf. Leighton 1993; Manly et al. 2002). Fallback foods are used in inverse proportion to the availability of preferred foods (Altmann 1988; Wrangham et al. 1998). These two classes of foods differ in their quality (preferred foods are relatively high quality, fallback foods are relatively low quality) and distribution in space and time (fallback foods are generally more abundant and available than preferred foods; Lambert 2007; Marshall and Wrangham 2007).

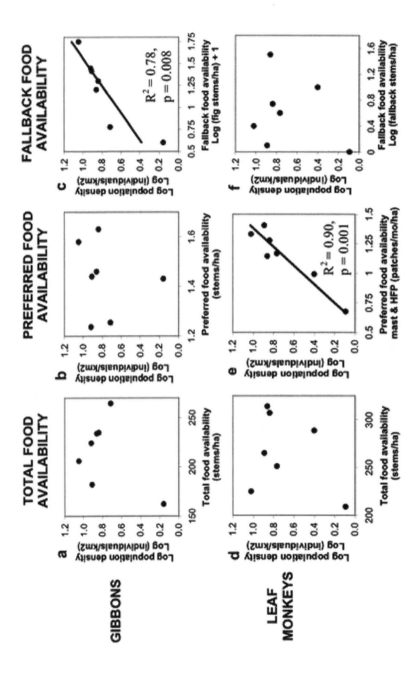

Fig. 9.2 Different classes of food have different effects on gibbons and leaf monkeys. Gibbon population density appears to be limited by the availability of fallback foods (**c**), but not total food availability (**a**) or preferred food availability (**b**). Leaf monkeys are limited by preferred foods (**e**), but not total foods (**d**) or fallback foods (**f**). Modified from Marshall et al. (2009b). For all figures (**a–f**), $n = 7$ forest types. Statistics and regression lines are provided for significant relationships only

There is some debate about whether preferred or fallback foods determine carrying capacity for primate populations. Some primatologists have suggested that because preferred foods are of high nutritional quality and provide energy necessary for reproduction, habitats with patches of preferred foods that are relatively large or abundant might be expected to maintain higher overall primate densities (e.g., Altmann et al. 1985; Balcomb et al. 2000). An alternative hypothesis suggests that fallback foods limit population density because they provide sustenance during periods of low food availability when competition for food is most intense (e.g., Cant 1980; Foster 1982; Marshall and Leighton 2006; Marshall et al. 2009c). Still other studies have suggested that alternative indices of food availability, such as total or maximum food availability or protein to fiber ratios in leaves, determine habitat quality (Chapman and Chapman 1999; Davies et al. 1988; Hanya et al. 2004; Mather 1992; McKey 1978; Wasserman and Chapman 2003).

Research on gibbons and leaf monkeys at CPRS suggests that consideration of foods in distinct classes (i.e., preferred vs. fallback) is warranted (Marshall et al. 2009b). Data from CPRS show that simple measures of total food abundance were not correlated with population density of either species across the seven habitats at the site (Fig. 9.2a, d). Furthermore, leaf monkey density was highly correlated with the availability of preferred foods during periods of high food abundance, whereas gibbon density was not (Fig. 9.2 b, e). In contrast, habitat-specific gibbon density was closely related to the availability of figs, their primary fallback food, while the availability of fallback foods did not explain any variation in leaf monkey density across the seven forest types (Fig. 9.2 c, f). These results are important because they suggest that only specific types of resources determine habitat quality (rather than general measures of food availability, as is often assumed) and that different species may be limited by distinct types of food resources, perhaps due to key differences in physiology, social system, or life history (Marshall et al. 2009b).

Effects of Variation in Habitat Quality

Primatologists assume that ecological factors (e.g., climate, food availability, disease, predation) drive macro-evolutionary processes (e.g., speciation, radiation, and extinction), and examination of the effects of variation in these factors over time and space has been the central goal of primate ecology for decades (e.g., Bourliére 1979; Chapman et al. 2002; Isbell 1991; Janson and Chapman 1999; Sterck et al. 1997; van Schaik 1983; Wrangham 1980). The unusual range of variation found among the seven forest types at CPRS provides an ideal setting in which to examine the consequences of variation in habitat quality on primate individuals, groups, and populations. Here, I provide examples of the effects of habitat quality on gibbons and leaf monkeys at each of these levels.

Basic ecological theory predicts that habitat quality will have important influences on individual fitness, but that these effects will be mediated by social system. In polygynous systems where females are relatively sedentary (e.g., leaf monkeys), the quality of the territory that can be defended by a male will dictate the number of females that he is able to attract. The logic behind this "polygyny threshold model" (Orians 1969; Verner and Willson 1966) is that in a heterogeneous landscape, a female may gain access to more resources by joining an existing pair in a high quality habitat than by establishing a new pair with a male in a habitat of lower quality. Females should therefore assort themselves in accordance with the "ideal free distribution" (Fretwell and Lucas 1969), and female fitness is equalized across the landscape. In reality, most primate females may not be optimally distributed among groups because a variety of factors might limit their ability to move freely between groups (e.g., dispersal costs, benefits of remaining near kin, the need to secure protection against predation). Nevertheless, the basic prediction is that variation in female fitness among polygynous groups will be small.

In monogamous territorial species (e.g., gibbons), social constraints prohibit females from freely assorting themselves (*sensu* Fretwell and Lucas 1969). Mated pairs typically defend territories to the exclusion of all other individuals. Therefore, fitness is not equalized across the landscape, and female reproductive success should be correlated with habitat quality, unless territory size is inversely correlated with habitat quality (see below). Male reproductive success should be correlated with habitat quality regardless of social system since in both systems, high quality males will outcompete low quality males for access to the best territories (e.g., Owen-Smith 1977). In polygynous systems, these higher quality territories allow a given male to attract more females, whereas in monogamous species, they provide the mated pair with additional resources for use in reproduction and may allow a male to attract a higher quality mate.

I used the average number of offspring (the sum of all infants, juveniles, and subadults) per adult female in each group as a proxy of reproductive success. This serves as a crude approximation of reproductive success as it does not include any effects of differential female life spans or differences in maturation age among groups, but is the best index presently available for populations at GPNP. As predicted, habitat quality had a strong influence on this measure of reproductive success in gibbons and leaf monkeys, but the effects of habitat quality differed between the two taxa. Since they live in monogamous pairs, the reproductive success of both male and female gibbons was positively correlated with habitat quality (Fig. 9.3a). In contrast, among polygynous leaf monkeys, males' reproductive success was higher in high quality habitats, while females' reproductive success is unaffected by habitat quality (Fig. 9.3b). This result confirms that social systems can fundamentally alter the way in which ecology affects individuals.

In high quality habitats (i.e., those supporting high population densities), primates might be expected either to live in larger groups or occupy smaller territories, or both. At CPRS, group size is positively correlated with population density in both gibbons (Fig. 9.4a, $r^2 = 0.58$, $n = 33$, $p < 0.0001$) and leaf monkeys (Fig. 9.4b, $r^2 = 0.77$, $n = 13$, $p < 0.0001$). Because of limited sampling, particularly of groups in

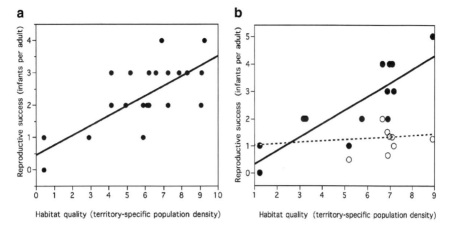

Fig. 9.3 An index of reproductive success (number of infants per adult) plotted vs. habitat quality (for groups whose entire territory was contained within one forest type this is equal to the habitat-specific density for that forest type; for groups whose territory spanned multiple forest types, this is the sum the habitat-specific density of all forest types occupied, scaled by the proportion of the territory in each, see text). Reproductive success of male and female gibbons is positively correlated with habitat quality (**a**; $r^2=0.58$, $p<0.0001$, $n=33$). Reproductive success for male (**b**; *dots, solid line*, $r^2=0.70$, $p=0.0004$, $n=13$) but not female leaf monkeys (**b**; *circles, dashed line*, $r^2=0.04$, $p=0.53$, $n=13$) is positively correlated with habitat quality

low quality habitats, the home range size of only a subset of groups was assessed. In this sample, a proxy of home range residuals (HRR) was unrelated to habitat quality for gibbons (Fig. 9.4c: $r^2=0.005$, $n=12$, $p=0.81$) or leaf monkeys (Fig. 9.4d: $r^2=0.004$, $n=10$, $p=0.86$). These results suggest that a given area of high quality habitat can support larger group sizes than the same area of low quality habitat, implying that feeding competition need not be positively correlated with group size. I return to this topic in the discussion below.

Animal taxa generally occupy a number of distinct habitats (Pulliam 1988), and understanding the effects of habitat quality on population growth rates has been a topic of considerable interest for animal ecologists for decades (Begon et al. 1996; Krebs 2001). The quality of a habitat influences the birth, death, immigration, and emigration rates of a population living there. In populations distributed across heterogeneous landscapes, a proportion of individuals can be found in habitats in which births exceed deaths and emigration exceeds immigration (i.e., they are demographic sources with the natural rate of population increase, r, >0; Pulliam 1996). In contrast, sink habitats, in which deaths exceed births and immigration exceeds emigration, have net negative population growth rates (r < 0; Pulliam 1988). In the absence of immigration from sources, populations in sink habitats will inevitably decline to extinction (Holt 1997). Although there are theoretical (e.g., Holt 1985; Watkinson and Sutherland 1995) and practical (e.g., Dias 1996; Doncaster et al. 1997) difficulties that complicate attempts to empirically demonstrate source-sink population dynamics in wild populations, there are mounting data

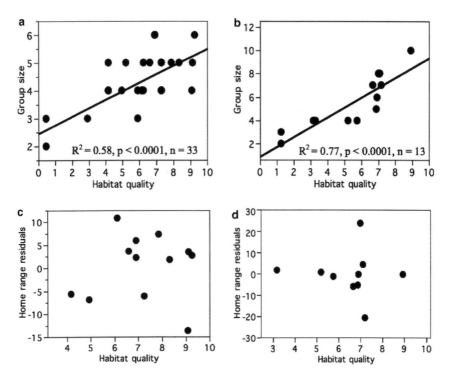

Fig. 9.4 Group size plotted against habitat quality (defined as in Fig. 9.3) for gibbons (**a**) and leaf monkeys (**b**). Habitat quality was unrelated to an index of home range size (HRR) for gibbons (**c**: $r^2 = 0.005$, $n = 12$, $p = 0.81$) or leaf monkeys (**d**: $r^2 = 0.004$, $n = 10$, $p = 0.86$). Statistics and regression lines on plots are provided for significant relationships only

(e.g., Kreuzer and Huntly 2003; others reviewed in Pulliam 1996 and Diffendorfer 1998) to suggest that these dynamics characterize at least some animal species. Since a wide range of mammalian taxa (including primates) exhibit substantially depressed population densities with increased altitude, determining whether montane forests might be demographic sinks at CPRS is particularly relevant.

Population density of both gibbons and leaf monkeys was negatively correlated with altitude (Fig. 9.5 a, b). At CPRS, the number of gibbon groups ($n = 33$) was sufficiently large to produce a simple demographic model, which indicated that montane forests are likely to be demographic sinks for this species, and suggests that montane forests could not support gibbon populations in the absence of continued input of individuals from higher quality lowland forests (Fig. 9.5a; Marshall 2009). The more limited sample of leaf monkey groups ($n = 13$) has precluded population modeling, but population density is clearly negatively related to altitude (Fig. 9.5b). This relationship is strongly significant when peat swamp forests (located in the lowlands but still of poor quality for leaf monkeys due to their very limited productivity) are removed from the analysis, and merely a statistical trend

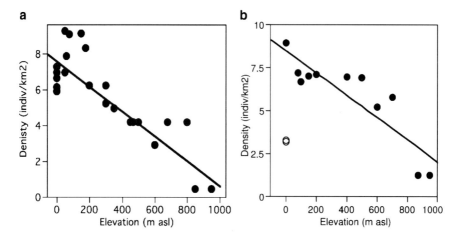

Fig. 9.5 Territory-specific population density (individuals/km², defined as the territory specific habitat-quality, as in Fig.9.3) of gibbons (**a**) and leaf monkeys (**b**) plotted against altitude (meters asl). Statistics: (**a**) $r^2 = 0.82$, $p < 0.0001$, $n = 33$, from Marshall 2009; (**b**) including two peat swamp groups (*open circles*): $r^2 = 0.29$, $p < 0.06$, $n = 13$; excluding peat swamp groups: $r^2 = 0.77$, $p < 0.0004$, $n = 11$. A simple demographic model using these cross-sectional data suggested that montane forests are sink habitat for gibbons (Marshall 2009); data are insufficient to estimate habitat-specific population growth rates for leaf monkeys

when the two peat swamp leaf monkey groups are retained. However, the implication of this result is similar to that found for gibbons: if lowland forests (most of which are of high quality for leaf monkeys) were destroyed, montane leaf monkey population densities might not be viable. These results have important conservation implications, which will be discussed at the end of the Discussion section.

Discussion

This chapter presents an overview of results that have emerged from studies of gibbons and leaf monkeys living in a range of distinct habitats. These results indicate that habitat quality (i.e., population density at carrying capacity) can vary substantially across forest types on relatively small spatial scales. These results also suggest that different classes of food resource (e.g., preferred and fallback foods) can have distinct effects on primate populations, that these effects may differ between primate taxa, and, therefore, that simple measures of food availability are inadequate to capture the ecological variation of most relevance to primates. Furthermore, habitat quality can have important implications for primate populations on the individual, group, and population level. For example, habitat quality can influence individual reproductive success, group size, and a population's probability of persistence. This suggests that observations and ecological inferences from one habitat should be extrapolated to other habitats with caution, and that a full understanding of a primate species' behavior, group and demographic structure, and population dynamics requires study of the species across the full range of habitats that it occupies.

As discussed earlier, most primate field studies are conducted in relatively high quality habitats, for understandable reasons. Therefore, the biological and demographic characteristics, such as reproduction (e.g., weaning age, inter-birth interval), density and demographic composition, and disease burden, observed in these habitats probably represent "best case scenarios" for these groups and are not representative of the natural range of variation in these species. Moreover, these biases likely affect our understanding of social and behavioral traits of certain primate taxa as well. It is hypothesized that long-term differences in habitat quality have led to a divergence of ecological adaptations in closely related species (e.g., Bornean vs. Sumatran orangutans: Delgado and van Schaik 2000; van Schaik et al. 2009; chimpanzees vs. bonobos: Wrangham 1986), and it is likewise reasonable to view some of the variation between different populations of a given primate species (e.g., chimpanzees: Doran et al. 2002; savanna baboons: Kamilar 2006) as a result of variation in habitat quality. Aspects of social behavior that are often considered to be hallmarks of a primate species' biology (e.g., intensity of food competition, presence of female bonding, degree of polygyny, hunting behavior) turn out to be quite variable within that species under different ecological conditions (e.g., rates of inter-group conflict and violent competition in gorillas living at high vs. low density: Yamagiwa 1999; female bonds in captive vs. wild chimpanzees: Baker and Smuts 1994; degree of polygyny in leaf monkeys: this study; hunting frequency and success rates of Kanayawara vs. Ngogo chimpanzees: Mitani and Watts 1999). This suggests that an increased awareness of the influence of habitat quality may improve our understanding of how ecological parameters have influenced the course of primate evolution.

This discussion of the determinants of habitat-specific carrying capacity focused on differences in indices of food availability among forest types, and did not consider the potentially important role of variation in habitat-specific costs. The role of ecological costs in limiting primate population density has generally received less attention from primate ecologists than has the role of food availability (but see, e.g., Caldecott 1980; Chapman et al. 2002; Davies et al. 1988; Dunbar 1992a, b; Ganzhorn 1992; McKey 1978; Milton 1979). This is notable given the extensive attention that has been paid to the costs of primate grouping (e.g., due to feeding competition, infanticide, and disease; Isbell 1991; Janson and Goldsmith 1995; Nunn 2003; van Schaik 1983; van Schaik and Janson 2000; Wrangham 1980). In principle, habitat-specific costs might differ among forest types at CPRS for several reasons, including thermoregulatory costs associated with elevation (Caldecott 1980; Hill et al. 2000; Iwamoto and Dunbar 1983); locomotor costs associated with differences in canopy structure (Cannon and Leighton 1994; 1996; Kappeler 1984); or costs of interspecific competition from other frugivorous vertebrates (Gautier-Hion 1978; Marshall et al. 2009a; Poulson et al. 2002). Ongoing work at CPRS is incorporating explicit consideration of habitat-specific costs, and will clarify how habitat-specific costs and benefits interact to determine primate carrying capacity.

Much theoretical and empirical work in primatology seeks to elucidate the factors underlying gregariousness in primates, and to understand the forces that influence group size (e.g., Wrangham 1980; van Schaik 1983). Factors such as within-group scramble competition and infanticide are thought to limit group size (Janson and Goldsmith 1995; van Schaik and Janson 2000), while predation risk and between-group contest competition are thought to increase the benefits of grouping (Wrangham 1980; van Schaik 1983). Owing to the presumed importance of within group scramble feeding competition, female fitness is predicted to be lower in larger groups (e.g., Borries et al. 2008; van Noordwijk and van Schaik 1999). This prediction is based on the tacit assumption that some variable other than food availability limits group size, and that females in larger groups experience more intense feeding competition. An alternative hypothesis is that fitness is equalized across groups of different size within a population because females distribute themselves according to an ideal free distribution (Fretwell and Lucas 1969). This perspective does not imply that feeding competition is unimportant, it simply suggests that the influence of feeding competition primarily occurs at the level of determining group size (e.g., affecting female decisions about which groups to join). Results from CPRS support the hypothesis that female leaf monkey reproductive success is independent of habitat quality (Fig 9.3b) and group size ($n = 13$, $R^2 = 0.03$, $p = 0.52$, Boyko, Boyko, and Marshall, unpublished analysis), implying that increased competition in larger groups is offset by the absolutely greater amounts of food available in higher quality habitats. Some other studies of primarily folivorous primates have also shown no effect of group size on reproductive success (e.g., gorillas: Stokes et al. 2003; Robbins et al. 2007; Thomas' langurs: Steenbeek and van Schaik 2001) although such results are not universal (e.g., Borries et al. 2008; Snaith and Chapman 2008).

Future Directions

The sampling methods employed at CPRS provide an unusually extensive sample of primate groups across a wide range of ecological conditions. This approach has provided a unique perspective on landscape-level processes, but has done little to improve understanding of how habitat quality influences the behavior of gibbons and leaf monkeys. Examination of the influence of habitat quality on behavior may improve our understanding of the range of behavioral flexibility exhibited in these species. For instance, residence times in food patches are predicted to vary as a function of the distribution and abundance of resources in the environment (Charnov 1976; Grether et al. 1992; Schoener 1971), and should vary systematically among forest types at CPRS. Assessment of whether and how preferred and fallback foods for gibbons and leaf monkeys are depleted would elucidate how food type influences foraging behavior, and would contribute to the recent reexamination of the long-held assumptions about the lack of feeding competition in folivorous primates (Koenig 2000; Snaith and Chapman 2005; 2007). Behavioral data collected across the mosaic of habitat types may also identify the proximate mechanisms that underlie some of the individual, group, and population effects reported here. For example, long-term monitoring of individual life histories (e.g., age at dispersal, offspring mortality, inter-birth intervals) would permit determination of which components of fitness are influenced by habitat quality. Finally, differences in rates of immigration and emigration among habitats would provide behavioral indications of source-sink population dynamics.

Understanding how habitat quality influences populations may improve our ability to protect and manage populations of threatened primate species in a number of ways. For example, if particular classes of foods are disproportionately important in limiting primate populations, special attention may be taken to spare these food resources during selective logging operations (Felton et al. 2003; Johns 1986; Leighton and Leighton 1983), and enrichment planting of these taxa may be used to raise the primate carrying capacity of degraded areas (Marshall et al. 2009b). Identifying and protecting habitats that are disproportionately used during periods of overall fruit scarcity also may be crucial for maintaining populations in heterogeneous landscapes (Cannon et al. 2007b; Curran and Leighton 2000; Furuichi et al. 2001; Johnson et al. 2005). In addition, understanding how species respond to natural variation in habitat quality may provide insight into their responses to future habitat alteration, through either human-induced habitat degradation or climate change (Marshall et al. 2006; Meijaard et al. 2008).

Explicit consideration of habitat-specific carrying capacity and an understanding of source-sink population dynamics suggest that higher elevation forests contribute relatively little to maintaining viable populations of some primate species (e.g., gibbons at CPRS, Marshall 2009; Bornean orangutans, Husson et al. 2009). Similarly, conservation plans that include population estimates based on remaining habitat area without regard to habitat quality will likely substantially overestimate the size of primate populations remaining in forest fragments, and lead to unrealistically optimistic estimates of their long term stability and viability (Chapman and Lambert 2000; Cowlishaw and Dunbar 2000; Marshall et al. 2009d). For example, the population densities of primates and other vertebrates may be extremely low in the large tracts of forest in Central Kalimantan (McConkey and Chivers 2004). Therefore, although these areas are attractive targets for conservation due to their high diversity and relatively limited disturbance, they may not support viable populations of threatened vertebrates unless the areas protected are very large (McConkey and Chivers 2004).

Acknowledgments I thank J. Supriatna and S. Gursky-Doyen for inviting me to contribute to this volume, and R. Garvey and two anonymous reviewers for helpful comments that substantially improved this chapter. Permission to conduct research at Gunung Palung National Park was kindly granted by the Indonesian Institute of Sciences (LIPI), the State Ministry of Research and Technology (RISTEK), the Directorate General for Nature Conservation (PHKA) and the Gunung Palung National Park Bureau (BTNGP). I gratefully acknowledge the financial support of the J. William Fulbright Foundation, the Louis Leakey Foundation, a Frederick Sheldon Traveling Fellowship, and the Department of Anthropology at Harvard University. I thank Universitas Tanjungpura (UNTAN), my counterpart institution in Indonesia since 1996, and M. Leighton and

R. W. Wrangham for support and encouragement. I gratefully acknowledge the assistance and support of the many students, researchers, and field assistants who worked at the Cabang Panti Research Station over the past two decades, particularly Albani, M. Ali A. K., Busran A. D., Edward Tang, Hanjoyo, J. R. Harting, Rhande, and J. R. Sweeney.

References

Altmann J, Hausfater G, Altmann SA (1985) Demography of Amboseli baboons, 1963–1983. Am J Primatol 8:113–125

Altmann SA (1988) Foraging for survival. Chicago University Press, Chicago

Baker KC, Smuts BB (1994) Social relationships of female chimpanzees: diversity between captive groups. In: Wrangham RW, McGrew WC, de Waal FBM, Heltne PG (eds) Chimpanzee cultures. Harvard University Press, Cambridge, pp 227–242

Balcomb SR, Chapman C, Wrangham RW (2000) Relationship between chimpanzee (*Pan troglodytes*) density and large, fleshy-fruit tree density: conservation implications. Am J Primatol 51:97–203

Begon M, Harper JL, Townsend CR (1996) Ecology: individuals, populations, and communities. Sinauer Associates, Inc, Sunderland, MA

Borries C, Larney E, Lu A, Ossi K, Koenig A (2008) Costs of group size: lower developmental and reproductive rates in larger groups of leaf monkeys. Behav Ecol 19:1186–1191

Bourliére F (1979) Significant parameters of environmental quality for nonhuman primates. In: Bernstein IS, Smith EO (eds) Primate ecology and human origins: ecological influences on social organization. New York, Garland Press, pp 23–46

Buckland ST, Anderson DR, Burnham KP, Laake JL, Borchers DL, Thomas L (2001) Introduction to distance sampling. Oxford University Press, Oxford

Caldecott JO (1980) Habitat quality and populations of two sympatric Gibbons (Hylobatidae) on a mountain in Malaya. Folia Primatol 33:291–309

Cannon CH, Curran LM, Marshall AJ, Leighton M (2007a) Long-term reproductive behavior of woody plants across seven Bornean forest types in the Gunung Palung National Park (Indonesia): suprannual synchrony, temporal productivity, and fruiting diversity. Ecol Lett 10:956–969

Cannon CH, Curran LM, Marshall AJ, Leighton M (2007b) Beyond mast-fruiting events: community asynchrony and individual dormancy dominate woody plant reproductive behavior across seven Bornean forest types. Curr Sci 93:1558–1566

Cannon CH, Leighton M (1994) Comparative locomotor ecology of gibbons and macaques: selection of canopy elements for crossing gaps. Am J Phys Anthro 93:505–524

Cannon CH, Leighton M (1996) Comparative locomotor ecology of gibbons and macaques: does brachiation minimize travel costs? Trop Biodiversity 3:261–267

Cant JGH (1980) What limits primates? Primates 21:538–544

Chapman CA, Chapman LJ (1999) Implications of small scale variation in ecological conditions for the diet and density of red colobus monkeys. Primates 40:215–231

Chapman CA, Chapman LJ, Bjorndal KA, Onderdonk DA (2002) Application of protein-to-fiber ratios to predict colobine abundance on different spatial scales. Inter J Primatol 23:283–310

Chapman CA, Lambert JE (2000) Habitat alteration and conservation of African primates: case study of Kibale National Park, Uganda. Am J Primatol 50:169–185

Charnov EL (1976) Optimal foraging, the marginal value theorem. Theor Pop Biol 9:129–136

Charnov EL, Berrigan D (1993) Why do female primates have such long lifespans and so few babies? or life in the slow lane. Evol Anthro 1:191–194

Cowlishaw G, Dunbar RIM (2000) Primate conservation biology. University of Chicago Press, Chicago, Illinois

Curran LM, Leighton M (2000) Vertebrate responses to spatiotemporal variation in seed production of mast-fruiting Dipterocarpaceae. Ecol Mono 70:101–128

Davies AG (1994) Colobine populations. In: Davies AG, Oates JF (eds) Colobine monkeys: their ecology, behaviour, and evolution. Cambridge University Press, Cambridge, pp 285–310

Davies AG, Bennett EL, Waterman PG (1988) Food selection by two south-east Asian colobine monkeys (*Presbytis rubicunda* and *Presbytis melalophos*) in relation to plant chemistry. Biol J Linn Soc 34:33–56

Delgado RA Jr, van Schaik CP (2000) The behavioral ecology and conservation of the orangutan (*Pongo pygmaeus*): A tale of two islands. Evol Anthro 9:201–218

Dias PC (1996) Sources and sinks in population biology. Trend Ecol Evol 11:326–330

Diffendorfer JE (1998) Testing models of source-sink dynamics and balanced dispersal. Oikos 81:417–433

Doncaster CP, Clobert J, Doligez B, Gustafsson L, Danchin E (1997) Balanced dispersal between spatially varying local populations: an alternative to the source-sink model. Am Nat 150:425–445

Doran DM, Jungers WL, Sugiyama Y, Fleagle JG, Heesy C (2002) Multivariate and phylogenetic approaches to understanding chimpanzee and bonobo behavioral diversity. In: Boesch C, Hohmann G, Marchant LF (eds) Behavioural diversity in Chimpanzees and Bonobos. Cambridge University Press, Cambridge, pp 14–34

Dunbar RIM (1992a) Time: a hidden constraint on the behavioural ecology of Baboons. Behav Ecol Sociobiol 31:35–49

Dunbar RIM (1992b) A model of the gelada socio-ecological system. Primates 33:69–83

Felton AM, Engström LM, Felton A, Knott CD (2003) Effects of selective hand-logging on orang-utan population density, forest structure and fruit availability in a peat swamp forest in West Kalimantan. Indonesia Biol Cons 114:91–101

Foster RB (1982) The seasonal rhythm of fruitfall on Barro Colorado Island. In: Leigh EGJ, Rand AS, Windsor DM (eds) The ecology of a tropical forest: seasonal rhythms and long-term changes. Smithsonian Inst. Press, Washington, DC, pp 151–172

Fretwell SD, Lucas HL Jr (1969) On territorial behavior and other factors influencing habitat distribution in birds. Acta Biotheor 19:16–36

Furuichi T, Hashimoto C, Tashiro Y (2001) Fruit availability and habitat use by chimpanzees in the Kalinzu Forest, Uganda: examination of fallback foods. Inter J Primatol 22:929–945

Ganzhorn JU (1992) Leaf chemistry and biomass of folivorous primates in tropical forests: test of a hypothesis. Oecologia 91:540–547

Gautier-Hion A (1978) Food niches and co-existence in sympatric primates in Gabon. In: Chivers DJ, Herbert CA (eds) Recent advances in primatology. Academic, London

Grether GF, Palombit RA, Rodman PS (1992) Gibbon foraging decisions and the marginal value model. Inter J Primatol 13:1–17

Hanya G, Yoshihiro S, Zamma K, Matsubara H, Ohtake M, Kubo R, Noma N, Agetsuma N, Takahata Y (2004) Environmental determinants of the altitudinal variations in relative group densities of Japanese macaques on Yakushima. Ecol Res 19:485–493

Hill RA, Lycett JE, Dunbar RIM (2000) Ecological and social determinants of birth intervals in Baboons. Behav Ecol 11:560–564

Holt RD (1985) Population dynamics in two-patch environments: some anomalous consequences of an optimal habitat distribution. Theor Popul Biol 28:181–208

Holt RD (1997) On the evolutionary stability of sink populations. Evol Ecol 11:723–731

Husson SJ, Wich SA, Marshall AJ, Dennis RD, Ancrenaz M, Brassey R, Gumal M, Hearn AJ, Meijaard E, Simorangkir M, Singleton I (2009) Orangutan distribution density abundance and impacts of disturbance. In: Wich SA, Utami S, Mitra Setia T, van Schaik CP (eds) Orangutans: geographic variation in behavioral ecology and conservation. Oxford University Press, Oxford, pp 77–96

Isbell LA (1991) Contest and scramble competition: patterns of female aggression and ranging behavior among primates. Behav Ecol 2:143–155

Iwamoto T, Dunbar RIM (1983) Thermoregulation, habitat quality and the behavioural ecology on Gelada Baboons. J Anim Ecol 52:357–366

Janson CH, Chapman CA (1999) Resources and primate community structure. In: Fleagle JF, Janson C, Reed KE (eds) Primate communities. Cambridge University Press, Cambridge, pp 237–267

Janson CH, Goldsmith ML (1995) Predicting group size in primates: foraging costs and predation risks. Behav Ecol 6:326–336

Johns AD (1986) Effects of selective logging on the behavioral ecology of West Malaysian primates. Ecology 67:684–694

Johnson AJ, Knott CD, Pamungkas B, Pasaribu M, Marshall AJ (2005) A survey of the orangutan population in and around Gunung Palung National Park, West Kalimantan, Indonesia based on nest counts. Biol Conserv 121:495–507

Kamilar JM (2006) Geographic variation in Savanna Baboon (Papio) ecology and its taxonomic and evolutionary implications. In: Lehman SM, Fleagle JG (eds) Primate biogeography: progress and prospects. Springer, New York, pp 169–200

Kappeler M (1984) The gibbon in Java. In: Preuschoft H, Chivers DJ, Brockelman WY, Creel N (eds) The lesser apes: evolutionary and behavioural biology. Edinburgh University Press, Edinburgh, pp 19–31

Knott CD (1998) Changes in orangutan diet, caloric intake, and ketones in response to fluctuating fruit availability. Inter J Primatol 19:1061–1079

Koenig A (2000) Competitive regimes in forest-dwelling Hanuman langur females (*Semnopithecus entellus*). Behav Ecol Sociobiol 48:93–109

Krebs CJ (2001) Ecology: the experimental analysis of distribution and abundance. Benjamin Cummings, San Francisco

Kreuzer MP, Huntly NJ (2003) Habitat-specific demography: evidence for source-sink population structure in a mammal, the pika. Oecologia 134:343–349

Laden G, Wrangham RW (2005) The rise of the hominids as an adaptive shift in fallback foods: plant underground storage organs (USOs) and australopith origins. J Hum Evol 49:482–498

Lambert JE (2007) Seasonality, fallback strategies, and natural selection: a chimpanzee and Cercopithecoid model for interpreting the evolution of the hominin diet. In: Ungar PS (ed) Evolution of the human diet: the known, the unknown, and the unknowable. Oxford University Press, Oxford, pp 324–343

Lambert JE, Chapman CA, Wrangham RW, Conklin-Brittain NL (2004) Hardness of cercopithecine foods: implications for the critical function of enamel thickness in exploiting fallback foods. Am J Phys Anthro 125(4):363–368

Leighton M (1993) Modeling diet selectivity by Bornean orangutans: evidence for integration of multiple criteria for fruit selection. Inter J Primatol 14:257–313

Leighton M, Leighton D (1983) Vertebrate responses to fruiting seasonality within a Bornean rain forest. In: Sutton SL, Whitmore TC, Chadwick AC (eds) Tropical rain forest: ecology and management. Blackwell Scientific Publications, Boston, pp 181–196

Manly BJF, McDonald LL, Thomas DL, McDonald TL, Erickson WP (2002) Resource selection by animals. Kluwer Academic, Dordrecht

Marshall AJ (2004) The population ecology of gibbons and leaf monkeys across a gradient of Bornean forest types. Ph.D. Dissertation, Department of Anthropology. Cambridge, MA, Harvard University

Marshall AJ (2009) Are montane forests demographic sinks for Bornean white-bearded Gibbons (*Hylobates albibarbis*)? Biotropica 41:257–267

Marshall AJ, Leighton M (2006) How does food availability limit the population density of white-bearded gibbons? In: Hohmann G, Robbins MM, Boesch C (eds) Feeding ecology of the Apes and other Primates. Cambridge University Press, Cambridge, pp 311–333

Marshall AJ, Wrangham RW (2007) The ecological significance of fallback foods. Int J Primatol 28:1219–1235

Marshall AJ, Nardiyono LM, Engström B, Pamungkas J, Palapa J, Meijaard E, Stanley SA (2006) The blowgun is mightier than the chainsaw in determining population density of Bornean orangutans (*Pongo pygmaeus morio*) in the forests of East Kalimantan. Biol Cons 129:566–578

Marshall AJ, Cannon CH, Leighton M (2009a) Competition and niche overlap between gibbons (*Hylobates albibarbis*) and other frugivorous vertebrates in Gunung Palung National Park, West Kalimantan, Indonesia. In: Lappan S, Whittaker DJ (eds) The Gibbons: new perspectives on small ape socioecology and population biology. Springer, New York, pp 161–188

Marshall AJ, Boyko CM, Feilen KL, Boyko RH, Leighton M. (2009b). Defining fallback foods and assessing their importance in primate ecology and evolution. *Am J Phys Anthro*, 140: 603–614

Marshall AJ, Ancrenaz M, Brearley FQ, Fredriksson GM, Ghaffar N, Heydon M, Husson SJ, Leighton M, McConkey KR, Morrogh-Bernard HC, Proctor J, van Schaik CP, Yeager CP, Wich SA (2009c) The effects of habitat quality, phenology, and floristics on populations of Bornean and Sumatran orangutans: are Sumatran forests more productive than Bornean forests? In: Wich SA, Utami S, Mitra Setia T, van Schaik CP (eds) Orangutans: Geographic variation in behavioral ecology and conservation. Oxford University Press, Oxford, pp 97–117

Marshall AJ, Lacy R, Ancrenaz M, Byers O, Husson S, Leighton M, Meijaard E, Rosen N, Singleton I, Stephens S, Traylor-Holtzer K, Utami Atmoko S, van Schaik CP, Wich SA (2009d) Orangutan population biology, life history, and conservation: Perspectives from PVA models. In: Wich SA, Utami S, Mitra Setia T, van Schaik CP (eds) Orangutans: Geographic variation in behavioral ecology and conservation. Oxford University Press, Oxford, pp 311–326

Mather RJ (1992) A field study of hybrid Gibbons in Central Kalimantan, Indonesia. PhD Dissertation, Department of Anthropology, Cambridge, Cambridge University, UK

McConkey KR, Chivers DJ (2004) Low mammal and hornbill abundance in the forests of Barito Ulu, Central Kalimantan, Indonesia. Oryx 38:439–447

McKey DB (1978) Soils, vegetation, and seed-eating by black colobus monkeys. In: Montgomery GG (ed) The Ecology of Arboreal Folivores. Smithsonian Institution Press, Washington DC, pp 423–437

Meijaard E, Sheil D, Marshall AJ, Nasi R (2008) Phylogenetic age is positively correlated with sensitivity to timber harvest in Bornean mammals. Biotropica 40:76–85

Milton K (1979) Factors influencing leaf choice by howler monkeys: a test of some hypotheses of food selection by generalist herbivores. Am Nat 114:362–378

Mitani JC (1990) Demography of Agile Gibbons (*Hylobates agilis*). Inter J Primatol 11:411–424

Mitani J, Watts DP (1999) Demographic influences on the hunting behavior of chimpanzees. Amer J Phys Anthro 109:439–454

Morrogh-Bernard H, Husson SJ, Knott CD, Wich SA, van Schaik CP, van Noordwijk MA, Lackman-Ancrenaz I, Marshall AJ, Kanamori T, Kuze N, bin Sakong R (2009) Orangutan activity budgets and diet: a comparison between species, populations, and habitats. In: Wich SA, Utami S, Mitra Setia T, van Schaik CP (eds) Orangutans: Geographic variation in behavioral ecology and conservation. Oxford University Press, Oxford, pp 119–133

Nunn CL (2003) Sociality and disease risk: a comparative study of leukocyte counts in primates. In: De Waal FBM, Tyack PL (eds) Animal social complexity. intelligence, culture, and individualized societies. Harvard University Press, Cambridge MA, pp 26–31

Orians GH (1969) On the evolution of mating systems in birds and mammals. Am Nat 103:589–603

Owen-Smith N (1977) On territoriality in ungulates and an evolutionary model. Quart Rev Biol 52:1–38

Poulson JR, Clark CJ, Connor E, Smith TB (2002) Differential resource use by primates and hornbills: implications for seed dispersal. Ecol 83:228–240

Pulliam HR (1988) Sources, sinks, and population regulation. Am Nat 132:652–661

Pulliam HR (1996) Sources and sinks: empirical evidence and population consequences. In: Rhodes OEJ, Chesser RK, Smith MH (eds) Population dynamics in ecological space and time. University of Chicago Press, Chicago, IL, pp 45–69

Robbins M, Robbins A, Gerald-Steklis N, Steklis H (2007) Socioecological influences on the reproductive success of female mountain gorillas (*Gorilla beringei beringei*). Behav Ecol Sociobiol 61:919–931

Rosenberger AL (1992) Evolution of feeding niches in new world monkeys. Am J Phys Anthro 88:545–562

Schoener TW (1971) Theory of feeding strategies. Ann Rev Ecol Syst 2:369–404

Schultz AH (1968) The recent hominoid primates. In: Washburn SL, Jay PC (eds) Perspectives on human evolution. New York, Holt, Rinehart and Winston, pp 122–191

Smith HB (1989) Dental development as a measure of life history in primates. Evolution 43:683–688

Snaith TV, Chapman CA (2005) Towards an ecological solution to the folivore paradox: patch depletion as an indicator of within-group scramble competition in red colobus. Behav Ecol Sociobiol 59:185–190

Snaith TV, Chapman CA (2007) Primate group size and interpreting socioecological models: Do folivores really play by different rules? Evol Anthro 16:94–106

Snaith TV, Chapman CA (2008) Red colobus monkeys display alternative behavioral responses to the costs of scramble competition. Behav Ecol 19:1289–1296

Steenbeek R, van Schaik CP (2001) Competition and group size in Thomas's Langurs (*Presbytis thomasi*): the folivore paradox revisited. Behav Ecol Sociobiol 49:100–110

Sterck EHM, Watts DP, van Schaik CP (1997) The evolution of female social relationships in nonhuman primates. Behav Ecol Sociobiol 41:291–309

Stokes E, Parnell R, Olejniczak C (2003) Female dispersal and reproductive success in wild western lowland gorillas (*Gorilla gorilla gorilla*). Behav Ecol Sociobiol 54:329–339

Strum SC, Western JD (1982) Variations in fecundity with age and environment in olive baboons (*Papio anubis*). Am J Primatol 3:61–76

Thomas L, Laake JL, Strindberg MS, Marques FFC, Buckland ST, Borchers DL, Anderson DR, Burnham KP, Hedley SL, Pollard JH, Bishop JRB, Marques TA (2006) *Distance* 5.0 Release 2, Research Unit for Wildlife Population Assessment, University of St. Andrews, UK. http://www.ruwpa.st-and.ac.uk/distance/

van Noordwijk MA, van Schaik CP (1999) The effects of dominance rank and group size on female lifetime reproductive success in wild long-tailed macaques, *Macaca fascicularis*. Primates 40:105–130

van Schaik CP (1983) On the ultimate causes of primate social systems. Behaviour 85:91–117

van Schaik CP, Janson CH (2000) Infanticide by males and its implications. Cambridge University Press, Cambridge

van Schaik CP, Marshall AJ, Wich SA (2009) Geographic variation in orangutan behavior and biology: its functional interpretation and its mechanistic basis. In: Wich SA, Utami S, Mitra Setia T, van Schaik CP (eds) Orangutans: Geographic variation in behavioral ecology and conservation. Oxford University Press, Oxford, pp 351–361

Verner J, Willson MF (1966) The influence of habitats on mating systems of North American birds. Ecol 47:143–147

Vogel ER, van Woerden JT, Lucas PW, Utami Atmoko SS, van Schaik CP, Dominy NJ (2008) Functional ecology and evolution of hominoid molar enamel thickness: *Pan troglodytes schweinfurthii* and *Pongo pygmaeus wurmbii*. J Hum Evol 55:60–74

Wasserman MD, Chapman CA (2003) Determinants of colobine monkey abundance: the importance of food energy, protein, and fibre content. J Anim Ecol 72:650–659

Watkinson AR, Sutherland WJ (1995) Sources, sinks and pseudo-sinks. J Anim Ecol 64:126–130

Wrangham RW (1980) An ecological model of female bonded primate groups. Behaviour 75:262–300

Wrangham RW (1986) Ecology and social relationships in two species of chimpanzee. In: Rubenstein DI, Wrangham RW (eds) Ecology and social evolution: birds and mammals. Princeton University Press, Princeton, pp 352–378

Wrangham RW, Conklin-Brittain NL, Hunt KD (1998) Dietary response of chimpanzees and Cercopithecines to seasonal variation in fruit abundance: I. Antifeedants. Int J Primatol 19:949–970

Yamagiwa J (1999) Socioecological factors influencing population structure of gorillas and chimpanzees. Primates 40:87–104

Part II
Indonesia's Monkeys

Chapter 10
Predator Recognition in the Absence of Selection

Jessica L. Yorzinski

Introduction

Animals are frequently confronted with changing environmental conditions (Houston and McNamara 1992; Komers 1997). When they are no longer exposed to the sources of selection that their ancestors once faced, they experience relaxed selection on these sources (Coss 1999). They may still retain behavior that was shaped to cope with the past selective forces, even though it no longer serves a specific function (Blumstein et al. 2000; Rothstein 2001).

Relaxed selection for predator recognition abilities occurs when animals live in environments which lack predators that previously preyed upon their ancestors. The effects of relaxed predation pressure have been studied in a wide taxonomic range of animals (Curio 1966; Kelley and Magurran 2003; Messler et al. 2007; Peckarsky and Penton 1988; Fullard et al. 2004; Blumstein et al. 2000; Hollén and Manser 2007). Some animals still retain specific antipredator behavior even though they do not coexist with their ancestral predators (Blumstein et al. 2000). The amount of time that has lapsed since animals were exposed to certain predators (Coss 1999; Berger et al. 2001) as well as whether they currently experience predation (Blumstein 2006) are potential factors that may influence the retention of appropriate antipredator behavior.

Compared to other taxa, primates have less often been the focus of studies on relaxed predation pressure (reviewed in Table 10.1). The majority of the studies on naïve primates have investigated the antipredator behavior of captive animals that have been isolated from their predators for only a few generations (van Schaik and van Noordwijk 1985; Takahashi 1997). We know little about predator recognition abilities of wild primates that have been isolated from their ancestral predators for thousands of years.

Indonesian primates offer a unique opportunity to explore this topic because many of them inhabit isolated islands and experience different predation pressures

J.L. Yorzinski (✉)
Animal Behavior Graduate Group, University of California, Davis, CA 95616, USA
e-mail: jyorzinski@ucdavis.edu

S. Gursky-Doyen and J. Supriatna (eds.), *Indonesian Primates*,
Developments in Primatology: Progress and Prospects,
DOI 10.1007/978-1-4419-1560-3_10, © Springer Science+Business Media, LLC 2010

Table 10.1 Review of the ability of naïve primates to recognize predators when they experience relaxed predation pressure. NA=the study did not provide the relevant information

	Common name	Scientific name	Current predator(s)	Novel predator type(s) tested[a]	Time passed (yr)[b]	Antipredator behavior?[c]	Novel control	Novel predator recognized?[d]	Compared to experienced conspecific?[e]	Reference(s)
Auditory	Mantled howler monkey	*Alouatta palliata*	None	Bird	50–100	Y	Bird	N[g]	Milder	Gil-da-Costa et al. (2003)
	Ringtailed lemur	*Lemur catta*	Mammal, bird	Bird	–[f]	Y	Bird	Y	NA	Macedonia and Young (1991)
	Cotton-top tamarin	*Saguinus oedipus*	None	Bird, mammal	–[f]	Y	Bird, mammal	N	NA	Friant et al. (2008)
	Pig-tailed langur	*Simias concolor*	Human, bird, snake	Mammal	0.5 mil-lion	Y	Mammal	N	NA	Yorzinski and Ziegler (2007)
Olfactory	Gray mouse lemur	*Microcebus murinus*	None	Bird, mammal	–[f]	Y	Bird, mammal	Y	NA	Sündermann et al. (2008)
	Red-bellied tamarin	*Saguinus labiatus*	None	Mammal	–[f]	Y	Mammal	Y	NA	Caine and Weldon (1989)
	Cotton-top tamarin	*Saguinus oedipus*	None	Mammal	–[f]	Y	Mammal	Y	NA	Buchanan-Smith et al. (1993)
Visual	Common marmoset	*Callithrix jacchus*	None	Snake	–[f]	Y	NA	NA	NA	Clara et al. (2008)
	Black tufted-eared marmoset	*C. penicillata*	None	Bird, mammal, snake	–[f]	Y	Toy	N	NA	Barros et al. (2002)
	Tufted capuchin	*Cebus apella*	None	Snake	–[f]	Y	NA	NA	NA	Vitale et al. (1991)

Vervet monkey	Cercopithecus aethiops	None	Bird, mammal, snake	$-^f$	Y	Bird	Y	NA	Brown et al. (1992)
Sooty mangabey	Cercocebus atys	None	Mammal	$-^f$	Y	NA	NA	NA	Davis et al. (2003)
Greater galago	Galago crassicaudatus	None	Mammal, snake	$-^f$	Y	NA	NA	NA	Jaenicke and Ehrlich (1972)
Crab-eating macaque	Macaca fascicularis	None	Snake	$-^f$	Y	NA	NA	NA	Vitale et al. (1991)
Rhesus macaque	M. mulatta	None	Snake	$-^f$	Y	Tube	Y	Milder	Joslin et al. (1964), Mineka et al. (1984), Nelson et al. (2003)
Rhesus macaque	M. mulatta	None	Mammal	$-^f$	Y	NA	NA	NA	Davis et al. (2003)
Pig-tail macaque	M. nemestrina	None	Mammal	$-^f$	Y	NA	NA	NA	Davis et al. (2003)
Bonnet macaque	M. radiata	None	Mammal, snake	$-^f$	Y	NA	NA	Less specific	Coss et al. (2007)
Mandrill	Mandrillus sphinx	Bird, snake	Mammal	30	Y	NA	NA	NA	Yorzinski and Vehrencamp (2008)
Slow loris	Nycticebus coucang	None	Mammal, snake	$-^f$	Y	NA	NA	NA	Jaenicke and Ehrlich (1972)
Cotton-top tamarin	Saguinus oedipus	None	Snake	$-^f$	Y	Mammal	N	Milder	Hayes and Snowdon (1990)

(continued)

Table 10.1 (continued)

Common name	Scientific name	Current predator(s)	Novel predator type(s) tested[a]	Time passed (yr)[b]	Antipredator behavior?[c]	Novel control	Novel predator recognized?[d]	Compared to experienced conspecific?[e]	Reference(s)
Cotton-top tamarin	*Saguinus oedipus*	None	Bird	—[f]	Y	NA	NA	NA	Moodie and Chamove (1990)
Squirrel monkey	*Saimiri sciureus*	None	Snake	—[f]	Y	Fish	Y	Milder	Murray and King (1973), Levine et al. (1993)

[a]The type of predator(s) that the naïve primate was shown in the experiment but had never experienced before.

[b]The minimum amount of time that passed since the primate was last exposed to the novel predator.

[c]An indication (Y=yes or N=no) of whether the naïve primate exhibited general antipredator behavior (avoidance, alarm and mobbing calls, piloerection, and/or changes in vigilance) toward the novel predator.

[d]An indication (Y=yes or N=no) of whether the naïve primate treated the novel predator differently than it treated a novel control.

[e]An indication of how the naïve primate responded to the novel predator compared to how an experienced conspecific that still lives with that predator responds.

[f]Because the exact date that this individual was brought into captivity (or moved to an urban setting) is unknown, it is likely that it was last exposed to its natural predators within the past 150 years.

[g]The authors conclude that the predator-naïve howler monkey no longer recognize the vocalizations of a native predator that it has not been exposed to for 50–100 years. However, the monkeys exhibited a greater response toward the vocalizations of a novel, native predator (harpy eagle) compared to the vocalizations of a novel, non-native predator (bald eagle) during the playback period.

Table 10.2   List of Indonesian monkeys and whether they live in environments with felid predators. Y = they live with felid predators (although some populations may no longer live with them due to relatively recent declines in felid populations or due to isolation from the main population) and N = they have not lived with felid predators for over 0.5 million years

Common name	Scientific name	Felid predators?
Muna-Butung macaque	*Macaca brunescens*	N
Heck's macaque	*M. hecki*	N
Moor macaque	*M. maura*	N
Sulawesi macaque	*M. nigra*	N
Gorontalo macaque	*M. nigriscens*	N
Ochre macaque	*M. ochreata*	N
Mentawai macaque	*M. pagensis*	N
Siberut macaque	*M. siberu*	N
Tonkean macaque	*M. tonkeana*	N
Mentawai langur	*P. potenziani*	N
Pig-tailed langur	*Simias concolor*	N
Crab-eating macaque	*M. fascicularis*	Y
Pigtailed macaque	*M. nemestrina*	Y
Proboscis monkey	*Nasalis larvatus*	Y
Grizzled langur	*Presbytis comata*	Y
Banded langur	*P. femoralis*	Y
White fronted langur	*P. frontata*	Y
Hose's langur	*P. hosei*	Y
Mitered langur	*P. melalophos*	Y
Maroon langur	*P. rubicunda*	Y
Thomas's langur	*P. thomasi*	Y
Javan langur	*Trachypithecus auratus*	Y
Silvered langur	*T. cristatus*	Y

than those experienced by their ancestors. In particular, many primates are no longer exposed to felid predation (Table 10.2). The study described here capitalizes on this fact and asks whether wild, naïve primates that have been isolated from ancestral felid predators for over 0.5 million years are still able to recognize them (Yorzinski and Ziegler 2007).

Case Study: Relaxed Predation Pressure in a Wild Primate

Pig-tailed Langurs

The pig-tailed langur (*Simias concolor*) is endemic to the Mentawai islands in Indonesia, which are located about 150 km off the west coast of Sumatra. Belonging within an Asian colobine clade (also consisting of species within the genera *Nasalis*, *Pygathrix*, and *Rhinopithecus*), it is thought to be most closely related to the proboscis monkey (N. larvatus; Groves, 1970; Delson, 1975; Whittaker et al., 2006). Two subspecies of the pig-tailed langur are recognized: *S. c. siberu* on Siberut Island (Chasen and Kloss 1927) and *S. c. concolor* on Sipora, North Pagai, South Pagai, and a few small islets off of South Pagai (Miller 1903; Mittermeier et al. 2007); the difference between these two subspecies is based on pelage coloration. Pig-tailed langurs are critically endangered (IUCN Red List 2008) and are considered one of the 25 most endangered primates (Mittermeier et al. 2007).

Because only a handful of studies have systematically documented their behavior, we know very little about these rare primates. They are medium-sized leaf monkeys that commonly live in one-male one-female or one-male multifemale groups (Tilson 1977; Watanabe 1981; Tenaza and Fuentes 1995; Hadi et al. 2009). They share their habitat with up to three other primate species: Siberut or Mentawai macaques (*Macaca siberu* or *Macaca pagensis*, respectively), Mentawai langurs (*Presbytis potenziani*), and Kloss gibbons (*Hylobates klossii*). Males emit long-distance calls that may function as intergroup communicative signals (Tenaza 1989; Erb 2006).

Pig-tailed langurs have likely been separated from their mainland predators for over 0.5 million years (Rohling et al. 1998; Abegg and Thierry 2002). No dangerous felids currently live in their environment (World Wildlife Fund 1980). However, related langur species living on the mainlands experience high rates of predation by felids (Seidensticker 1983; Rabinowitz et al. 1987; Karanth and Sunquist 1995; Støen and Wegge 1996; Sankar and Johnsingh 2002) and exhibit antipredator behavior when seeing these predators or models of these predators (Thapar 1986; Ramakrishnan and Coss 2000b; Wich and Sterck 2003). Humans are their primary and only confirmed predator; serpent eagles (*Spilornis cheela sipora*) and reticulated pythons (*Python reticulatus*) are probably predators (Whitten and Whitten 1982; C. Abegg pers. comm.), but predation events have never been documented.

Hypotheses and Predictions

A series of auditory playbacks was conducted to investigate the predator-recognition abilities of the pig-tailed langur. The reactions of langurs to the vocalizations of different animals were evaluated to test two hypotheses regarding their acoustic predator-recognition abilities. The first hypothesis is that pig-tailed langurs recognize the vocalizations of dangerous felids. If this hypothesis is supported, then these langurs will exhibit antipredator behavior toward the calls of felids and humans because they recognize both as predators (the human voices are presumed to convey information about human predators because only nonhabituated monkeys were tested). Their responses to the felid calls will be different from their responses to the vocalizations of elephants (novel animals but not predators) and pigs (familiar animals but not predators) because these latter two mammals are not predators of primates (pigs freely roamed the rainforest but were not common).

The second hypothesis is that langurs are afraid of novel vocalizations that they have never heard before. If this hypothesis is supported, then they will respond similarly to the felid and elephant calls because both of these vocalizations are novel. Because they will still exhibit fear towards the vocalizations, their response to the novel sounds should have some similarities to their response toward human voices.

The null hypothesis is that langurs are not afraid of the felid or elephant vocalizations. If this is the case, then their response to the calls of felids and elephants will be different from their response to the voices of their known human predator. We would expect their reactions to all novel vocalizations (felid and elephant) to be the same and also be similar to their responses toward known and nonpredatory animals (pig and bird).

Preliminary visual presentations were also conducted to determine whether the predator-recognition abilities the monkeys exhibited in response to the auditory stimuli were similar to their response to visual stimuli (Yorzinski unpublished data). Two-dimensional visual models of a felid (*Panthera tigris*) and rhinoceros (*Rhinoceros unicomis*) were used. The rhinoceros represented a nonpredatory, novel animal that was not present on the island but exists on the mainland (similar to the elephant vocalizations in the auditory experiments).

Field Site and Experimental Procedure

The langurs were studied at the Siberut Conservation Project (SCP) field site in northeast Siberut Island. The Siberut Conservation Project collaborates with local people to protect the rainforest from logging and hunting activities. The field site encompasses 10.7 km^2 of primary and secondary mixed lowland rainforest. An extensive trail system allows researchers to navigate through the dense understory. Even though the monkeys were not hunted for two years prior to the onset of this study, they were not habituated to the presence of humans.

Over 300 h were spent searching the rainforest for pig-tailed langur groups. When a group was found, I randomly chose an adult langur that was relatively still (i.e., it was resting, grooming, or eating), hid within the understory, and began filming this focal individual. Meanwhile, the field assistant placed the speaker in a concealed spot on the ground at about 35 m from the closest individual of the group and initiated the playback. A 10 s segment of a felid, elephant, person, pig or bird vocalization was broadcast (the particular vocalization that was played was randomized across trials and only one trial was conducted within a given observation period). I continued filming the focal animal until it left its original position (in which case visual contact was usually lost). The video recordings were analyzed frame-by-frame to quantify the behavior of the focal individual.

Sound levels of the playback stimuli were adjusted to a mean of 80–85 dB at 1 m from the speaker. Most of the felid vocalizations were recorded by Gustav Peters and obtained from the Animal Sound Archives at the Zoological Research Museum Alexander Koenig. The elephant and other felid vocalizations were purchased from the Wildlife Section of the British Library Sound Archive. I recorded the human, pig, and bird vocalizations on Siberut Island.

Because the estimated home range of the langur is 3–5 ha (Watanabe 1981), we tested groups that were about 600 m (mean 600 ± 50 m; range: 300–1,100) away from groups that were previously tested with the same stimulus type. It is therefore unlikely that the same group was tested on multiple occasions with the same stimulus type. However, it is possible that some of the same individuals were repeatedly tested with different stimuli; even so, this type of resampling would have minimal effects on the statistical analyses (Coss et al. 2005). Planned comparisons were made to investigate differences in the behavior of the langurs in response to the felid vocalizations and the other treatments.

Two preliminary visual experiments were conducted. The two-dimensional visual models were based on copies of high quality photographs (tiger: Whittaker 2002; rhinoceros: McHugh 2003) that were enlarged to approximate the actual size of the animals (tiger: 95 cm length, 70 cm height; juvenile rhinoceros: 141 cm length, 86 cm height (Stankowich and Coss 2007)). Two blinds were built about 1,100 m apart and only one experiment was conducted at each blind. Based on their home range size (Watanabe 1981), it is likely that two different langur groups were tested. The field assistant and I waited inside of the blinds for over 100 h. When a group of langurs (at least two individuals) randomly passed in front of the blind, a model was displayed for 90 s. One individual in each group was filmed and the video was later analyzed frame-by-frame.

Results and Discussion

The results supported the second hypothesis (langurs are afraid of novel vocalizations). Langurs that heard felid vocalizations spent similar amounts of time looking in the direction of the speaker compared to the langurs that heard the elephant vocalizations (both novel vocalizations); in contrast, langurs that heard the felid vocalizations spent less time looking in the direction of the speaker compared to langurs hearing human voices. Langurs spent similar amounts of time looking at the speaker in response to the felid and pig vocalizations. The langurs likely spent a substantial amount of time looking at the speaker in response to the pig vocalizations because pigs were present in the forest but not abundant enough to ensure frequent interactions between the two species (Fig. 10.1).

The langurs that heard felid vocalizations fled more slowly than those hearing human voices. Langurs fled at similar latencies for both of the novel vocalizations (felid and elephant vocalizations; Fig. 10.2). As indicated by this flight behavior, the langurs appeared to quickly recognize the human vocalizations, while their delayed responses to novel playbacks indicated sensitivity to novel sounds. Similar numbers of individuals fled in response to the novel and/or dangerous stimuli (felid, elephant, and human), but none fled in response to the familiar and nondangerous stimuli (pig and bird). The total number of monkeys that fled did not differ among

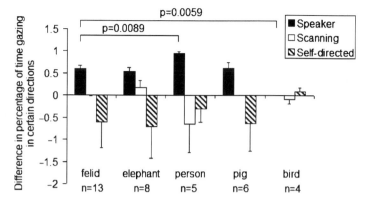

Fig. 10.1 Difference in the percentage of time pig-tailed langurs spent gazing in certain directions before and after different playback treatments: looking in the direction of the speaker (*speaker*), scanning in different directions (*scanning*), and looking at their own body (resting, grooming, or feeding; self-directed). The percentage of time the langur was engaged in each of the three categories in the pre-playback period was subtracted from the percentage of time gazing in each category in the post-playback period. Positive values indicate that the monkeys spent more time gazing in specific directions after the playback compared to before the playback. Means ± SE are displayed

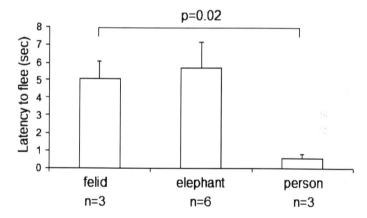

Fig. 10.2 Latency to flee in response to different playback treatments. Means ± SE are displayed

the felid, elephant, and person playbacks (Fig. 10.3). Further experiments that evaluate the responses of langurs to playbacks of a wider range of novel vocalizations (i.e., not only broadcasting felid and elephant vocalizations) would indicate the extent to which their responses to novel vocalizations can be generalized across different types of sound stimuli.

The preliminary experiments with visual models also supported the second hypothesis (although the results are speculative because only two experiments were conducted).

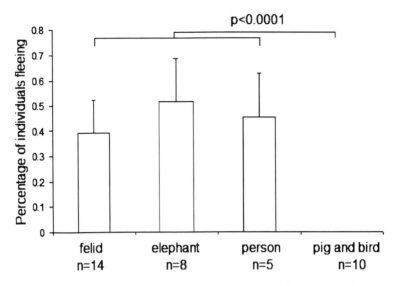

Fig. 10.3 Total number of individuals fleeing in response to different playback treatments. Means ± SE are displayed

Langurs exhibited fear toward the felid and rhinoceros visual models by alarm calling (emitted over 25 calls to each model) and fleeing (waited at least 15 s before fleeing). Although no visual models of humans were presented, the response of the langurs to the felid and rhinoceros models was qualitatively different from their response to actual encounters with humans. When langurs encounter humans, they tend to alarm call less frequently or not at all (pers. obs.); this behavior is adaptive because humans can easily kill langurs with their bow and arrows if the monkeys remain conspicuous. In contrast, a group of pig-tailed langurs was observed mobbing a snake by gathering around it and alarm calling frequently (Pak Tarzan, pers. comm.). Although different from their response toward humans, their response to the model felid and rhinoceros was more similar to their response toward the snake and may be better suited to countering the attacks of nonhuman predators.

Antipredator behavior may persist in populations under relaxed selection that are still exposed to at least one predator (i.e., the multipredator hypothesis; Blumstein et al. 2004). The persistence of these behaviors would probably only occur if the remaining predator elicits the same type of antipredator response (e.g., fleeing or mobbing) as the historical predator. However, a remaining predator may not even be necessary for some aspects of predator recognition. For example, perceptual features indicating danger, such as two facing eyes, are shared by felid predators and conspecifics. Aggressive social contexts might maintain the provocative aspects of these perceptual features (Coss et al. 2005).

Because the pig-tailed langur has been heavily hunted by humans for centuries (Tenaza and Tilson 1985) and is the likely prey of native eagles and pythons (Whitten and Whitten 1982; C. Abegg pers. comm.), it may be particularly sensitive to potentially dangerous sights and sounds. Indeed, the langurs often reacted with a generalized fear response toward novel animals (felid and elephant calls; felid and rhinoceros models) but did not react strongly to familiar, nondangerous animals (pigs and birds). Because the felid vocalizations were not treated differently than other novel vocalizations (elephant), the langurs did not appear to retain specific recognition of felid predators.

Evaluating Predator Recognition

Studies that conduct predator presentation experiments (see above case study), monitor reintroductions, and track developmental changes can all contribute to our knowledge of naïve animals' responses to predators. While these types of studies have been conducted on diverse species, relatively little is known about primates.

Predator Presentations

We can investigate predator recognition abilities by presenting predator-naïve animals with predators. When presented with a predator, some naïve animals appear to recognize it (e.g., Hollén and Manser 2007). In contrast, other naïve animals do not identify it as a predator (e.g., Blumstein et al. 2006).

Few studies have explored the abilities of predator-naive primates to recognize predators. Primates often exhibit generalized antipredator behavior (avoidance, alarm and mobbing vocalizations, piloerection, and/or changes in vigilance) in response to novel olfactory, auditory, and visual predatory stimuli. However, we often do not know whether primates are specifically responding to predators or simply responding to novelty. Very few studies have presented both predator and novel stimuli to primates in order to evaluate their predator recognition abilities. Experiments that present both of these stimuli can help us fill this gap in our knowledge (Table 10.1).

In the few studies that investigated naïve primates and their ability to recognize predators by olfactory cues, the primates demonstrate that they are able to recognize the predators (Caine and Weldon 1989; Buchanan-Smith et al. 1993; Sündermann et al. 2008). In contrast, naïve primates that hear the vocalizations of novel animals fail to make distinctions between predatory species and novel, nonpredatory species (e.g., Yorzinski and Zeigler 2007 but see Macedonia and Young 1991). Lastly, when primates see novel animals, they are sometimes able to differentiate between the novel and predatory animals (Brown et al. 1992), but not always (Hayes and Snowdon 1990). The limited number of studies investigating this topic makes it difficult to draw general conclusions.

Reintroductions

We can learn about predator recognition abilities when naïve animals are reintroduced into predator-rich environments. Predation can account for a significant percentage of mortality in reintroduced animals (Short et al. 1992; Miller et al. 1994;

Pietsch 1994; Kuehler et al. 1996) but is not always the main cause of death (Wolf et al. 1998). This variation in mortality rate due to predation may reflect differences in the abilities of animals to recognize predators and respond appropriately to them. Because animals can have higher survival rates if they have experience with live predators before they are released into the wild (van Heezik et al. 1999), some naïve animals may lack detailed predator recognition abilities.

Among primates, predation is also a major cause of mortality in reintroductions. An inability to recognize predators as well as inappropriate antipredator responses may explain this high mortality rate. Relative to other causes of mortality in reintroductions, predation ranks as one of the highest (22% mortality due to predation in *Leontopithecus rosalia*, Beck et al. 1991; 57% in *Callithrix geoffroyi*, Passamani and Passamani 1995; 71% in *Varecia variegata*, Britt et al. 2003). However, the offspring of reintroduced parents suffer reduced predation (0% mortality due to predation in *Leontopithecus rosalia*; Beck et al. 1991). Because primates are able to learn about predators from their conspecifics (Custance et al. 2002; Griffin 2004), it may only take a few generations for naïve primates to become knowledgeable about predators. Wild primates have even been shown to learn appropriate antipredator behavior within their lifetimes after being exposed to a new predator (Gil-da-Costa et al. 2003).

Ontogeny

Immature animals living in environments with predators are also relatively naïve and can teach us about predator recognition abilities. For example, immature California ground squirrels (*Spermophilus beecheyi*) react more intensely to snakes than novel stimuli. This suggests that the young are predisposed to recognizing and responding appropriately to these predators (Owings and Coss 1977). In contrast, great tit fledglings (*Parus major*) do not appear to recognize predators – they respond similarly to dangerous and nondangerous stimuli (Kullberg and Lind 2002).

Only several studies have explored the development of antipredator behavior in immature primates. Immature vervet monkeys, bonnet macaques, and spectral tarsiers emit alarm calls in response to a wider range of potentially dangerous stimuli than adults (Seyfarth and Cheney 1986; Ramakrishnan and Coss 2000a; Gursky 2003). Because they receive feedback from conspecifics after they make these alarm calls (e.g., conspecifics emit further alarm calls or flee if a real danger exists), they can use this information to learn which stimuli are in fact dangerous (Seyfarth and Cheney 1986).

Conclusions

The results from the above types of studies (predator presentation experiments, reintroductions, and developmental changes) provide us with important information about the responses of naïve primates to predators. In general, these studies suggest that primates are often fearful of novel stimuli and can learn to react appropriately to them. When naïve primates are relying on auditory and visual assessments, they may not know whether a novel animal is dangerous or not. However, when they rely on olfactory assessments, they can better make this distinction.

Although we are rapidly accumulating knowledge about the responses of animals to relaxed predation pressures, we still have much to discover. Naïve animals learn to fear certain types of animals faster than others (Öhman and Mineka 2001). For example, naïve rhesus macaques can quickly learn to associate fear with snakes and crocodiles (but not with nondangerous rabbits) when conspecifics are fearful of them (Cook and Mineka 1989). Future studies that explored whether animals learn more quickly with respect to other predator types (not just snakes and crocodiles) would provide us with a better understanding of naive primates' responses toward predators.

We also know little about the features that are salient for predator recognition in naïve primates. Because animals must first recognize predators before they can respond appropriately, it is critical to understand how naïve animals categorize dangerous and nondangerous animals. When naïve animals recognize predators based on olfactory cues, they may be relying on specific metabolites in the feces that indicate whether the animal was carnivorous (Blumstein et al. 2006; Sündermann et al., 2008). When they categorize novel predators based on visual cues, they may be relying on the relative size of the animal as well as the presence of forward-facing eyes (Coss and Goldthwaite 1995; Coss et al. 2005). And when they make assessments based on auditory cues, the acoustic properties of the calls (e.g., low-pitched vocalizations) may provide information about the size or motivation of the potential threat (Owings and Morton 1998). While all of these factors may influence a naive primate's ability to recognize a predator, there is little systematic research pinpointing the exact features that are salient for recognition.

Acknowledgments I thank Thomas Ziegler, Keith Hodges, Christophe Abegg, Muhammad Agil, and Bogor Agricultural University for allowing me to conduct research at SCP, providing logistical support, and offering useful suggestions on the case study. Pak Nauli and Risel were excellent field guides. Daniel Blumstein, Richard Coss, Peter Klopfer, Mark Laidre, Gail Patricelli, Thomas Ziegler, and two anonymous reviewers provided useful comments on this chapter. JLY was funded by a Morley Student Research Grant, a National Science Foundation Graduate Research Fellowship, and the German Primate Center, Göttingen.

References

Abegg C, Thierry B (2002) Macaque evolution and dispersal in insular south-east Asia. Biol J Linn Soc 75:555–576

Barros M, Boere V, Mello EL, Tomaz C (2002) Reactions to potential predators in captive-born marmosets (*Callithrix penicillata*). Int J Primatol 23:443–454

Beck B, Kleiman DG, Dietz JM, Castro I, Carvalho C, Martins A, Rettberg-Beck B (1991) Losses and reproduction in reintroduced golden lion tamarins *Leontopithecus rosalia*. Dodo J Jersey Wildl Preserv Trust 27:50–61

Berger J, Swenson JE, Persson I (2001) Recolonizing carnivores and naïve prey: conservation lessons from Pleistocene extinctions. Science 291:1036–1039

Blumstein DT (2006) The multi-predator hypothesis and the evolutionary persistence of anti-predator behavior. Ethology 112:209–217

Blumstein DT, Daniel JC, Griffin AS, Evans CS (2000) Insular tammar wallabies (*Macropus eugenii*) respond to visual but not acoustic cues from predators. Behav Ecol 11:528–535

Blumstein DT, Daniel JC, Springett BP (2004) A test of the multi-predator hypothesis: rapid loss of antipredator behaviour after 130 years of isolation. Ethology 110:919–934

Blumstein DT, Mari M, Daniel JC, Ardron JG, Griffin AS, Evans CS (2006) Olfactory predator recognition: wallabies may have to learn to be wary. Anim Conserv 5:87–93

Britt A, Welch C, Katzb A (2003) Can small, isolated primate populations be effectively reinforced through the release of individuals from a captive population? Biol Conserv 115:319–327

Brown MM, Kreiter NA, Maple JT, Sinnott JM (1992) Silhouettes elicit alarm calls from captive vervet monkeys (*Cercopithecus aethiops*). J Comp Psychol 106:350–359

Buchanan-Smith HM, Anderson DA, Ryan CW (1993) Responses of cotton-top tamarins (*Saguinus oedipus*) to faecal scents of predators and non-predators. Anim Welf 2:17–32

Caine NG, Weldon PJ (1989) Responses by red-bellied tamarins (*Saguinus labiatus*) to fecal scents of predatory and non-predatory neotropical mammals. Biotropica 21:186–189

Chasen FN, Kloss CB (1927) Spolia Mentawiensia – mammals. Proc Zool Soc Lond 53:797–840

Clara E, Tommasi L, Rogers LJ (2008) Social mobbing calls in common marmosets (*Callithrix jacchus*): effects of experience and associated cortisol levels. Anim Cogn 11:349–358

Cook M, Mineka S (1989) Observational conditioning of fear to fear-relevant versus fear-irrelevant stimuli in rhesus monkeys. J Abnorm Psychol 98:448–459

Coss RG (1999) Effects of relaxed natural selection on the evolution of behavior. In: Foster SA, Endler JA (eds) Geographic variation in behavior: perspectives in evolutionary mechanisms. Oxford University Press, Oxford, pp 180–208

Coss RG, Goldthwaite RO (1995) The persistence of old designs for perception. Perspect Ethol 11:83–148

Coss RG, Ramakrishnan U, Schank J (2005) Recognition of partially concealed leopards by wild bonnet macaques (*Macaca radiata*) the role of the spotted coat. Behav Process 68:145–163

Coss RG, McCowan B, Ramakrishnan U (2007) Threat-related acoustical differences in alarm calls by wild bonnet macaques (*Macaca radiata*) elicited by python and leopard models. Ethology 113:352–367

Curio E (1966) How finches react to predators. Animals 9:142–143

Custance DM, Whiten A, Fredman T (2002) Social learning and primate reintroduction. Int J Primat 23:479–499

Davis JE, Parr L, Gouzoules H (2003) Response to naturalistic fear stimuli in captive old world monkeys. Ann NY Acad Sci 1000:91–93

Delson E (1975) Evolutionary history of the Cercopithecidae. Contrib Primatol 5:167–217

Erb WM (2006) Patterns and variation in long-distance communication of simakobu monkeys (*Simias concolor*) on Siberut Island, Indonesia – a pilot study. Am J Phys Anthropol 129(S42):87

Friant SC, Campbell MW, Snowdon CT (2008) Captive-born cotton-top tamarins (*Saguinus oedipus*) respond similarly to vocalizations of predators and sympatric nonpredators. Am J Primatol 70:707–710

Fullard JH, Ratcliffe JM, Soutar AR (2004) Extinction of the acoustic startle response in moths endemic to a bat-free habitat. J Evol Biol 17:856–861

Gil-da-Costa R, Palleroni A, Hauser MD, Touchton J, Kelley JP (2003) Rapid acquisition of an alarm response by a neotropical primate to a newly introduced avian predator. Proc R Soc Lond B 270:605–610

Griffin AS (2004) Social learning about predators: a review and prospectus. Learn Behav 32:131–140

Groves CP (1970) The forgotten leaf-eaters and the phylogeny of Colobinae. In: Napier JP, Napier PR (eds) Old world monkeys. Academic, New York

Gursky S (2003) Predation experiments on infant spectral tarsiers (*Tarsius spectrum*). Folia Primatol 74:272–284

Hadi S, Zeigler T, Hodges JK (2009) Group structure and physical characteristics of simakobu monkeys (*Simias concolor*) on the Mentawai Island of Siberut, Indonesia. Folia Primatol 80:74–82

Hayes SL, Snowdon CT (1990) Predator recognition in cottontop tamarins (*Saguinus oedipus*). Am J Primatol 20:283–291

Hollén LI, Manser MB (2007) Persistence of alarm-call behaviour in the absence of predators: a comparison between wild and captive-born meerkats (*Suricata suricatta*). Ethology 113:1038–1047

Houston AI, McNamara JM (1992) Phenotypic plasticity as a state-dependent life-history decision. Evol Ecol 6:243–253

IUCN Redlist (2008) IUCN red list of threatened species. www.iucnredlist.org. Accessed 17 Nov 2008

Jaenicke C, Ehrlich A (1972) Effects of animate vs. inanimate stimuli on curiosity behavior in greater galago and slow loris. Primates 23:95–104

Joslin J, Fletcher H, Emlen J (1964) A comparison of the responses to snakes of lab- and wild-reared rhesus monkeys. Anim Behav 12:348–352

Karanth KU, Sunquist ME (1995) Prey selection by tiger, leopard, and dhole in tropical forests. J Anim Ecol 64:439–450

Kelley JL, Magurran AE (2003) Effects of relaxed predation pressure on visual predator recognition in the guppy. Behav Ecol Sociobiol 54:225–232

Komers PE (1997) Behavioural plasticity in variable environments. Can J Zool 75:161–169

Kuehler C, Kuhn M, Kuhn JE, Lieberman A, Harvey N, Rideout B (1996) Artificial incubation, hand-rearing, behavior, and release of Common `Amakihi (*Hemignathus virens virens*): surrogate research for restoration of endangered Hawaiian forest birds. Zoo Biol 15:541–553

Kullberg C, Lind J (2002) An experimental study of predator recognition in great tit fledglings. Ethology 108:429–441

Levine S, Atha K, Wiener SG (1993) Early experience effects on the development of fear in the squirrel monkey. Behav Neural Biol 60:225–233

Macedonia JM, Young PL (1991) Auditory assessment of avian predator threat in semi-captive ringtailed lemurs (*Lemur catta*). Primates 32:169–182

McHugh T (2003) In: Hutchins M, Keiman DG, Geist V, McDade MC (eds) Grzimek's animal life encyclopedia. 2nd edn. vol 12. Gale Group, Farmington Hills, MI, p 22

Messler A, Wund MA, Baker JA, Foster SA (2007) The effects of relaxed and reversed selection by predators on the antipredator behavior of the threespine stickleback, *Gasterosteus aculeatus*. Ethology 113:953–963

Miller GS (1903) Seventy new Malayan mammals. Smithson Misc Coll 45:1–73

Miller BD, Biggins D, Hanebury L, Vargas A (1994) Reintroduction of black-footed ferret (*Mustela nigripes*). In: Olney PJS, Mace GM, Feistner ATC (eds) Creative conservation: interactive management of wild and captive animals. Chapman and Hall, London, pp 45–464

Mineka S, Davidson M, Cook M, Keir R (1984) Observational conditioning of snake fear in rhesus monkeys. J Abnorm Psychol 93:355–372

Mittermeier RA, Ratsimbazafy J, Rylands AB, Williamson L, Oates JF, Mbora D, Ganzhorn JU, Rodríguez-Luna E, Palacios E, Heymann EW, Cecília M, Kierulff M, Yongcheng L, Supriatna J, Roos C, Walker S, Aguiar JM (2007) Primates in Peril: the world's 25 most endangered primates, 2006–2008. Primate Conserv 22:1–40

Moodie EM, Chamove AS (1990) Brief threatening events beneficial for captive tamarins? Zoo Biol 9:275–286

Murray SG, King JE (1973) Snake avoidance in feral and laboratory reared squirrel monkeys. Behaviour 47:281–288

Nelson EE, Shelton SE, Kalin NH (2003) Individual differences in the responses of naïve rhesus monkeys to snakes. Emotion 3:3–11

Öhman A, Mineka S (2001) Fears, phobias, and preparedness: toward an evolved module of fear and fear learning. Psychol Rev 108:483–522

Owings DH, Coss RG (1977) Snake mobbing by California ground squirrels: adaptive variation and ontogeny. Behaviour 62:50–68

Owings DH, Morton ES (1998) Animal vocal communication: a new approach. Cambridge University Press, Cambridge

Passamani M, Passamani JA (1995) Losses of reintroduced Geoffroy's marmoset. Australian Primatol 10:12–13

Peckarsky BL, Penton MA (1988) Why do Ephemerella nymphs scorpion posture: a "ghost of predation past?". Oikos 53:185–193

Pietsch RS (1994) The fate of urban common brushtail possums translocated to sclerophyll forest. In: Serena M (ed) Reintroduction biology of Australian and New Zealand fauna. Surrey Beatty and Sons, Chipping Norton, New South Wales, Australia, pp 239–246

Rabinowitz A, Andau P, Chai PPK (1987) The clouded leopard in Malaysian Borneo. Oryx 21:107–111

Ramakrishnan U, Coss RG (2000a) Age differences in the responses to adult and juvenile alarm calls by bonnet macaques (*Macaca radiata*). Ethology 106:131–144

Ramakrishnan U, Coss RG (2000b) Recognition of heterospecific alarm vocalizations by bonnet macaques (*Macaca radiata*). J Comp Psychol 114:3–12

Rohling EJ, Fenton M, Jorissen FJ, Bertrand P, Ganssen G, Caulet JP (1998) Magnitudes of sea-level lowstands of the past 500,000 years. Nature 394:162–165

Rothstein SI (2001) Relic behaviours, coevolution and the retention versus loss of host defenses after episodes of avian brood parasitism. Anim Behav 61:95–107

Sankar K, Johnsingh AJT (2002) Food habits of tiger (*Panthera tigris*) and leopard (*Panthera pardus*) in Sariska Tiger Reserve, Rajasthan, India, as shown by scat analysis. Mammalia 66:285–289

Seidensticker J (1983) Predation by Panthera cats and measures of human influence in habitats of South Asian monkeys. Int J Primatol 4:323–326

Seyfarth R, Cheney D (1986) Vocal development in vervet monkeys. Anim Behav 34:1640–1658

Short J, Bradshaw SD, Giles J, Prince RIT, Wilson GR (1992) Reintroduction of macropods (Marsupialia: Macropodoiden) in Australia: a review. Biol Cons 62:189–204

Stankowich T, Coss RG (2007) The re-emergence of felid camouflage with the decay of predator recognition in deer under relaxed selection. Proc R Soc B 274:175–182

Støen OG, Wegge P (1996) Prey selection and prey removal by tiger (*Panthera tigris*) during the dry season in lowland Nepal. Mammalia 60:363–373

Sündermann D, Scheumann M, Zimmermann E (2008) Olfactory predator recognition in predator-naïve gray mouse lemurs (*Microcebus murinus*). J Comp Psychol 122:146–155

Takahashi H (1997) Huddling relationships in night sleeping groups among wild Japanese macaques in Kinkazan Island during winter. Primates 38:57–68

Tenaza RR (1989) Intergroup calls of male pig-tailed langurs (*Simias concolor*). Primates 30:199–206

Tenaza R, Tilson RL (1985) Human predation and Kloss's gibbon (*Hylobates klossii*) sleeping trees in Siberut Island, Indonesia. Am J Primatol 8:299–308

Tenaza RR, Fuentes A (1995) Monandrous social organization of pigtailed langurs (*Simias concolor*) in the Pagai Islands, Indonesia. Int J Primatol 16:295–310

Thapar V (1986) Tiger: portrait of a predator. Facts on File, New York, pp 139–145

Tilson RL (1977) Social-organization of simakobu monkeys (*Nasalis concolor*) in Siberut Island, Indonesia. J Mammal 58:202–212

van Heezik Y, Seddon PJ, Maloney RF (1999) Helping reintroduced houbara bustards avoid predation: effective anti-predator training and the predictive value of pre-release behaviour. Animal Conserv 2:155–163

van Schaik CP, van Noordwijk MA (1985) Evolutionary effect of the absence of felids on the social organization of the macaques on the island of Simeulue (*Macaca fascicularis fusca*, Miller 1903). Folia Primatol 44:138–147

Vitale AF, Visalberghi E, De Lillo C (1991) Responses to a snake model in captive crab-eating macaques (*Macaca fascicularis*) and captive tufted capuchins (*Cebus apella*). Int J Primatol 12:277–286

Watanabe K (1981) Variation in group composition and population density of the two sympatric Mentawaian leaf-monkeys. Primates 22:145–160

Whittaker T (2002) In: Sunquist M, Sunquist F (eds) Wild cats of the world (plate 46). University of Chicago Press, Chicago

Whittaker DJ, Ting N, Melnick DJ (2006) Molecular phylogenetic affinities of the simakobu monkey (*Simias concolor*). Mol Phylogenet Evol 39:887–892

Whitten AJ, Whitten JEJ (1982) Preliminary observations of the Mentawai macaque on Siberut Island, Indonesia. Int J Primatol 3:445–459

Wich SA, Sterck EHM (2003) Possible audience effect in Thomas langurs (Primates; *Presbytis thomasi*): an experimental study on male loud calls in response to a tiger model. Am J Primatol 60:155–159

Wolf CM, Garland T Jr, Griffith B (1998) Predictors of avian and mammalian translocation success: reanalysis with phylogenetically independent contrast. Biol Conserv 86:243–255

World Wildlife Fund (1980) Saving Siberut: a conservation master plan. World Wildlife Fund, Bogor, Indonesia

Yorzinski JL, Vehrencamp SL (2008) Preliminary report: antipredator behaviors of mandrills. Primate Rep 75:11–18

Yorzinski JL, Ziegler T (2007) Do naïve primates recognize the vocalizations of felid predators? Ethology 113:1219–1227

Chapter 11
The Relationship Between Nonhuman Primate Densities and Vegetation on the Pagai, Mentawai Islands, Indonesia

Lisa M. Paciulli

Introduction

Trees serve countless functions for animals such as supplying food, offering cover from predators and the elements, providing substrates for locomotion, furnishing places to rest and sleep, etc. It is not surprising then, that the densities of many primary consumers such as primates, are related to the plant resources in their environment (Brown 1981), and specifically, to tree variables. These include the density and/or basal area of important tree species such as figs (*Ficus* spp.), palms (Palmae), and lianas, as well as forest structure indices such as total stem density; tree species richness, diversity, and equitability; stand basal area, mean patch size, and percent canopy cover (Table 11.1).

There are four endemic and endangered nonhuman primates inhabiting the Mentawai Islands of West Sumatra, Indonesia – Kloss's gibbon (*Hylobates klossii*), the Mentawai pig-tailed macaque (*Macaca pagensis*), the Mentawai Island leaf langur/ sureli (*Presbytis potenziani*), and the simakobu monkey (*Simias concolor*). Information on the way various factors, including forest vegetation, affect the population densities of these endemic and endangered primates is much needed. However, published data on the feeding behavior of habituated individuals exist for only one of the four species – Kloss's gibbon (Whitten 1982).

While known food species can undoubtedly be good indicators of primate population numbers (*Ficus* spp.: Skorupa 1986; *Eschweilera* spp.: Stevenson 2001), general forest structure variables appear to be just as closely related to primate population patterns (total stem density: Skorupa 1986; canopy cover: Peres 1997). Therefore, in this study, the presence/absence, density, and basal areas of important food resources, as well as forest structure indices were computed from measures taken in nine forests on the Pagai, Mentawai Islands. In addition, primate surveys

L.M. Paciulli (✉)
Department of Anthropology, University of West Georgia, Carrollton, GA 30118, USA
e-mail: lpaciull@westga.edu

S. Gursky-Doyen and J. Supriatna (eds.), *Indonesian Primates*,
Developments in Primatology: Progress and Prospects,
DOI 10.1007/978-1-4419-1560-3_11, © Springer Science+Business Media, LLC 2010

Table 11.1 Studies that report a positive relationship between various vegetation variables and primate densities, abundances, and/or biomass

Vegetation variables	Species	References
Total stem density	*Colobus guereza, Lophocebus albigena*	Skorupa (1986)
Plant species richness	*L. albigena*, New World Monkeys	Skorupa (1986) and Stevenson (2001)
Total tree stand basal area	*Piliocolobus badius, Procolobus rufomitratus, Alouatta palliata Mexicana*	Skorupa (1986), Mbora and Meikle (2004), and Arroyo-Rodríguez et al. (2007)
Tree species diversity	*Microcebus murinus, Cheirogaleus medius, Phaner furcifer, Lepilemur ruficaudatus, Eulemur fulvus rufus, Propithecus verreauxi, Alouatta* spp.	Ganzhorn et al. (1997) and Peres (1997)
Percent canopy cover	*Pan troglodytes, Cercopithecus l'hoesti, Pongo pygmaeus pygmaeus, Saguinus bicolor*	Skorupa (1986), Felton et al. (2003), and Vidal and Cintra (2006)
Overall plant productivity	New World Monkeys (species richness)	Kay et al. (1997)
Food tree density	*P. badius, P. rufomitratus*	Chapman et al. (2002a) and Mbora and Meikle (2004)
Basal area of food trees	*P. rufomitratus, A. p. mexicana*	Mbora and Meikle (2004) and Arroyo-Rodríguez et al. (2007)
Basal area density of food trees	*C. mitis*	Worman and Chapman (2006)
Food resource abundance	*A. caraya, Callithrix jacchus*	De A. Moura (2007)
Large food trees	*P. p. pygmaeus*	Felton et al. (2003)
Moraceae basal area	New World Monkeys	Stevenson (2001)
Moraceae tree diversity	*Ateles geoffroyi*	Weghorst (2007)
Ficus spp. tree density	*Hylobates* spp., *C. ascanius*	Leighton and Leighton (1983) and Skorupa (1986)
Palm (Palmae) density	*Cebus albifrons, C. apella, Saguinus fuscicollis, S. imperator, Saimiri*, New World Monkeys	Terborgh (1986) and Stevenson (2001)
Liana density	*E. fulvus* spp.	Johnson and Overdorff (1999)
Herbaceous plant family density (Marantaceae and Zingerberaceae)	*Gorilla g. gorilla*	Tutin et al. (1991) and White et al. (1995)
Legumes	Colobinae subfamily	Davies et al. (1988) and Waterman et al. (1988)
Eschweilera spp. basal area	Pithecinae subfamily	Stevenson (2001)
Eschweilera spp. tree seeds	*Chiropotes satanas*	van Roosmalen et al. (1988)
Leaf protein-to-fiber ratio	*L. mustelinus, P. verreauxi, C. guereza, P. badius*	Ganzhorn (1995) and Chapman et al. (2002a)
Large tree density	*A. p. mexicana*	Arroyo-Rodríguez et al. (2007)
Basal area of persistent tree species	*A. p. mexicana*	Arroyo-Rodríguez et al. (2007)
Height of tree, first bough	*Presbytis hosei*	Nijman (2004)
Abundance of forest logs, snags	*S. bicolor*	Vidal and Cintra (2006)

were conducted in the same forests and species' densities were calculated. Subsequently, the relationship between primate densities and vegetation measures was examined to determine which variables, if any, were related to primate densities. Several basic hypotheses were tested.

Hypotheses and Predictions

Hylobates klossii: For 20 years, it has been speculated that gibbon densities are limited by *Ficus* densities (Leighton and Leighton 1983; Johns and Skorupa 1987). Sumatran gibbons feed on figs more than other hylobatids (60–70% of feeding time vs. 35–50%: Palombit 1997) and *H. klossii* seems to be one of the most avid fig eaters (72% fruit, 23% figs: Whitten 1982). Thus, it is predicted that Kloss gibbons will follow the patterns of other gibbons and have densities that are significantly positively correlated with (1) fig densities and (2) the presence, frequency, density, and basal area of known *H. klossii* food species.

Macaca pagensis: Macaques are known for their broad diet, high intelligence, extreme adaptability, and their propensity for taking advantage of unpredictable food resources at all levels of the canopy (Richard et al. 1989; Caldecott et al. 1996; Nakayama et al. 1999; Mastripieri 2007). In short, it seems that macaques have no trouble finding food wherever they are. Therefore, it is predicted that Mentawai pig-tailed macaque densities will not vary across sites.

Presbytis potenziani: P. melalophus, the species with which *P. potenziani* is most often grouped (Wilson and Wilson 1975; Tilson 1976; Brandon-Jones 1993), forages for food in all strata of the canopy (Curtin 1980; MacKinnon and MacKinnon 1980; Bennett and Davies 1994). In addition, many *Presbytis* species consume large amounts of seeds, as well as young leaf and fruit parts from common tree species (Bennett and Davies 1994). Thus, *Presbytis* may be able to maintain similar densities in varied forests by feeding on the most common species in those forests. Conversely, some colobine densities fluctuate with forest structure variables such as total stem density (Skorupa 1986), total tree stand basal area (Skorupa 1986; Mbora and Meikle 2004), and height of tree (Nijman 2004). *P. potenziani* densities may vary with these variables as well.

Simias concolor: Habituated simakobus have been observed spending roughly equal amounts of time feeding on young leaves, fruits, and flowers (Paciulli unpub. data). However, during the Indonesian drought of 1997, simakobus spent approximately 80% of their feeding time consuming keruing flowers (*Dipterocarpus haselthii*: Paciulli unpub. data). Thus, it seems that keruing flowers serve as a fallback food for simakobus, and it is predicted that their densities will vary with keruing densities. In addition, *Simias* densities may also vary with forest structure variables such as total stem density (Skorupa 1986), total tree stand basal area (Skorupa 1986; Mbora and Meikle 2004) and height of tree (Nijman 2004), as these affect other colobines.

Methods

I conducted this study on the two southern Mentawai Islands – North Pagai (2°42′S 100°5′E) and South Pagai (3°00′S 100°20′E) (Fig. 11.1). The Pagai Islands span a length of approximately 110.5 km and a width of about 57.5 km (WWF 1980; Eudey 1987). Annual precipitation ranges from 2,655 to 6,383 mm (Tenaza and Fuentes 1995) and the maximum height above sea level is 368 m (Nelles Verlag 1999). A total of nine dipterocarp or mixed dipterocarp forests were selected for this study. The names of the nine forests are Kilometer 28, Kilometer 60, Kinumbu, Manganjo, Saumanganyak, Simpang G, Area I, Area III, and Area IV.

The four primate species inhabiting the Pagai Islands are Kloss's gibbon (*Hylobates klossii*), the Mentawai pig-tailed macaque (*Macaca pagensis*), the Mentawai Island langur/sureli (*Presbytis potenziani*), and the simakobu monkey (*Simias concolor*) (Eudey 1987; WWF 1980). *H. klossii* and *P. potenziani* are endangered while *M. pagensis* and *S. concolor* are critically endangered due to habitat loss, logging, and hunting (IUCN 2009).

I conducted primate and vegetation surveys between May 1999 and July 2000 in the nine forests. I used established line-transect methodology (Burnham et al. 1980; NRC 1981; Defler and Pintor 1985; Janson and Terborgh 1986; Brockelman and Ali 1987; Whitesides et al. 1988; Buckland et al. 1993). In each area, assistants cut three 4 km long transects, 300 m apart (Brockelman and Ali 1987), along cardinal directions. Each transect was surveyed randomly on three separate days during

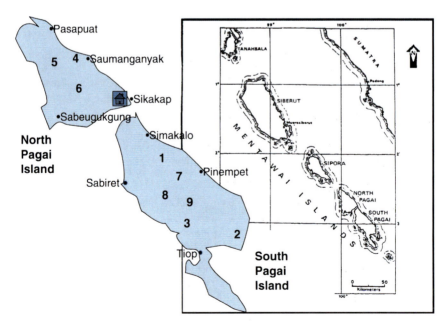

Fig. 11.1 Map of the Mentawai Islands (WWF 1980) showing the nine areas surveyed on the Pagai Islands

a 9 day period (Sterling and Rakotoarison 1998; Johnson and Overdorff 1999) at a rate of ~0.9 km/h. The total line length sampled was 324 km.

During the surveys, I recorded all sights and sounds of primates. In addition, I wrote down the angle and distance of the first individual detected, as well as the time and location of the detection. I also noted the number of individuals and their age and sex. For a more detailed description of the methodology see Paciulli (2004).

I collected data on vegetation at 20 places along the primate survey trails. A 25 m by 5 m quadrat was demarcated every 200 m. Within each quadrat, I recorded the tree species, estimated total height, and diameter at breast height (DBH) for each tree ≥3 cm DBH to include more trees than the usual measurement of trees greater than 10 cm DBH would yield. I also gauged canopy coverage at 60 points per site (20 points×3 trails) by estimating the percent of sky visible directly above the center of each quadrat. The total area I examined in each forest was 7,500 m^2 or 0.75 ha, covering a total of 6.75 ha across all nine forests.

Data Analysis

I conducted analyses on primate densities using all detections and Kings' method (observer-to-animal: King 1929 cited in Leopold 1933; Janson and Terborgh 1986). I used JMP 3.2.2 (SAS Institute 1989–1997) and SPSS 11.0.1 (SPSS 1989–2001) statistical programs for analyses. I used correspondence analysis (Sokal and Rohlf 1995) and hierarchical cluster analysis (using the average linkage method) (Everitt 1980) to identify co-occurring assemblages of trees, on the basis of species presence/absence, frequencies, and basal areas. First, the presence/absence, frequencies, and basal areas of all trees were organized by site and used in three separate correspondence analyses (CA). CA uses a graphical technique to show which rows or columns of tables have similar patterns. In addition, the CA output provides a plot and the coordinates (i.e., values) that were used to position the data points in the plot. The coordinates represent the variation of a single datum point in relation to all other data points in the plot. The three coordinates for each CA datum point were used in hierarchical cluster analyses (HCA). HCA groups species into clusters that have values that are close to each other relative to those of other clusters, with respect to their occurrence in distinct sampling sites.

I used multiple regressions (Sokal and Rohlf 1995) to establish which, if any, of the variables could predict primate densities. Multiple regressions were run on the densities of each species at the nine sites and the six presence/absence clusters, the six frequency clusters, and the six basal area clusters. To determine in which multiple regressions the variables from the other vegetation categories (i.e., the composite vegetation variables, the important tree variables, and the *H. klossii* food variables), could be included, I used Kendall's correlations (Sokal and Rohlf 1995). I conducted the correlations to see which variables were correlated with one another and therefore, could not be used as independent variables in the same regression analysis. Furthermore, linear regressions between species' densities and *D. haselthii* variables were run independently because this was thought of post-hoc. Significance levels were set at $p \le 0.05$ and tests were two-tailed unless noted.

Results

All four of the primate species were found in all of the forests surveyed in this study. Primate densities, forest structure variables, percent canopy cover, and the densities and basal areas of important tree species are listed in Appendix.

There were no significant relationships between any of the primate densities and the forest structure variables. However, two of the important tree variables predicted one of the species' densities. When *Ficus* spp. densities were low, Mentawai Island leaf langur densities were high ($r = -0.1390$, $F = 9.42$, $p = 0.037$). In addition, when liana densities were high, leaf langur densities were high as well ($r = 1.0755$, $F = 35.03$, $p = 0.004$).

In addition, Kloss's gibbon food variables seemed predictive of two species' densities and overall primate densities. For instance, when the basal area of known gibbon food species was high, *Presbytis* densities were high too ($r = 1.2155$, $F = 11.00$, $p = 0.045$). Likewise, when the basal area ($r = 0.6148$, $F = 19.05$, $p = 0.012$) or presence ($r = 0.9237$, $F = 29.50$, $p = 0.006$) of known gibbon food species and genera were high, so were *Presbytis* densities. In contrast, when the presence of *H. klossii* food species was high, simakobu densities were low ($r = -3.3052$, $F = 23.25$, $p = 0.017$). Moreover, although primate densities were low when the presence or frequency of gibbon food species were high ($r = -3.8941$, $F = 33.13$, $p = 0.01$; $r = -0.1959$, $F = 16.54$, $p = 0.027$, respectively), primate densities were high when either the basal areas of known gibbon food species were high ($r = 3.5982$, $F = 41.96$, $p = 0.008$), or the presence of gibbon food genera was high ($r = 3.2424$, $F = 43.40$, $p = 0.007$).

Only one cluster variable in the hierarchical cluster analysis significantly predicted a single species' density. When the presence of Cluster 6 trees was high, *Hylobates'* densities were high as well ($r = 1.4872$, $F = 19.41$, $p = 0.048$).

Discussion

Relationship Between Vegetation Variables and Specific Hypotheses and Predictions

Of the four hypotheses and corresponding predictions proposed, only one was supported, that Mentawai pig-tailed macaque densities would not be affected by any of the vegetation and forest structure variables used in this study. In contrast, the three hypotheses regarding the gibbons, leaf langurs, and simakobus were rejected. Upon closer examination, it became apparent that other researchers have found that many vegetation and forest structure indices did not always correlate with primate densities, abundances, and/or biomass.

It was predicted that Kloss gibbon densities would be related to fig densities and the presence, frequency, and basal area of *H. klossii* food species. Likewise, it was expected that simakobu densities would vary with keruing densities. Neither of these predictions were supported by the data. Although these results were surprising at first, other studies have had similar outcomes. Stevenson (2001) found that indices such as the basal areas of figs were not good predictors of total primate biomass or of any primate guild. He reasoned that *Ficus* trees often occur at low densities and/or in hemiepiphitic form. As such, it is difficult to get good sample sizes for low-density species. Second, measures such as the presence, density, and/or basal area of trees may miss vital information on plants that grow on other plants, such as lianas (Stevenson 2001). In addition, there are other examples of important foods not being related to the population parameters of the primates that feed on them (leguminous plants: *P. badius* and *C. polykomos*: Davies et al. 1999).

Contrary to expectations, indices such as the basal area of Moraceae were not related to gibbon densities either. Stevenson (2001) found the same for the biomass of *Alouatta* spp. This was unanticipated considering that howlers frequently feed on Moraceae plants (Estrada et al. 1999; Serio-Silva et al. 2002). Likewise, after collecting and analyzing the nutritional content of leaves from numerous tree species at four sites, Chapman et al. (2002a) found that the average protein-to-fiber ratio of species eaten by red colobus (*C. badius*) did not improve the predictive power of tests for either red colobus biomass or total colobine biomass. This was not unexpected as the mean protein-to-fiber ratios of red colobus food species did not differ significantly from the protein-to-fiber ratios of nonfood species.

At first, it was perplexing to find that there was no relationship between simakobu densities and the presence, frequency, or basal area of *Dipterocarpus haselthii*. As stated earlier, in a time of a severe drought, simakobus were observed feeding on *D. haselthii* flowers approximately 80% of the time, every day, for a 2-month period (Paciulli unpub. data). This lack of a relationship between a primate species density and a keystone resource was unexpected. It is, however, not unprecedented. For instance, Stevenson (2001) did not find any significant correlations between the basal areas of suggested keystone resources (e.g., figs, palms) and the abundance of any primate guild (except for a weak effect of the basal area of figs on the biomass of large atelines) (Stevenson 2001). Perhaps one of the reasons for this apparent nonrelationship between primate densities and keystone food resources is the fact that many primates have broad ecological tolerances and can readily "switch" to less preferred resources, when the preferred ones are unavailable (e.g., *P. b. rufomitratus* switching from young to mature leaves: Mowry et al. 1996). Thus, although a single resource comprises a significant portion of a primates' daily consumption, it does not necessarily mean that densities will be affected by this one species. For example, although Kloss gibbons on Siberut fed on figs 23% of the time, 77% of the time they ate other foods (Whitten 1982). This may be why low (or high) densities of figs do not appear to have much bearing on Kloss gibbon densities.

Although the case can be made for why the density of a primate species might not be strongly tied to its' main food item, it is more difficult to explain why a primates' density would not be strongly affected by all of the foods in its' diet. More specifically, why do Kloss gibbon densities not vary at all with their foods? One possible reason is that the diets of individual groups of primates fluctuate significantly over time. For instance, Chapman et al. (2002b) found that food intake of the same red colobus (*Procolobus badius*) groups showed significant and consistent changes over a 4-year period. Not only were different plant species consumed, but the time spent feeding on different types of foods (i.e., leaves vs. fruits) as well as plant parts, varied. In addition, there were dietary differences between groups whose ranges overlapped by as much as 49% (Chapman et al. 2002b). Thus, although Whitten (1982) collected data on *H. klossii* food intake for approximately 650 h over a 2-year period, a study of one or two demographically changing groups occupying a single habitat during a brief period of time may not adequately represent the overall diet of the species (Chapman et al. 2002b:349). Diet differs significantly over both small spatial, as well as short temporal, scales within a single group, never mind, a species. This may explain why *H. klossii* foods seemed unrelated to *H. klossii* densities.

Another result that is difficult to explain is the significant positive relationship between gibbon densities and the presence of trees in Cluster 6. Cluster 6 was comprised of only two tree species – *Trichosanthes* sp. and *Xylopia caudata* – neither are known food species of Kloss's gibbons. Therefore, the significant regression probably is spurious. Notwithstanding, it seems intriguing at the very least to mention that both *Trichosanthes* spp. and *Xylopia* spp. have curative value for humans (*T. kirilowii*: Akihisa et al. 2001; Campbell et al. 2002; Choi et al. 2002; Krishnan et al. 2002 and *Xylopia*: de Melo et al. 2001).

It was predicted that Mentawai Island leaf langur densities would be unrelated to vegetation and this was not the case. The significant negative relationship between *P. potenziani* densities and figs is difficult to explain because there have been no published accounts of the feeding habits of habituated Mentawai Island leaf langurs. Although some colobines consume a substantial amount of figs (e.g., *Semnopithecus entellus*: Hladik 1977; *P. b. rufomitratus*: Marsh 1986; *P. pileata*: Stanford 1991), others do not. For example, in some parts of Malaysia, banded leaf monkeys (*P. melalophus*) rarely eat figs (Lambert 1990). However, it is still unclear why Mentawai Island leaf langur densities would be significantly lower where *Ficus* spp. densities were high.

It is easier to explain the significant relationship between liana densities and leaf langur densities. Many colobines exploit liane resources (*C. polykomos*: Dasilva 1994; Oates 1994; Davies et al. 1999; *P. melalophus*: Bennett and Davies 1994; *P. b. rufomitratus*: Oates 1994; Mowry et al. 1996; *P. rubicunda*: Davies 1991). Analysis of protein, fiber, and energy values of foods selected versus items available (but not eaten), suggests that liana preference is related to protein and energy maximization (Dasilva 1994). This makes sense as in some areas, lianas generate young leaves more reliably and abundantly than trees do (Palombit 1997). In addition, legume vine shoots can contain up to 55% protein, whereas immature tree leaves usually contain only about 20% protein (Hladik 1978). Liana leaves also may not be as well defended chemically as tree-leaf shoots (Whitten 1982). It is probably for all of these reasons that some colobines, such as the Hanuman langurs (*P. entellus*) in southern Nepal, choose a climbing species, *Spatholobus parviflorus*, as their number one food resource, feeding on the leaves 42.2% of the time (Koenig et al. 1998). The Mentawai Island leaf langurs also may spend a significant proportion of their feeding time on liane species as well, and this may be the reason for the significant relationship between their densities and the densities of lianas.

When the basal area of known gibbon food species was high, overall primate and *Presbytis* densities were high too. In contrast, when the presence or frequency of gibbon food species was high, primate and simakobu densities were low. These seemingly conflicting results could be explained by the fact that the presence/absence of tree species is not a good predictor of tree size, which is more indicative of the potential resources a tree could provide. This also implies that *P. potenziani* and the other Pagai primates may be exploiting some of the same resources that Kloss's gibbons do.

Relationship Between Other Vegetation Variables and Primate Densities

There were no significant relationships between any of the primate densities and the forest structure variables. There could be several reasons for this including; (1) the measures used in this study were too simple, (2) the botanical measures are not related to primate densities and/or have yielded insignificant results in similar studies, (3) other variables are better predictors of primate densities, and/or (4) the area sampled was not large enough.

One possible reason for the composite botanical indices not yielding significant results is that they may have been too crude. For example, Chapman et al. (2002a) found that resource quality was key in understanding determinants of colobine abundance and cautioned that a simple quantification of resource abundance might not be a good indicator of primate abundances (Chapman et al. 2002a). In the present study, finer analyses of resource quality, such as calculating the mean protein-to-fiber ratios of trees, were not undertaken. Collecting primate foods from several sites and analyzing nutritional composition yields important information on associations with primate population parameters (Ganzhorn 1995; Chapman et al. 2002a). However, such practices are both time-consuming (Chapman et al. 2002a) and expensive. In addition, where the foods of as yet unstudied primates are still largely unknown, the task of determining the relationship between single species of plants and primate populations can be daunting.

In addition, although each of the composite vegetation variables has been found to be associated with various primate abundances, densities, and biomasses (see section "Introduction"), each also has been found to be nonassociated with just as many primate population parameters. For example, Stevenson (2001) found that there was no significant relationship between total tree density per hectare and primate biomass.

He proposed that one reason for this could be that measures such as density do not reflect size, which is an important determinant of reproductive output.

Likewise, basal area does not always turn out to be a good predictor of primate population sizes. For instance, in his broad analysis of New World sites, Stevenson (2001) found that total basal area per hectare was not correlated with the biomass of any primate guild. He extended his earlier observation that size is an important determinant of reproductive output by adding that other factors such as habitat quality and differences between plant families also affect plant reproduction. However, as Janson and Chapman (1999) point out, even measures of overall plant productivity can sometimes be negatively related to total primate biomass in some communities (Kay et al. 1997).

Species diversity is another less-than-perfect predictor of primate densities. For instance, Tutin (1999) compared fruit diversity and abundance in a forest fragment to a neighboring continuous forest. She found that despite the lower diversity and abundance of fruit in the fragment, the local density of primates in the fragment was equivalent to that of the adjoining continuous forest. Tutin (1999) proposed that the reason for the similar densities was that there were many benefits of living in such a small area (9 ha) such as having exceptional knowledge of the location and quality of food resources, reduced feeding competition between individuals through group fission, lower travel costs, etc.

Perhaps one of the clearest illustrations of how some vegetation indices at a single site can be correlated with some primate species, but not with others, comes from Skorupa (1986). Skorupa (1986) found that the strongest vegetation correlates were unique to each species. For instance, although the abundances of black-and-white colobus (*C. guereza*) and gray-cheeked mangabeys (*C. albigena*) were significantly related to total stem density, red colobus (*C. badius*) abundances were associated with stand basal area, and chimpanzee (*P. troglodytes*) and L'hoests' guenon (*C. l'hoesti*) abundances were correlated with percent canopy cover. In other words, most species' abundances covaried with a single vegetation variable (Skorupa 1986).

In addition, there may be variables other than those included in the current study that are better predictors of primate population parameters. For example, Peres (1997) found that latitude, total rainfall, and the degree of seasonality were all significantly related to howler (*Alouatta* spp.) densities. Stevenson (2001) also found that climatic variables were important in predicting total primate biomasses. Although it is not impossible, it is doubtful that variables such as latitude, total rainfall, and degree of seasonality affect Mentawai primate densities. These variables probably vary little across the tiny Mentawai archipelago, not to mention just the two Pagai Islands, which span a total length of approximately 110.5 km and a width of about 57.5 km.

Moreover, other biogeographic factors that affect primate populations such as fragment size were not examined in this study. Estimates of primate densities in southeastern Brazil were on the order of several hundreds to thousands of individuals/species in 20,000 ha size fragments and <50 individuals/species in 200 ha size fragments (Chiarello and de Melo 2001). Forest size could have affected primate densities in this study too.

An additional reason for the lack of significant results between primate densities and the botanical indices could be that the area sampled was not large enough. For instance, Stevenson (2001) found that basal areas reached stable values only after sampling at least 1.5 ha. In this study, 0.75 ha/site were sampled – half the area Stevenson (2001) reported was needed to reach a consistent value. Perhaps sampling a larger area would reflect the true basal areas at each of the sites and in turn, this might be more predictive of primate densities.

Conclusions

For conservation purposes, it is important to identify the general forest and tree indices, as well as the specific tree species that affect primate densities. The results of this study show that overall, none of the forest structure variables were consistently associated with primate densities. However, more meaningful vegetation measures such as liana densities and the plant species and genera consumed by at least one of the four primate species (*H. klossii*), were related to some primate densities. This makes sense because the forest structure indices were created from the overall data set from each site. In other words, there was no a priori reason to assume that one of the forest structure variables would be related to any one of the Mentawai primate densities, other than the fact that the same measures have been shown to be important indicators of some primate species' densities in other regions. In contrast, only the plant species and genera compiled for the gibbon food variables and keruing trees (*D. haselthii*) were selected because they are important plant species for Mentawai primates. Therefore, it is not surprising that *H. klossii* foods were the strongest predictors of primate densities.

Additional data on the feeding habits of habituated Mentawai primates are needed to reveal the associations between Mentawai primate densities and vegetation variables. Once these are known, logging companies that practice selective logging can be informed of the tree species that are valuable to the Mentawai primates, with the hope that they will be left standing. It also is important to remember that trees are not only used by primates and other animals for food, but for cover from predators and the elements, as substrates for locomotion, and as places to rest and sleep. If all of these factors were taken into account in this and similar studies, then perhaps the relationship between nonhuman primate population parameters and the trees they *use* (i.e., not just the species on which they feed), would be more conspicuous.

Acknowledgments Generous funding was provided by the Wildlife Conservation Society, Primate Conservation, Inc., and Conservation International/the Margo Marsh Biodiversity Foundation. Permits to conduct research were granted by the Indonesian Institute of Science (LIPI), the Indonesian Department of Forestry, and the project sponsor, Dr./Bapak Amsir Bakar of Universitas Andalas. Logistical help came from the Hutagalung family and Mr. Aurelius Napitupulu. Kathleen Donovan assisted with data collection and Amna Ali helped with the tables and appendices. Special thanks to Drs. Patricia C. Wright, Charles H. Janson, John G. Fleagle, Joshua Ginsberg, and two anonymous reviewers for commenting on and helping improve an earlier version of this manuscript.

Appendix. Primate Densities and Vegetation Variables. Key is at Bottom of Appendix

Variable/site	Km 28	Km 60	Kin	Mang	Smng	Simp	Area I	Area III	Area IV
Primate density	12.25	9.77	13.72	9.18	22.84	6.08	14.78	19.97	12.79
H. klossii density	2.558	0.83	1.501	1.03	1.129	1.12	1.258	1.276	0.762
M. pagensis density	4.832	6.84	9.026	5.73	17.98	5.69	13.73	14.45	8.633
P. potenziani density	1.153	0.58	1.332	0.52	0.731	0.69	2.208	5.148	4.401
S. concolor density	4.8	3.15	7.554	2.76	10.45	0.27	1.609	4.332	1.69
Total stem density	409.3	420	500	471	402.7	652	514.7	482.7	402.7
Species richness	59	49	65	88	56	73	74	78	70
Stand basal area	32.42	51.1	71.65	34.8	25.64	32.9	27.64	31.25	33.64
Mean patch size	0.106	0.16	0.191	0.1	0.085	0.07	0.072	0.086	0.111
Species diversity	4.641	4.39	4.884	5.78	5.042	5.03	5.364	5.361	5.288
Species equitability	0.81	0.76	0.824	0.98	0.883	0.81	0.901	0.91	0.926
Percent canopy cover	78.59	80.8	86.87	83.5	77.62	83.7	72.57	75.02	78.62
Ficus spp. density	1.333	2.67	0	8	6.667	24	5.333	0	5.333
Ficus spp. basal area	0.013	0.98	0	0.14	0.137	0.75	0.064	0	1.126
Moraceae density	1.333	2.67	5.333	8	9.333	28	5.333	0	8
Moraceae basal area	0.013	0.98	0.691	0.14	0.187	1.07	0.064	0	1.211
Palmae density	8	4	28	1.33	5.333	48	8	20	6.667
Palmae basal area	0.108	0.08	0.495	0.02	0.123	0.83	0.122	0.524	0.162
Liana density	0	1.33	1.333	1.33	0	5.33	1.333	4	5.333
Liana basal area	0	0.04	0.038	0.03	0	0.08	0.042	0.072	0.07
Pres. *H. klossii* fdsp	1	1	1	1	1	1	1	1	1
Pres. *H. klossii* fdgen	1	1	1	1	1	1	1	1	1
Pres. *H. klossii* fdsp + gen	1	1	1	1	1	1	1	1	1
No. *H. klossii* fdsp	5	5	5	5	3	5	6	5	6
No. *H. klossii* fdgen	10	6	9	9	9	9	11	10	9
No. *H. klossii* fdsp + gen	15	11	14	14	12	14	17	15	15

(continued)

(continued)

Variable/site	Km 28	Km 60	Kin	Mang	Smng	Simp	Area I	Area III	Area IV
Freq. *H. klossii* fdsp	27	58	20	30	42	66	46	60	18
Freq. *H. klossii* fdgen	41	33	35	31	45	37	38	31	56
Freq. *H. klossii* fdsp + gen	68	91	55	61	87	103	84	91	74
BA *H. klossii* fdsp	0.857	5.41	1.937	1.62	3.479	2.65	2.582	5.257	2.952
BA *H. klossii* fdgen	2.629	2.39	2.564	1.67	2.711	2.36	1.834	2.326	3.136
BA *H. klossii* fdsp + gen	3.485	7.8	4.501	3.29	6.19	5.02	4.416	7.582	6.089
D. haselthi density	30.67	69.33	97.33	20.00	10.67	28.00	2.667	13.33	4.00
D. haselthi basal area	8.782	22.48	26.76	4.54	1.668	3.687	0.457	1.352	1.325

Key

Variable	Definition and/or formula
Density	Number of stems/ha
Total stem density	Number of trees at a site
Species richness	Total number of tree species in an area
Basal area (BA)	$$\mathrm{BA} = \frac{\pi}{40{,}000} \times \frac{\sum \mathrm{DBH}^2}{a} = 0.0000785398 \times \frac{\sum \mathrm{DBH}^2}{a}$$
Stand basal area (G)	Sum of the basal area of all living trees at each site
Mean patch size	G/N (N = total number of stems)
Species diversity (Shannon–Wiener Index)	$$H' = -\sum_{i=1}^{s} (p_i)(\log_2 p_i)$$
Species equitability (Shannon–Wiener Evenness Index)	$H' = \log_2 S$
Percent canopy cover	Average of 60 canopy cover estimates at each site

Pres presence; *No* number; *Freq* frequency; *fd* food; *sp* species; *gen* Genera; *fdsp* Whitten (1982) observed *H. klossii* feeding on these species; *fdgen* Whitten (1982) observed *H. klossii* feeding on species in the same genera. These foods are either a different and/or an unknown species; *fdsp + gen* a combination of known *H. klossii* food species (Whitten 1982) and species belonging to the same genera as known food species.

References

Akihisa T, Tokuda H, Ichiishi E, Mukainaka T, Toriumi M, Ukiya M, Yasukawa K, Nishino H (2001) Anti-tumor promoting effects of multiflorane-type triterpenoids and cytotoxic activity of karounidol against human cancer cell lines. Cancer Lett 173:9–14

Arroyo-Rodríguez V, Mandujano S, Benítez-Malvido J, Cuende-Fanton C (2007) The Influence of Large Tree Density on Howler Monkey (*Alouatta palliata mexicana*) Presence in Very Small Rain Forest Fragments. Biotropica 39:760–766

Bennett EL, Davies AG (1994) Ecology of Asian colobines. In: Davies AG, Oates JF (eds) Colobine monkeys: their ecology, behaviour and evolution. Cambridge University Press, Cambridge, pp 129–171

Brandon-Jones D (1993) The taxonomic affinities of the Mentawai Islands Sureli, *Presbytis potenziani* (Bonaparte, 1856) (Mammalia: Primata: Cercopithecidae). Raffles Bull Zool 41:331–357

Brockelman WY, Ali R (1987) Methods of surveying and sampling forest primate populations. In: Marsh CW, Mittermeier RA (eds) Primate conservation in the tropical rain forest. Alan R. Liss, New York

Brown JH (1981) Two decades of homage to Santa Rosalia – Toward a general theory of diversity. Am Zool 21:877–888

Buckland ST, Anderson DR, Burnham KP, Laake JL (1993) Distance sampling: estimating abundance of biological populations. Chapman and Hall, London

Burnham KP, Anderson DR, Laake JL (1980) Estimation of density from line transect sampling of biological populations. Wildl Monogr 72:1–202

Caldecott JO, Feistner ATC, Gadsby EL (1996) A comparison of ecological strategies of pig-tailed macaques, mandrills and drills. In: Fa JE, Lindburg DG (eds) Evolution and ecology of macaque societies. Cambridge University Press, Cambridge, pp 73–97

Campbell MJ, Hamilton B, Shoemaker M, Tagliaferri M, Cohen I, Tripathy D (2002) Antiproliferative activity of Chinese medicinal herbs on breast cancer cells in vitro. Anticancer Res 22:3843–3852

Chapman CA, Chapman LJ, Bjorndal KA, Onderdonk DA (2002a) Application of protein-to-fiber ratios to predict colobine abundance on different spatial scales. Int J Primatol 23:283–310

Chapman CA, Chapman LJ, Gillespie TR (2002b) Scale issues in the study of primate foraging: red colobus of Kibale National Park. Am J Phys Anthropol 117:349–363

Chiarello AG, de Melo FR (2001) Primate population densities and sizes in Atlantic forest remnants of Northern Espírito Santo. Braz Int J Primatol 22:379–396

Choi JH, Choi JH, Kim DY, Yoon JH, Youn HY, Yi JB, Rhee HI, Ryu KH, Jung K, Han CK, Kwak WJ, Cho YB (2002) Effects of SKI 306X, a new herbal agent, on proteoglycan degradation in cartilage explant culture and collagenase-induced rabbit osteoarthritis model. Osteoarthritis Cartilage 10:471–478

Curtin SH (1980) Dusky and banded leaf monkeys. In: Chivers DJ (ed) Malayan forest primates. Plenum, New York, pp 107–145

Dasilva GL (1994) Diet of *Colobus polykomos* on Tiwai Island: selection of food in relation to its seasonal abundance and nutritional quality. Int J Primatol 15:1–26

Davies AG (1991) Seed-eating by red leaf monkeys (*Presbytis rubicunda*) in dipterocarp forest of northern Borneo. Int J Primatol 12:119–144

Davies AG, Bennett EL, Waterman PG (1988) Food selection by two South-east Asian colobine monkeys (*Presbytis rubicunda* and *Presbytis melalophos*) in relation to plant chemistry. Biol J Linn Soc 34:33–56

Davies AG, Oates JF, Dasilva GL (1999) Patterns of frugivory in three West African colobine monkeys. Int J Primatol 20:327–357

De A. Moura AC (2007) Primate group size and abundance in the Caatinga Dry Forest, Northeastern Brazil. Int J Primatol 28:1279–1297

Defler TR, Pintor D (1985) Censusing primates by transect in a forest of known primate density. Int J Primatol 6:243–259 de Melo AC, Cota BB, de Oliveira AB, Braga FC (2001) HPLC quantitation of kaurane diterpenes in *Xylopia* species. Fitoterapia 72:40–45

Estrada A, Juan-Solano S, Martinez TO, Coates-Estrada R (1999) Feeding and activity patterns of a howler monkey (*Alouatta palliata*) troop living in a forest fragment at Los Tuxtlas, Mexico. Am J Primatol 48:167–183

Eudey AA (1987) Action Plan For Asian Primate Conservation: 1987–1991. UNEP, IUCN, WWF, Riverside

Everitt BS (1980) Cluster analysis. Halsted Press, New York

Felton AM, Engström LM, Felton A, Knott CD (2003) Orangutan population density, forest structure and fruit availability in hand-logged and unlogged peat swamp forests in West Kalimantan, Indonesia. Biol Conserv 114:91–101

Ganzhorn JU (1995) Low-level forest disturbance effects on primary production, leaf chemistry, and lemur populations. Ecology 76:2084–2096

Ganzhorn JU, Malcomber S, Andrianantoanina O, Goodman SM (1997) Habitat characteristics and lemur species richness in Madagascar. Biotropica 29:331–343

Hladik CM (1977) A comparative study of the feeding strategies of two sympatric species of leaf monkeys: *Presbytis senex* and *Presbytis entellus*. In: Clutton-Brock TH (ed) Primate ecology: studies of feeding and ranging behaviour in lemurs, monkeys, and apes. Academic, London, pp 323–353

Hladik A (1978) Phenology of leaf production in rain forest of Gabon: distribution and composition of food for folivores. In: Montgomery GG (ed) The ecology of arboreal folivores. Smithsonian Institution Press, Washington, DC, pp 51–71

IUCN (2009) IUCN Red List of Threatened Species. Version 2009.1. <www.iucnredlist.org>. Downloaded on 15 June 2009

Janson CH, Chapman CA (1999) Resources and primate community structure. In: Fleagle JG, Janson CH, Reed KE (eds) Primate communities. Cambridge University Press, Cambridge, pp 237–267

Janson CH, Terborgh JW (1986) Censando primates en el bosque lluvioso. In: Rios MA (ed) Reporte Manu. Centro Datos para la Conservacion, Lima, Peru, pp 1–48

Johns AD, Skorupa JP (1987) Responses of rain-forest primates to habitat disturbance: a review. Int J Primatol 8:157–191

Johnson SE, Overdorff DJ (1999) Census of brown lemurs (*Eulemur fulvus* spp.) in Southeastern Madagascar: methods-testing and conservation implications. Am J Primatol 47:51–60

Kay RF, Madden RH, van Schaik C, Higdon D (1997) Primate species richness is determined by plant productivity: implications for conservation. Proc Natl Acad Sci USA 94:13023–13027

Koenig A, Beise J, Chalise MK, Ganzhorn JU (1998) When females should contest for food – Testing hypotheses about resource density, distribution, size, and quality with Hanuman langurs (*Presbytis entellus*). Behav Ecol Sociobiol 42:225–237

Krishnan R, McDonald KA, Dandekar AM, Jackman AP, Falk B (2002) Expression of recombinant trichosanthin, a ribosome-inactivating protein, in transgenic tobacco. J Biotechnol 97:69–88

Lambert FR (1990) Some notes on fig-eating by arboreal mammals in Malaysia. Primates 31:453–458

Leighton DR, Leighton M (1983) Vertebrate responses to fruiting seasonality within a Bornean rain forest. In: Sutton SL, Whitmore TC, Chadwick AC (eds) Tropical rain forest: ecology and management. Blackwell, Oxford, pp 181–196

Leopold A (1933) Game management. The University of Wisconsin Press, Madison

MacKinnon K, MacKinnon J (1980) Niche differentiation in a primate community. In: Chivers DJ (ed) Malayan forest primates. Plenum, New York, pp 167–190

Marsh CW (1986) A resurvey of Tana primates and their forest habitat. Primate Conserv 7:72–81

Mastripieri D (2007) Macachiavellian intelligence: how rhesus macaques and humans have conquered the world. University of Chicago Press, Chicago

Mbora DNM, Meikle DB (2004) Forest fragmentation and the distribution, abundance and conservation of the Tana river red colobus (*Procolobus rufomitratus*). Biol Conserv 118:67–77

Mowry CB, Decker BS, Shure DJ (1996) The role of phytochemistry in dietary choices of Tana River red colobus monkeys (*Procolobus badius rufomitratus*). Int J Primatol 17:63–84

Nakayama Y, Matsuoka S, Watanuki Y (1999) Feeding rates and energy deficits of juvenile and adult Japanese monkeys in a cool temperate area with snow coverage. Ecol Res 14:291–301

Nelles Verlag (1999) Indonesia 4. Nelles Verlag GmbH, München, Germany

Nijman V (2004) Effects of habitat disturbance and hunting on the density and the biomass of the endemic Hose's leaf monkey *Presbytis hosei* (Thomas, 1889) (Mammalia: Primates: Cercopithecidae) in east Borneo. Contrib Zool 73:283–291

NRC (1981) Techniques for the study of primate population ecology. National Academy Press, Washington, DC

Oates JF (1994) The natural history of African colobines. In: Davies AG, Oates JF (eds) Colobine monkeys: their ecology, behaviour and evolution. Cambridge University Press, Cambridge, pp 75–128

Paciulli LM (2004) The effects of logging, hunting, and vegetation on the densities of the Pagai, Mentawai Island primates (Indonesia). Dissertation thesis, State University of New York, Stony Brook

Palombit RA (1997) Inter- and intraspecific variation in the diets of sympatric siamang (*Hylobates syndactylus*) and lar gibbons (*Hylobates lar*). Folia Primatol 68:321–337

Peres CA (1997) Effects of habitat quality and hunting pressure on arboreal folivore densities in Neotropical forests: a case study of howler monkeys (*Alouatta* spp.). Folia Primatol 68:199–222

Richard AF, Goldstein SJ, Dewar RE (1989) Weed macaques: the evolutionary implications of macaque feeding ecology. Int J Primatol 10:569–594

SAS Institute (1989–1997) *JMP*, SAS Institute Inc

Serio-Silva JC, Rico-Gray V, Hernández-Salazar LT, Espinosa-Gomez R (2002) *Ficus* (Moraceae) in the diet and nutrition of a troop of Mexican howler monkeys, *Alouatta palliata mexicana*, released on an island in southern Veracruz, Mexico. J Trop Ecol 18(6):913–928

Skorupa JP (1986) Responses of rainforest primates to selective logging in Kibale Forest, Uganda: a summary report. In: Benirschke K (ed) Primates: the road to self-sustaining populations. Springer, New York, pp 57–70

Sokal RR, Rohlf FJ (1995) Biometry: the principles and practice of statistics in biological research. W. H. Freeman and Company, Stony Brook, NY

SPSS (1989–2001) SPSS for Windows, SPSS Inc

Stanford CB (1991) The diet of the capped langur (*Presbytis pileata*) in a moist deciduous forest in Bangladesh. Int J Primatol 12:199–216

Sterling EJ, Rakotoarison N (1998) Rapid assessment of richness and density of primate species on the Masoala Peninsula, eastern Madagascar. Folia Primatol 69:109–116

Stevenson PR (2001) The relationship between fruit production and primate abundance in Neotropical communities. Biol J Linn Soc 72:161–178

Tenaza RR, Fuentes A (1995) Monandrous social organization of pigtailed langurs (*Simias concolor*) in the Pagai Islands. Indones Int J Primatol 16:295–310

Terborgh J (1986) Keystone plant resources in the tropical forest. In: Soule ME (ed) Conservation biology: the science of scarcity and diversity. Sinauer, Sunderland, MA, pp 330–344

Tilson RL (1976) Infant coloration and taxonomic affinity of the Mentawai Islands leaf monkey, *Presbytis potenziani*. J Mammal 57:766–769

Tutin CEG (1999) Fragmented living: behavioural ecology of primates in a forest fragment in the Lopé Reserve, Gabon. Primates 40:249–265

Tutin CEG, Williamson EA, Rogers ME, Fernandez M (1991) A case study of a plant-animal relationship – *Cola lizae* and lowland gorillas in the Lope Reserve. Gabon J Trop Ecol 7:181–199

van Roosmalen MGM, Mittermeier RA, Fleagle JG (1988) Diet of the Northern Bearded Saki (*Chiropotes satanus chiropotes*): a neotropical seed predator. Am J Primatol 14:11–35

Vidal MD, Cintra R (2006) Effects of forest structure components on the occurence, group size and density of groups of bare-face tamarin (*Saguinus bicolor* – Primates: Callitrichinae) in Central Amazonia. Acta Amazonica 36:237–248

Waterman PG, Ross JAM, Bennett EL, Davies AG (1988) A comparison in the floristics and leaf chemistry of the tree flora in two Malaysian rain forests and the influence of leaf chemistry on populations of colobine monkeys in the Old World. Biol J Linn Soc 34:1–32

Weghorst JA (2007) High population density of black-handed spider monkeys (*Ateles geoffroyi*) in Costa Rican lowland wet forest. Primates 48:108–116

White LJT, Rogers ME, Tutin CEG, Williamson EA, Fernandez M (1995) Herbaceous vegetation in different forest types in the Lopé Reserve, Gabon: implications for keystone food availability. Afr J Ecol 33:124–141

Whitesides GH, Oates JF, Green SM, Kluberdanz RP (1988) Estimating primate densities from transects in a west African rain forest: a comparison of techniques. J Anim Ecol 57:345–367

Whitten AJ (1982) Diet and feeding behaviour of Kloss gibbons on Siberut Island, Indonesia. Folia Primatol 37:177–208

Wilson WL, Wilson CC (1975) Species-specific vocalizations and the determination of phylogenetic affinities of the *Presbytis aygula-melalophos* group in Sumatra. In: Kondo S, Kawai M, Ehara A (eds) Contemporary primatology. S. Karger, Basel

Worman COD, Chapman CA (2006) Densities of two frugivorous primates with respect to forest and fragment tree species composition and fruit availability. Int J Primatol 27:203–225

WWF (1980) Saving Siberut: a conservation master plan. World Wildlife Fund, Bogor

Chapter 12
Proboscis Monkey (*Nasalis larvatus*): Bio-ecology and Conservation

Muhammad Bismark

Introduction

Proboscis monkeys (*Nasalis larvatus*) are endemic to Borneo and primarily reside in peat swamp forest, mangrove, and riparian forest. Some smaller populations have also been found in upstream dipterocarp forest, as well as in rubber plantations, 300 km from coastal areas (Soendjoto 2003). Swamp forest along riverbank and riparian mangrove in the coast are also potential habitats for proboscis monkeys. The proboscis monkey population is very dependent on the quality of the wetland ecosystem, especially mangrove forest and riparian forest. This species' focus on quality habitat makes it relatively intolerant to habitat disturbance (Bennett and Gombek 1993; Yeager 1992). McNeely et al. (1990) reported that there are 29,500 km² of proboscis monkey habitat. Since 1990, 49% of this habitat has been lost, and only 4.1% of this habitat occurs within designated conservation areas. Undoubtedly, as village settlements and agricultural areas along the river's edge tend to increase, the proboscis monkey habitat will decline as will its population. The increasing frequency of river traffic, forest concession activity, forest fire, illegal logging, and conversion of swamp forest to plantation estate and agricultural land, or fishpond development at mangrove forest, all represent primary causes of proboscis monkey habitat destruction.

The degradation of proboscis habitat has occurred quite fast because most of it has high economic value to the people who reside in the surrounding riverbank. The community uses the river as a transportation line, while the riparian forest is used by people to open agricultural gardens and settlements. All of these behaviors cause destruction to proboscis monkey habitat. As a result, the population of proboscis has decreased, the distribution has become more spotty (increased distance between subpopulations) (Bismark and Iskandar 2002; Ma'ruf et al. 2005), and local people continue to view the proboscis monkey as a pest.

M. Bismark (✉)
Forest and Nature Conservation, Research and Development Center, Bogor, Indonesia
e-mail: bismark_forda@yahoo.com

S. Gursky-Doyen and J. Supriatna (eds.), *Indonesian Primates*,
Developments in Primatology: Progress and Prospects,
DOI 10.1007/978-1-4419-1560-3_12, © Springer Science+Business Media, LLC 2010

Habitat Distribution

The distribution of proboscis monkeys is indicated in Fig. 12.1. The distribution and habitat types used by proboscis monkeys in south Kalimantan has already been reported by Soendjoto et al. (2005) and Bismark (1995), while in east Kalimantan it has been reported by Bismark and Iskandar (2002) and Ma'ruf et al. (2005). The distribution of proboscis monkeys in Kalimantan has also been surveyed by Meijaard and Nijman (2000) in more than 30 different locations comprising mangrove forest, small islands, coastal deltas, riverbanks, and swamp forest. They observed that the proboscis monkey population is distributed from the coast to more inland areas. More than 20% of the population was observed in coastal areas, 18% of the population was located between 100 and 200 km from the coast, 16% of the population was located between 20 and 100 km, and 58% of the population was located 50 km from the coast. Smaller proportions of the population were also found between 300 and 750 km from the coast. Over 90% of the locations were at an altitudinal range below 200 m asl. The highest altitudinal distribution reported was 350 m asl.

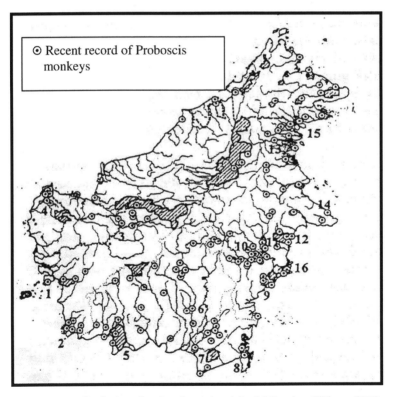

Fig. 12.1 The recent distribution of proboscis monkey (after Meijaard and Nijman 2000)

Approximately 49% of proboscis monkey habitat was lost between 1990 and 1995; the remaining habitat was estimated to be about 39%, of which only 15% is located within conservation areas (Meijaard and Nijman 2000). The rate of habitat loss is estimated to be approximately 2% per year. Habitat destruction occurred not only outside designated conservation areas, but also within conservation areas. Lowland forest of conservation areas in Kalimantan had been degraded by more than 56% in 2001 (Curran et al. 2004). For example, habitat degradation occurred in Pulau Kaget, where only 10% of forest remains; 90% of the island became agricultural land (Meijaard and Nijman 2000). Formerly, tree vegetation in the proboscis's habitat at Pulau Kaget, 20–55 m from the riverbank, had a density of 150 trees per ha (Bismark 1997). A major forest fire occurred in 1997 at Tanjung Puting National Park that destroyed 75% of the forest area. In 1998, a forest fire also occurred in Kutai National Park, destroying 95% of the forest area, including large swaths of proboscis monkey populations.

Mangrove forest in Kalimantan, one of the primary habitat sites for proboscis monkeys, only constitutes 7% of the forest. Long, wide (>10 m) and deep rivers form riparian mangrove forest. In such habitats, the mangrove trees grow relatively high, have large trunk diameters, and support the ecological needs and daily activity of proboscis monkeys. Mangrove riparian forest has higher productivity than other types of mangrove forest (Mitch and Gosselink 1984). The taller trees provide safety to groups of proboscis monkeys, especially when choosing sleeping trees to protect the group from predator attack (Bismark 1986; Yeager 1990).

Population Density

Researchers have conducted many population surveys. Some population survey results have also been reported by Yeager and Blondal (1992), Ruhiyat (1986), Yasuma (1989), and Bennett and Sebastian (1988). The population density in many locations has been reported to be between 8.3 and 58 individual/km^2. It has been reported that in Kutai National Park (Kutai NP), three groups of proboscis monkey were found downstream of the Sangata River and some groups had been found upstream of the Sangata River (Rodman 1978). In mangrove forest of Kutai NP, the population of proboscis monkey distributed only in mangrove forest of Sangkimah River, Teluk Kaba, Pemedas River, and Padang River. Along 2 km of Sangkimah River, there were 117 individuals of proboscis monkey (Bismark 1986). The proboscis monkey population located in mangrove forest can reach 60 individuals/km^2 (Bismark 1986). Based on the age composition of the group, which can consist of four infants per group, proboscis monkeys seem to have a high reproductive rate. Over our 9-year research project, we observed that the population of proboscis monkey in Kutai National Park has declined 28.2% or an average of 3.1% per year. This resulted from high intensity habitat destruction of the mangrove forest along the riverbank.

Yeager and Blondal (1992) analyzed the density of proboscis monkeys in degraded habitat. It was about 9 individuals/km^2 in the most severely degraded habitat,

Table 12.1 Average group size and age composition of proboscis monkeys in Sangata River (Bismark and Iskandar 2002)

Location	Group size	Group composition				
		Adult male	Adult female	Subadult	Young	Infant
Upstream	12.0	1.0	3.75	1.75	2.75	2.25
Downstream	17.4	1.4	6.4	2.8	3.4	1.8
Mangrove forest	21.0	2.75	7.0	3.5	3.25	4.0

25 individuals/km^2 in habitat with severe destruction, 33 individuals/km^2 in areas with moderate habitat degradation, and 62 individuals/km^2 in areas with low habitat degradation. Proboscis monkeys are very sensitive to habitat disturbance (Wilson and Wilson 1975); therefore, the population of proboscis monkey could be used as an indicator of habitat destruction level, especially that of mangrove forest and riparian forest. The differences in group size and population density based on ecosystem are presented in Table 12.1.

The population growth of proboscis monkey in degraded habitat at Pulau Kaget (267 ha) is positive. Its' habitat is dominated by *Sonneratia caseolaris,* which is distributed between 20 and 55 m from riverbank, with a density of 150 trees per ha. At that time, the population was around 300 individuals (Bismark 1997). In comparison to Kutai National park, the population in Pulau Kaget is better off than Kutai National Park in which only 400 individuals were located in 200,000 ha of this national park (Bismark and Iskandar 2002). The relatively high population density in Pulau Kaget results from the ability of this population to utilize water plants such as *Limnocharis flava, Agapanthus africanus, Hymenachne amplicaulis* dan *Vittis trifolia* as food resources. Water plant species contain higher mineral contents than do many terrestrial plants (Oates 1978). To maintain population numbers, proboscis monkeys need specific amounts of minerals (Bennett and Sebastian 1988). They need 179.9 mg of K per kg of body weight per day (Bismark 1995). Moreover, *Vittis trifolia* contains high concentrations of K, about 1.06 %. The high mineral contents of water plants are also supported by chemical analysis of the aerial roots of *S. Caseolaris.* The result detected that aerial roots have mineral concentration of Zn, Cu, and Al two to eight times higher than those same plants grown on soil, and the mineral concentration of Al was 6–17 times higher than the concentration of Al in soil.

In Sangata River, there were three populations of proboscis monkeys distributed along a distance of 18–40 km. This population consisted of one to four groups within 1–2 km of riparian forest (Bismark and Iskandar, 2002). Previously, Suzuki (1986) reported that a population of proboscis monkey in Kutai National Park was distributed between 4 and 25 km (an average distance of 10.6 km). At present, the population of proboscis monkeys in Kutai is distributed within an average distance of 30 km between groups (Bismark and Iskandar 2002) and now an average distance between population was 50 km in mouth of Mahakam river (Ma'ruf 2004). The increase in distance between proboscis monkey groups is an indication of declining habitat quality.

Table 12.2 Total population of proboscis monkey in Kutai National Park (Bismark and Iskandar 2002)

Location	Square of survey area (km^2)	The number of proboscis monkey that was observed	Density/km^2	Estimated total population
Upstream	60	50	0.8	122
Downstream	10	89	8.9	107
Mangrove forest	12	84	7.0	81
Total population		224		310

Habitat type and quality affects the number of individuals in a group. Upstream, a group of proboscis monkeys was between 6 and 15 individuals. However, only 10 km downstream, a larger group was found containing 10–25 individuals. In the disturbed mangrove forest, group size was between 6 and 10 individuals, whereas in the good vegetation coverage of mangrove forest, group size could reach between 17 and 25 individuals. This study, which was conducted at many locations, showed a difference in the number of individuals based on the location. The differentiation showed a data's correction value in analyzing of proboscis population is 1.8. Yasuma (1989) considered a correction data of 2.46, based on correction value, so the total population of proboscis monkey in Kutai National Park estimated 400 individuals (Table 12.2).

Bennett and Sebastian (1988), Salter et al. (1985), and Ma'ruf (2004) also reported the frequency they encountered proboscis monkeys in mangrove forest and nipah. The proboscis monkeys that live in mangrove habitat have a high conservation threat because only 8% of mangrove habitat is located in protected forest. In 1990s, it was estimated that there were 2,000 proboscis monkeys in the peat swamp forest of Tanjung Puting National Park. In Sarawak, the total population of proboscis monkey was estimated to be approximately 1,000 individuals, where 300 individuals are found in conservation areas (Yeager and Blondal 1992). Previously, MacKinnon (1986) suggested the population of proboscis monkey about 250,000 individuals, and a part of this population was 25,000 individual in conservation area, whereas Yeager and Blondal (1992) estimated that population of proboscis monkey in conservation areas was less than 5,000 individual. In 1994, population of proboscis monkey in Kalimantan was predicting about 114,000 individuals (Bismark 1997).

Population Threats

The population of proboscis monkey in Tanjung Puting National Park was 62.9 individual/km^2 in 1985, in 1989, the population decreased to about 27.7 individuals/km^2. Subsequently, in 1991, the population of proboscis monkeys was 41 individuals/km^2. It means that within 6 years the population of proboscis monkey has decreased by 35% or 6% per year. Decreases in the population of proboscis monkeys are primarily caused by increasing river pollution, gold mining activity, and river traffic (Yeager 1992). The population decrease may also be caused in part by increased predator pressure resulting from forest clearing. For example, in the riparian forest in Sarawak, populations of salvator's lizard (*Varanus salvator*), a known predator of proboscis monkeys, was high (Rodman 1978; Yeager 1992). Other aspects that are also contributing to the decline in the proboscis monkey population include, parasites (Rijksen 1978; Freeland 1976), geographic distribution (Happel et al. 1987; Chivers 1974), social system (Happel et al. 1987), and a different of potential food availability in term of intestine physiology (Bennett 1983).

Primate hunting by indigenous peoples generally occurs to fulfill their protein requirements. For example, in Siberut Island, primate hunting is a part of their culture and tradition. It might have begun as a way to control the primate populations because there were no predators that preyed on the mammalians. Over the last 35 years, hunting and cultivation are the major problems correlated with the declining population of proboscis monkey. This is especially true since indigenous peoples now utilize guns for hunting and speedboats for transportation (Meijaard and Nijman 2000). Subsequently, the cultivation was planted with fruits that are also attractive to proboscis monkey as a source of food. The presence of proboscis monkey is regarded as a pest and therefore the local people have effort to hunt them (Soendjoto et al. 2006). Besides that, proboscis monkey also hunted to be bait for catching monitor lizard (*Varanus salvator*) as added income.

Undoubtedly, forest fires are another threat to the proboscis monkey. Forest fire in Tanjung Puting NP (1997) destroyed 75% of the wetland forest and in Kutai NP only 5% of the forest remains following another intensive forest fire (Meijaard and Nijman 2000). Moreover, for the populations that survived the fire, there has been an increase in mortality due to food scarcity, due to a loss of habitat, and due to increased prevalence of disease (Manangsang et al. 2005).

The habitat destruction evident through Kalimantan makes it easier for predators to attack proboscis monkeys. As mentioned earlier, the salvator's lizard (*Varanus salvator*) is abundant in riparian forest and is a known predator of proboscis monkeys (Rodman 1978; Yeager 1992). Another reptile that preys on proboscis monkeys is the cobra snake (*Ophiophagus hannah*). The problem with habitat destruction is not only due to the shrinkage of forest area, but also the change in the quality of river water as drinking water, bathing and swimming areas for the proboscis monkey. The development of village settlements and industry in upstream areas reduces river water quality by spreading parasitic pollution. This was shown based on the *Ascaris* and *Trichiuris* egg worms found in fecal of proboscis monkey. *Trichiuris* egg worm is generally found in other primate such as *M. fasicularis* (Matsubayashi and Sayuthi 1981), orangutan, and chimpanzee (Rijksen 1978).

Based on all the problems mentioned above, in the PHVA for Proboscis monkeys, 12 locations were identified as having proboscis monkeys. These locations and the number of individuals observed at each of these locations are presented in Table 12.3. The total number of proboscis monkey living in Kalimantan estimated by Bismark (1997) was 114,000 individuals. Based on the PHVA conducted in 2004, proboscis monkeys presently number about 9,200 individuals because of the major threats that come from anthropogenic activities (Manangsang et al. 2005). The conservation of

Table 12.3 Estimated carrying capacity of population of proboscis monkey (Manangsang et al. 2005)

No	Location/population	Carrying capacity (individual)
1.	Rivers in Central Kalimantan	500
2.	Sentarum Lake, West Kalimantan	700
3.	Gunung Palung National Park, West Kalimantan	500
4.	Kutai National Park, East Kalimantan	1,300
5.	Kendawangan Nature Reserve, West Kalimantan	1,000
6.	South Barito, South Kalimantan	1,700
7.	Delta Mahakam, East Kalimantan	300
8.	Sambas Paloh Nature Reserve, West Kalimantan	200
9.	Sangkurilang, East Kalimantan	100
10	Sesayap, Sebulu, Sebakung, East Kalimantan	700
11.	South Mahakam, East Kalimantan	200
12.	Tanjung Puting National Park, Central Kalimantan	2,000
	Total 12 locations	9,200

proboscis monkeys require the prevention of habitat destruction and prevention of declining habitat quality that result from illegal logging, forest fire, hunting, village settlement and agricultural land, and develop of fish pond at mangrove forest.

Geometric and Biomass

Proboscis monkeys are sexually dimorphic in terms of their body shape, nose shape, and body weight. The comparison between shape and body weight according to sex and age classes and height sitting (length of body to head) is presented in Table 12.4. Body geometrical parts that can be utilized as a parameter of body weight were sitting-height and basal body (Table 12.5). However, sitting-height is easier to observe and measure in the field. Another parameter that is closely correlated with body weight is width of body surface (body, head, and part of gesture). The width of the body surface of an animal is between 0.02 and 1,400 kg, is equivalent with body weight 3/4 degree or closely 2/3 degree (Montheith and Unsworth 1991). Using measurement taken of proboscis monkeys, in conjunction with an exponential regression model (Bismark 1994), I calculated the relationship between sitting height and body weight, separated for age and sex classes. Regression analysis correlated with SH (cm) and W (m^2) of male and female, produced the following formulas:

$$W\ (\male) = 0.0514e^{0.0395SH}\ (r = 0.90)$$

$$W\ (\female) = 0.1048e^{0.0662SH}\ (r = 0.87)$$

As a comparison of the sitting-height (SH) of male and female proboscis monkeys, the formula to estimate weight from body surface (W), for males and females was different.

Table 12.4 The comparison of sitting-height and basal body of proboscis monkey (Bismark 1994)

Sex/age's class	Sitting-height (SH)	Body surface width (W)
Adult female/adult male	4/5	>1/2
Subadult male/adult male	3/4	>1/2
Subadult female/adult female	>4/5	3/4
Subadult male/adult female	>4/5	>4/5

Table 12.5 Geometric body parts affected by body weight in proboscis monkey

	Geometric body		
Age's class	SH (cm)	W (cm²)	Body weight (BW) (kg)
Adult male	65.50	7,204.26	25.17
Adult female	56.25	4,135.22	12.50
Subadult male	51.67	3,904.91	6.67
Subadult female	50.00	320.75	5.00
Young female	38.00	2,828.11	3.50

Therefore, body surface (W) is used to estimate body weight of proboscis monkey through an indirect measurement of sitting-height. Table 12.5 shows that basal body square (W in m²) was correlated with body weight of proboscis monkey (BW kg). The formula was $W = 0.1324\,BW^{0.67}$ Thus, I systematically estimated the body weight of male and female proboscis monkeys based on sitting-height (SH) and basal body square (L), as follows:

1. The measurement of sitting-height (SH in cm) in the field
2. The calculation of weight of body surface (W in m²)

   (a) $W(\male) = 0.514 e^{0.039 SH}$
   (b) $W(\female) = 0.1048 e^{0.0662 SH}$

3. The estimated of body weight (BW)

   (c) $W = 0.1324\,BW^{0.67}$
   (d) $SH = 33.03\,BW^{0.25}$ ($r = 0.91$)

The estimation result is based on the accuracy of the proboscis monkey's sitting-height in the field. To observe sitting-height position, we only recorded sitting-height when the monkey was observed sitting on the branch of a tree of less than 10 cm diameter and a height of less than 5 m.

Daily Activities and Energy Requirement

In high-quality mangrove forest, proboscis monkeys occur at relatively high population density. Table 12.6 illustrates the group size; day range, home range of three groups, as well as the percent overlap of the home range. Yeager (1989)

Table 12.6 The home range of proboscis monkeys in mangrove forest (Bismark 1994)

Groups	Total individuals	Day range (m)	Home range (ha)	Overlapping (%)
K	25	500	18.25	43.84
O	17	516	19.44	62.82
U	20	475	20.50	20.73

Table 12.7 Daily activity budget of proboscis monkeys (Bismark 1994)

Activities	Observation (%)	Time budget (hours)
Feeding	23.2	3.01
Moving	25.2	3.27
Resting	42.3	5.50
Playing	8.2	1.07
Grooming	1.1	0.15

reported that the home range of proboscis monkey was 125–137.5 ha ($x = 130.3$ ha) with 95.9% of the home ranges overlapping. The adjusted home range was 19.3 ha of per group. Later studies (Bismark 1994) have replicated these results. To minimize competition for food resources and sleeping trees, proboscis monkeys move to a new sleeping trees each day. The distance between sleeping trees was 50–400 m with a mean of 180 m. The use of sleeping trees along the riverbank may represent an adaptation of proboscis monkey to diseases and predator.

The proboscis monkey populations that reside in mangrove forest have an average biomass of 194.7 kg per group with an average of 21 individuals per group. Thus, the average biomass of an individual proboscis monkey is 9.27 kg (the estimate weight of adult male is 25.2 kg; adult female of 12.5 kg, subadult of 6.7 kg, young of 3.5 kg, and infant of 1.5 kg).

The caloric needs of proboscis monkeys depend on each individual's body weight and daily activities. Based on this study, the average of body weight was 8.84 kg and daily activities as shown in Table 12.7. A group of proboscis monkeys travels an average distance of 497.2 m/day. When the group travels, they walk quadrupedally ($N = 247$) 59.92% of the time, climb 11.34% of the time, jump 23.89% of the time, and brachiate 4.85% of the time.

Based on the formulation of Moen (1973) and Wheatley (1982), the caloric needs for each individual with a body weight of 8.84 kg, traveling arboreally are 133.76 kcal. Jumping is the most expensive locomotors activity requiring 93.55 kcal, whereas quadruped locomotion, walking, climbing, and hanging only require 4.92 kcal. Vertical movements with a 50% gradient require 35.29 kcal. There was 358.89 kcal used during basal metabolism, 14.38 kcal expended during feeding, 92.89 kcal expended during resting, and 47.91 kcal expended during playing.

According to Wheatley (1982), when arboreal activity occurs in an area that is not flat, the kcal needed for the motion double. In this situation, the total number of calories needed for 8.84 kg of proboscis monkey is 781.60 kcal. Wheatley (1982) said that the supply of calorie needed for 5 kg *Macaca fascicularis* was 855 kcal.

Table 12.8 Dietary composition of proboscis monkeys residing in mangrove habitat (Bismark 1994)

Food types	Proportion (%)	Fresh weight (g)	Dry weight (g)
Leaves	81.14	767.50	218.53
Fruit	8.38	101.50	41.16
Flowers	7.68	16.30	4.56
Bark, insect, crabs, etc.	2.80	15.00	6.00
Total		900.00	270.25

This disparity result because of the difference in the total distance traveled per day by *Macaca* and *Nasalis*. *M. fasicularis* walks 1,869 m/day and needs 65.61% of the energy requirement; however, *N. larvatus* needs 41.3% of the energy requirement. Undoubtedly, habitat destruction has been declined potency of food resources and population of proboscis monkey.

Therefore, for supply of food calorie, proboscis monkeys with 8.84 kg of body weight consume 900 g of food per day (Table 12.8). The content of food calorie that consumed by an average of proboscis monkey was 3,947.5 kcal/g of dry weight. Therefore, proboscis monkey food of 270.25 g of dry weight equivalent with 30.57 g/kg of body weight was contained 1,066.8 kcal or 120.68 kcal/kg of body weight. Conversely, *Macaca fuscata* with 8 kg of body weight needed food requirement of 254 g of dry weight and 1,050 kcal of food calorie (Iwamoto 1982).

Subsequently, carrying capacity of mangrove forest for proboscis monkey was 84 individual/km^2,778.68 kg of biomass and required 93,971.1 kcal of calorie, while habitat productivity was 570,000 kcal/km^2 (Bismark et al. 1994), and so the consumption of food energy for optimal population of proboscis monkey was 16.5% of habitat productivity.

Food and Mineral Requirement

The diet of proboscis monkeys is 98.25% leaves, shoots, fruit, and flowers. The mineral concentration of the foods consumed by proboscis monkeys living in mangrove forest is shown in Table 12.9. Some foods were consumed, in very small quantities, due to their having an essential mineral content such as 0.87% of *Alophyllus cobbe*, which has higher P and Zn values than other leaves possess. Although leaves are a major part of the diet, fruit and flowers are added due to other mineral requirements. For example, the flower of *R. apiculata* and *A. officinalis* contained a higher Cu value than do the leaves of *A. officinalis*. The P content of leaves and fruit of *A. officinalis* is higher than *Rhizophora* and *Bruguiera*, while the bark of *R. apiculata* contains higher Ca values than other foods consumed by this primate species (Table 12.9).

The density of proboscis monkeys, which live along the riverbank, was affected by the changing vegetation types from downstream wetland forest to upstream dipterocarp forest. Upstream forest generally has relatively poor nutrient composition, while proboscis monkeys as a folivorous primate requires enough

Table 12.9 Mineral content of food consumed by proboscis monkey in mangrove forest (mg/day) (Bismark 1994)

Food resources (species)	(%) eaten	Minerals (mg)									
		P	Ca	K	Na	Mg	Cl	Fe	Mn	Cu	Zn
R. apiculata	Leaf (64.22)	265.3	1,227.5	1,211.3	2,003.3	919.6	3,109.3	83.0	6.0	2.6	1.7
R. apiculata	Flower (7.68)	4.7	0.4	17.1	12.7	16.9	39.7	0.4	0.1	0.0	0.0
R. apiculata	Bark (1.05)	2.1	38.3	3.3	18.4	10.9	2.6	0.1	0.2	0.0	0.0
A. officinalis	Leaf (14.31)	34.6	12.5	159.3	149.5	50.8	194.5	1.9	0.6	0.0	0.0
A. officinalis	Flower (8.38)	118.9	17.2	183.1	51.8	42.8	391.0	2.0	0.8	0.0	0.0
B. gymnorhiza	Leaf (0.87)	3.5	12.7	7.3	19.7	8.9	36.4	2.1	0.0	0.0	0.0
B. parviflora	Leaf (0.87)	0.7	3.9	4.7	19.3	1.7	2.2	0.0	0.0	0.0	–
Alophyllus cobbe	Leaf (0.87)	1.2	2.0	2.1	0.7	2.1	1.3	0.0	0.0	0.0	0.0
Total		431.4	1,314.9	1,588.5	2,275.6	1,053.9	3,777.1	89.8	7.9	2.6	1.7

minerals to support the food fermentation process in its intestine (Hladik 1978). For example, optimal amounts of Na are required for the fermentation process in rumen and Cu is needed for protein synthesis (Durand and Kawashima 1980). To supply these minerals, colobine primates also consume swamp plants and clay soil (Oates 1978). Proboscis monkeys living in riparian forest in Samboja Kuala, East Kalimantan have been observed consuming *Mangifera caesia*, *Garcinia mangostana*, *Durio zibethinus*, *Sondaricum koetjapi*, *Hevea brasiliensis*, as well as *Sonneratia caseolaris* (Alikodra et al. 1995).

Proboscis monkeys adapt to their habitat type and food sources within the habitat by adjusting their population density and group size within each habitat. The population of proboscis monkey located upstream of the Sangata River, a habitat that is dominated by Dipterocarps, was 12 individuals per group. In mangrove forest, the population density of proboscis monkeys was 17.4 and 21 individuals per group (Table 12.1).

In Situ Conservation

The South East Asia Zoo Association (SEAZA) and Indonesian Zoo Association (IZA) have identified that proboscis monkeys have the highest priority for the development of in situ and *ex situ* conservation. To address these conservation needs, an international conservation workshop was organized in Bogor by the Conservation Breeding Specialist Group of the IUCN, the World Conservation Union (CBSG). In addition, Indonesia Proboscis Monkey Population and Habitat Viability Assessment (PHVA) workshop was conducted in December 2004 (Manangsang et al. 2005).

The high priority for proboscis monkey conservation is caused by awareness of how rapidly the population is declining. In 1994, the total population of proboscis monkey was 114,000 individuals. The main population, scattered in 12 locations throughout Kalimantan, was traditionally believed to comprise 25,000 individuals. Supriatna (2004) suggested the population contained only 15,000 individuals based on the habitat deforestation rate of 2.5%. Manangsang et al. (2005) estimated that only 9,200 individuals occur in these locations. If the population of proboscis monkey was 114,000 individuals in 1994 and 15,000 individuals in 2004, then within the last of 10 years, the population declined on average 10% per year.

According to Bismark and Iskandar (2002), the decline in the proboscis monkey population is in large part due to the loss of habitat which now averages 3.1% yearly (Bismark and Iskandar 2002). This does not take into account habitat degradation, which is reported to be approximately 3.49% a year (Supriatna et al. 2001). Bismark and Iskandar (2002) note that proboscis monkeys are very sensitive to habitat fragmentation. Habitat fragmentation along riverbank causes increased hunting pressure, higher parasite loads, higher predator pressure, and higher stress loads, and thus directly affects the proboscis monkey population.

On average, proboscis monkeys traverse 500 m of riparian forest daily, between 300 and 800 m in the rubber plantation, and 800–2,000 m in mangrove forest. Therefore, a minimum of 500 m of the riverbank forest needs to be protected forest for the successful conservation of one proboscis monkey group. This will also minimize the rate of surface runoff, which carries soil particle erosion. Therefore, a forest concession (practicing of Reduced Impact Logging) bordering the habitat of proboscis monkeys should be implemented to preserve forest quality for the proboscis monkeys.

Ex Situ Conservation

The maintenance of proboscis monkeys in zoos has already been achieved in habitat countries as well as foreign countries. The first successful captive breeding of proboscis monkeys was reported in 1993. Prior to this, the majority of captive born proboscis monkeys died due to high levels of stress. In early 1998, a proboscis monkey maintained at the Singapore Zoological Garden (SZG) gave birth to a baby.

The diet of captive breed proboscis monkeys include instant noodles, fruit of *Dillenia* sp., banana, and papaya. A proboscis monkey with 8 kg of body weight consumes 100 g of instant noodles, 30 g of leaf, 40 g of banana, and 5 g of papaya per day. The average weight of dry food given to captive proboscis monkeys is 157.93 g/day or 27.85 g/kg of body weight/day. At the Singapore Zoological Garden, proboscis monkeys eat 25 g of apple; 25 g of banana; three bunches of long nourishing bean; ½ boiled egg; and 25 g of rice with meat, formed into a rounded mass. In addition, five proboscis monkeys were also given supreme primate dry food with fish-oil, 3–4 gain of neotroplek, and three bundles or 4 kg of leaves. At the Indonesian Safari Garden in Bogor, proboscis monkeys eat 37% vegetables, 4% protein, 50% leaves, and 9% of fruits. Proboscis monkey with 6 kg of body weight eat 1.8 kg or 30.18% of body weight (Bismark 1994). The leaf species for proboscis monkey in the Indonesian Safari Garden were *Cinnamomum* sp., *Ficus benyamina*, *Pterocarpus indicus*, *Paraserianthes falcataria*, *Terminalia cattapa*, *Artocarpus heterophylla*, and *Ficus* sp. These leaves contained 6–30% of Ca, while *Ficus* sp. has 30.5 ppm of Ca, 0.1–2.3% of P, and 11.3–22.8% of protein. Our knowledge of food composition and nutrient content will support an acceleration of captive breeding program or *ex situ* conservation for the proboscis monkey.

Ex situ conservation through the development of captive breeding in zoos can be used as conservation education to increase awareness regarding the conservation of proboscis monkeys in nature, especially preventing hunting and encouraging rehabilitation of the riverbank, lake, and forest plantation.

Conservation Programs

To solve the problems of habitat loss and degradation, as well as the rapidly declining population of proboscis monkeys, many programs are needed including an inventory of the distribution, habitat use and population density of proboscis monkeys; the rehabilitation and restoration of potential habitat for population development; the development of community awareness for the conservation of the riverbank and wildlife; control of river use to minimize pollution and limit village settlement along the riverbank; the development of *ex situ* conservation; and the development of nature recreation with proboscis monkey as an object to increase economic value of local community and wildlife use.

The increase in the use of land along the riverbank and the resulting river traffic has resulted in tremendous habitat fragmentation and the decline of proboscis monkey populations. Nevertheless, to determine the conservation status of this species, additional data are needed on this species' population distribution.

At present, the majority of research on the proboscis monkey has generally focused on their distribution and population density, while behavioral research to support conservation is very limited due. Soendjoto et al. (2005) conducted the newest behavioral research about the adaptation of proboscis monkeys to rubber plantations.

The river and water quality of riparian forest are important to conserve as they represent prime habitat for proboscis monkeys. The 500 m of riparian forest containing high tree species diversity and good sleeping trees are also important to conserve as habitat for the proboscis monkey. Conversely, in areas with low biodiversity such as cultivated land, the daily movements of proboscis monkey increase around 800–1000 m. Nevertheless, the uses of riparian forest for cultivation and village settlement have to be 500 m away from riverbank. This distance is necessary to protect the area of cultivation from disturbance by the proboscis monkey, which is very depended to water resources.

The conservation of proboscis monkey outside designated conservation areas is necessary to prevent the declining population from falling below the effective minimum population size. Even though hunting of proboscis monkey is rare, the declining population is caused by the development area of village settlement, cultivation, and fishpond.

Wildlife management is the responsibility of the Forestry Department. However, some habitats are the responsibility of local government such as the buffer zone of a national park. The buffer zones of national parks represent a strong habitat for endangered species including the proboscis monkey. However, the buffer zones are often adjacent to village settlements that support cultivation development and forest farming for economic efforts. Therefore, the community surrounding the buffer zone needs to be informed about the endangered status of the proboscis monkey and why we need to conserve this species.

Finally, the coordination programs of watershed management, swamp forest, mangrove forest, village settlements and cultivation have to be synergic each other in order to conserve habitat and proboscis monkey populations, both within and external to designated conservation areas. Therefore, a Steering Institution for conservation coordinated by Regional Development Planning Agency (*Bappeda*) with members of related local institutions, such as Forestry office, Estate Office, and others, needs to be developed.

Conclusions

The declining population of the proboscis monkey is accelerating simultaneously as habitat quality and habitat availability decrease. Over the last 10 years, proboscis monkeys have been found in scattered populations averaging 18–40 km from other populations. Nowadays, an average distance between scattered populations has increased to 50 km.

The group size of proboscis monkey varied based on habitat types. The group size was 10–25 individuals in mangrove forest downstream, six to ten individuals in riparian forest, 6–15 individuals in forest upstream, an average of 12.3 individuals in small islands. Group size was also affected by the habitat's carrying capacity and essential mineral content in the food resources. Daily movement of proboscis monkey in mangrove forest was about 500 m for foraging where they consumed 900 g/day of food or 270.25 g of dry weight, which contained 81.14% of leaf, 8.38% of fruit, and 7.68% of flower. The food choice was relatively close with mineral consumption, especially P, K, Ca, and Zn.

The carrying capacity habitat of proboscis monkey was the highest in mangrove forest, which supported 84 individual/km^2 of population with 778.68 kg of biomass. The estimated biomass of proboscis monkey in habitat could be done with study of height sitting (td) and width of body surface formulated as $W = 0.1324\ BW^{0.67}$, $W\ (\male) = 0.0514e^{0.0395SH}\ (r = 0.90)$, and $W\ (\female) = 0.1048e^{0.0662SH}\ (r = 0.87)$. The supply of food calorie was 120.68 kcal/kg of body weight, which was obtained by 16.5% of primary habitat productivity.

The adaptation of proboscis monkey to habitat disturbance was shown as the change in food species composition. Proboscis monkeys still choose foods with high mineral content and formed a strategic foraging of little group size around one to three adult individuals.

Based on the sensitivity of proboscis monkeys to the decline in habitat quality, evidenced on the low population density and smaller group size in lower quality habitat, implementation of conservation efforts in areas outside conservation areas, such as production forest and buffer zone development, are needed.

References

Alikodra HS, Mustari AH, Santosa N, Yasuma S (1995) Social interaction of proboscis monkey (*Nasalis larvatus*) group at Samboja Koala, East Kalimantan. Annual Report of Pusrehut

Bennett EL (1983) The banded langur: ecology of a colobinae in West Malaysian rain forest. Ph.D. Dissertation, Cambridge University, Cambridge

Bennett EL, Gombek F (1993) Proboscis monkey of Borneo. Natural History Publication (Borneo) Sdn.Bhd & Koktas Sabah Berhad, Kuala Lumpur

Bennett EL, Sebastian AC (1988) Social organization and ecology of proboscis monkeys (*Nasalis larvatus*) in mixed coastal forest in Sarawak. Int J Primatol 9(3):233–255

Bismark M (1986) Perilaku bekantan (Nasalis larvatus) dalam memanfaatkan lingkungan hutan bakau di Taman Nasional Kutai, Kalimantan Timor. Thesis Magister Sains, Progam Pascasarjana IPB, Bogor

Bismark M (1994) Analisis geometri tubuh bekantan (*Nasalis larvatus*). Bull Penelitian Hutan 561:41–52

Bismark M (1995) Kandungan mineral dalam pakan bekantan (*Nasalis larvatus*) di habitat hutan bakau. Kongres Nasional Biologi XI, Jakarta

Bismark M (1997) Pengelolaan habitat dan populasi bekantan (*Nasalis larvatus*) di Cagar Alam Pulau Kaget. Kalimantan Selatan. Diskusi hasil Penelitian, Pusat Litbang Hutan dan Konservasi Alam

Bismark M, Iskandar S (2002) Kajian total populasi dan struktur sosial bekantan (*Nasalis larvatus*) di Taman Nasional Kutai, kalimantan Timur. Bull Penelitian Hutan 631:17–29

Bismark M, Soerianegara I, Sastradipradja D, Suratmo FG, Alikodra HS, Pawitan H (1994) The potency of mangrove forest habitat to the proboscis monkey's source at Kutai National Park. East Kalimantan. International Primatological Society Congress, Bali, Indonesia

Chivers DJ (1974) The siamang in Malaya: a field study of a primate in tropical rain forest. Contrib Primatol 4:1–335

Curran LM, Trigg SN, McDonald AK, Astini D, Hardiono YM, Siregar P, Caniago T, Kasischke E (2004) Lowland forest loss in protected areas of Indonesian Borneo. Science 303:1000–1003

Durand M, Kawashima R (1980) A general account of the mangove of Princess Charlotte Bay with particular reference to zonation of the open shoreline. In: Teas HJ (ed) Tasks of vegetation science. W. Junk Publishers, The Hague, pp 37–46

Freeland WJ (1976) Pathogens and the evolution of primate sociality. Biotropica 8(1):12–24

Happel RE, Noss JF, Marsh CW (1987) Distribution, abundance, and endangerment of primates. In: Marsh CW, Mittermeier RA (eds) Primate conservation in the tropical rain forest. Alan R. Liss. Inc., New York, pp 63–82

Hladik CM (1978) Adaptive strategies of primate in relation of leaf eating. In: Mongomery GG (ed) The ecology of arboreal folivores. Smithsonian Institution Press, Washington, DC, pp 373–395

Iwamoto T (1982) Food and nutritional condition of free ranging Japanese monkeys on Koshima Islet during winter. Primates 23(2):153–170

Ma'ruf A (2004) Studi perilaku bekantan (*Nasalis larvatus*) di daerah Balik Papan dan Sekitarnya. Lap. Penelitian Loka Litbang Primata, Samboja

Ma'ruf A, Triatmoko, Syahbani I (2005) Studi populasi bekantan (*Nasalis larvatus*) di Muara Sungai Mahakam, Kalimantan Timur. Lap. Penelitian Loka Litbang Primata, Samboja

MacKinnon K (1986) Alam Asli Indonesia: Flora, Fauna dan Keserasian. Gramedia, Jakarta

Manangsang J, Traylor-Holzer K, Reed D, Leus K (2005) Indonesian proboscis monkey population and habitat viability assessment: final report. IUCN/SSC Conservation Breeding Specialist Group, Apple Valley, MN

Matsubayashi K, Sayuthi D (1981) Microbiological and clinical examination of cynomolgus monkeys in Indonesia, In: Kyoto University Overseas Report of Studies on Indonesian Macaque. Kyoto Univ. Primate Center Institute, pp 47–56

McNeely JA, Miller KR, Reid WV, Mittermeier RA, Werner TB (1990) Conserving the world's biological diversity. IUCN, Gland, Switzerland

Meijaard E, Nijman V (2000) Distribution and conservation of proboscis monkey (*Nasalis larvatus*) in Kalimantan Indonesia. Biol Conserv 92:15–24

Mitch WJ, Gosselink JG (1984) Wetlands. Van Nostrand Reinhold Comp, New York

Moen AN (1973) Wildlife ecology: an analytical approach. W.H. Freeman and Company, San Francisco

Montheith M, Unsworth M (1991) Principles of environmental physics, 2nd edn. Edward Arnold, London

Oates JF (1978) Water-plant and soil consumption by guereza monkeys (*Colobus guezera*): a relationship with mineral and toxic in diet. Biotropica 10(4):241–253

Rijksen HD (1978) A field study of Sumatra: ecology, behavior and conservation. Vienman, Wageningen

Rodman PS (1978) Diets, densities and distribution of Bornean primates. In: Montgomery GG (ed) The ecology of arboreal folivore. Smithsonian Institution Press, Washington, DC, pp 465–478

Ruhiyat Y (1986) Preliminary study of proboscis monkey (*Nasalis larvatus*) in Gunung Palung Nature Reserve, West Kalimantan. In: Kyoto University Overseas Research Report of Studies on Asian non-human primates, No. 5. Kyoto Univ. Primates Research. Inst. Kyoto, pp 59–69

Salter RE, MacKenzie NA, Nightingale N, Aken KM, Chai PK (1985) Habitat use, ranging behavior and food habits of proboscis monkey *Nasalis larvatus*. Primates 26(4):436–451

Soendjoto MA (2003) Adaptasi bekantan (*Nasalis larvatus*) terhadap hutan karet: Studi kasus di Kabupaten Tabalong, Kalimantan Selatan. *Usulan Penelitian*. Progam Pasca Sarjana, Institut Pertanian Bogor

Soendjoto AM, Alikodra HS, Bismark M, Setijanto H (2005) Vegetasi tepi baruh pada habitat bekantan (*Nasalis larvatus*) di hutan karet Kabupaten Tabalong, Kalimantan Selatan. Biodiversitas 6(1):40–44

Soendjoto AM, Alikodra HS, Bismark M, Setijanto H (2006) Jenis dan komposisi pakan bekantan (*Nasalis larvatus*) di hutan karet Kabupaten Tabalong, Kalimantan Selatan. Biodiversitas 7(1):34–38

Supriatna J, Manangsang J, Tumbelaka L, Andayani N, Indrawan M, Darmawan L, Leksono S, Djuwantoko M, Seal U, Bryers O (2001) Conservation assessments and management plan for the primates of indonesia: final report. Conservation Breeding Specialist Group (SSC/ IUCN), Apple Valley, MN

Supriatna J. (2004) Primates of Sundaland and Wallacea: A review of threats and conservation efforts. Folia Primatologica 75:206

Suzuki A (1986). The socio-ecological study on the orangutan in the Mentoko-BT Sinara Study Area, in Kutai National Park, East kalimantan. In: Kyoto University Overseas Research Report of Studies on Asia non-human primates No. 5. Kyoto Univ. Primate Research. Institute, Kyoto, pp 23–28

Wheatley BP (1982) Energetics of foraging in *Macaca fascicularis* and *Pongo pygmaeus* and selective advantage of large body size in the orangutan. Primates 23(3):348–363

Wilson CC, Wilson WL (1975) The influence of selective logging on primates and some other animals in East Kalimantan. Folia Primatol 23:245–274

Yasuma S (1989) The present situation of proboscis monkey *Nasalis larvatus*, concern with it's distribution in and around the Bukit Soeharto protection Forest. Tropical Rain Forest Research Project. JICA Report

Yeager CP (1989) Feeding ecology of the proboscis monkey (*Nasalis larvatus*). Int J Primatol 10(6):497–529

Yeager CP (1990) Proboscis monkey (*Nasalis larvatus*) social organization group structure. Am J Primatol 20:95–106

Yeager CP (1992) Changes in proboscis monkey (*Nasalis larvatus*) group size and density at Tanjung Puting National Park, Central Kalimantan, Indonesia. Trop Biodivers I 1:49–55

Yeager CP, Blondal TK (1992) Conservation status of proboscis monkey (*Nasalis larvatus*) at Tanjung Puting National park, Kalimantan tengah Indonesia. Forest Biology and Conservation in Borneo. Center for Borneo Studies Publication 2:133–137

Chapter 13
Pests, Pestilence, and People: The Long-Tailed Macaque and Its Role in the Cultural Complexities of Bali

Kelly E. Lane, Michelle Lute, Aida Rompis, I. Nengah Wandia, I.G.A. Arta Putra, Hope Hollocher, and Agustin Fuentes

Introduction

Bali's unique religious tradition established the foundations of a system of island-wide rice agriculture that is organized around interconnected water temples, known as *subaks* (Lansing 2007). This temple-oriented rice agricultural system was well established at the time of the Dutch Colonization of Indonesia, approximately 500 years ago, and has remained relatively stable since that time with *subaks* and traditional rice agriculture practices enjoying renewed success within the last 30 years since Indonesia's Green Revolution (Wheatley 1999; Lansing 2007). Historically, the island's temples have effectively governed and coordinated the timing of planting, irrigating, and harvesting rice, preventing water shortages and disease outbreaks in doing so. As the site for religious, agricultural, and cultural events, the temple system acts as the cohesive and organizing power for whole villages. It is this interwoven framework of temples, rice fields, villages, and associated forest patches that forms the complex cultural and physical landscape in which Balinese long-tailed macaques (*Macaca fascicularis*) exist (Fuentes et al. 2005).

Bali macaques and their close association with humans have intrigued biologists, anthropologists, ecologists, and artists since the beginning of the twentieth century (Wheatley 1999). Early anthropological work focused largely on macaque behavior and the interconnectedness of macaques in the traditional workings of the Balinese culture. The multifaceted and intimate nature of the human-macaque relationship on Bali has led researchers now to take a more interdisciplinary approach and consider religion, culture, and biology simultaneously to provide additional insights into important issues that arise uniquely at the intersection between these two species. Current research on macaques in Bali includes work on behavior, human-macaque interactions, population genetics, reproductive and dietary endocrinology, obesity, the role of macaques in the human social context of the Balinese, and most recently

A. Fuentes (✉)
Department of Anthropology, University of Notre Dame, Notre Dame, IN 46556-5611, USA
e-mail: afuentes@nd.edu

S. Gursky-Doyen and J. Supriatna (eds.), *Indonesian Primates*,
Developments in Primatology: Progress and Prospects,
DOI 10.1007/978-1-4419-1560-3_13, © Springer Science+Business Media, LLC 2010

pathogens and infectious diseases (Fuentes 2006; Fuentes et al. 2005; Jones-Engel et al. 2005; Lane et al. in review a, b; Wandia 2007).

Macaque Geographical Distribution

Macaques are one of the most successful primate radiations, second only to humans. As many as 21 species of macaque have been described to date, clustering into four main groups: the sinica group, the arctoides group, the silenus-sylvanus group, and the fascicularis group. The long-tailed macaque (*Macaca fascicularis*) and its closest relative the rhesus macaque (*Macaca mulatta*) are classified within the fascicularis group (Thierry 2007). Although there is ongoing dispute about the classification of certain macaque species, the placement of *Macaca fascicularis* is well established (Fooden 1995; Fa and Lindburg 1996).

The long-tailed macaque has one of the largest distributions of nonhuman primates, second only to rhesus macaques. The range extends north through Burma, Thailand, and Vietnam and south throughout the Indonesian archipelago, forming hybrid zones with rhesus macaques at its northern boundaries (Fooden 1995; Tosi et al. 2002). Long-tailed macaques thrive in all available habitat zones throughout the region, and it is this ability to persist across wide-ranging environmental variation that has contributed to the evolutionary success of the species throughout Southeast Asia. Populations of long-tailed macaque are often comprised of one to four matrifocal groups, consisting of between approximately 30 and 50 individuals per group (Fooden 1995).

The Bali Macaque

On the island of Bali (Fig. 13.1), at least 43 well-established monkey forest populations exist (Southern 2002; Fuentes et al. 2005). Variation between the populations exists in local habitat, climate as well as water and food resources, group size, responsiveness to humans, arboreality, provisioning, and, to some extent, morphology (see Table 13.1 for monkey forest details). Regional and local distinctions in human socioeconomic status and education affect the human interface with local ecologies probably contributing to differences in the characteristics of macaque populations. The largest, most established populations on the island, Padangtegal-Ubud, Uluwatu, Sangeh, Pulaki, and Alas Kedaton, are all established tourism sites and have routine, large-scale provisioning of food by a paid staff. The smaller, more transient populations, such as Selumbung and Kuning, are located in areas where provisioning occurs only during temple ceremonies and macaques there are often considered pests by local villagers. These smaller populations are often along the island's periphery – to the north, east, and west – while the largest temple populations are located predominantly in the central core of the island, along with most other tourist destinations and the largest human populations. The macaques of Bali live in forest patches in disparate ecological contexts from coastal, dry scrub habitats and seasonal rainforest to isolated beaches and the summit of an active volcano (Whitten et al. 1997; Fuentes et al. 2005).

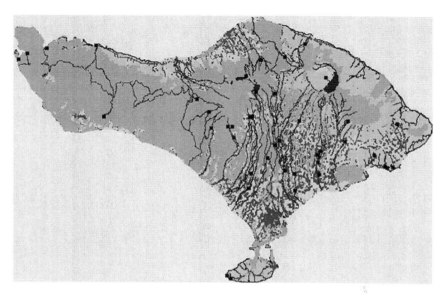

Fig. 13.1 In this GIS map, forest land, wet rice agricultural areas, city and village areas and riversand lakes are represented. The dark squares represent the monkey temple forest sites. Adapted from Southern (2002)

An Anthropogenic Ecology

Physiological and behavioral plasticity make macaques successful in a variety of habitats, but especially in human-altered habitats. Macaque behavior and demographic patterns on Bali are modified in many ways by human activity with macaque groups exploiting human food sources, becoming less arboreal, manipulating nonfood objects more, and experiencing higher population sizes, birth rates, and lower mortality rates as compared to macaques found in other parts of the range (Fuentes et al. 2005). While other primates are more limited in their distributions, including other macaque species (i.e. *M. assamensis* or *M. munzala*), due to their specific and limited habitat and/or dietary requirements (Thierry 2007), the gut of the long-tailed macaque is simple, capable of digesting many foods rapidly, and in many ways, reminiscent of the human gut. Further, Bali macaque morphology varies with both East-West and altitudinal gradients (Southern 2002). Variation in morphology is most notable in the pelage of the macaques, with populations living in the easternmost stretches of the island as well as at high elevations often displaying a darker, thicker coat than their counterparts throughout the remainder of the island (Southern 2002). While intriguing, this variation in pelage, and any further variations in morphology, has yet to be fully explored.

Provisioning has enhanced the growth of monkey forest populations overall. In large temple sites with organized management teams, provisioning has transitioned from simple offerings, comprised of flowers, eggs, and locally available fruits and nuts, to large scale, routine feedings of carrots, potato, and sweet potato. Bananas and peanuts from tourists make up a substantial portion of the diet of these macaques.

Table 13.1 Details describing the geophysical location and environmental components of monkey forests across Bali. NDC indicates no data collected for this site. Adapted from Lane (unpublished data)

Population	N	Geophysical location	Habitat type	Proximity to humans	Provisioning rate	Proximal water source
Pedangtegal	350	8.519 S, 115.258 E	Forest, rice	High	Daily, large	Stream, sawah
Sangeh	300	8.479 S, 115.206 E	Forest, rice	High	Daily, large	Stream, sawah
Pelaga	NDC	8.289 S, 115.230 E	NDC	NDC	NDC	NDC
Uluwatu	300	8.829 S, 115.084 E	Dry scrub	High	Daily, large	Ocean, temple pool
Batur	25	8.240 S, 115.386 E	Volcano apex	High/medium	Daily, small	Vent steams
Cekik	50	8.189 S, 114.444 E	Dry forest	Low	NDC	NDC
Penyaringan	NDC	8.360 S, 114.706 E	NDC	NDC	NDC	NDC
Muncan	NDC	8.432 S, 115.445 E	NDC	NDC	NDC	NDC
Angseri	40	8.356 S, 115.161 E	Bamboo, rice	Medium/low	Ceremonial	Stream, sawah
Bedugul	200	8.238 S, 115.143 E	Roadside	Medium	Daily, small	Vendors, channels
Batu Pageh	45	8.838 S, 115.093 E	Dry scrub	Low	Ceremonial	Ocean
Kuning	45	8.468 S, 115.361 E	Forest ravine	Low	Ceremonial	Stream, sawah
Tegal Bangli	NDC	8.491 S, 115.359 E	NDC	NDC	NDC	NDC
Pulaki	200	8.146 S, 114.682 E	Dry	Medium	Daily, mid	Sea, vendors
Teluk Terima	35	8.156 S, 114.530 E	Dry scrub	Low	Ceremonial	Temple pool
Tejakula	50	8.140 S, 115.320 E	Dry forest	Medium	Daily, small	Temple pools
Tegalalang	NDC	8.370 S, 115.030 E	NDC	NDC	NDC	NDC
Gilimanuk	NDC	8.171 S, 114.471 E	NDC	NDC	NDC	NDC
Besakih	NDC	8.368 S, 115.454 E	NDC	NDC	NDC	NDC
Selumbung	NDC	8.470 S, 115.532 E	NDC	NDC	NDC	NDC
Alas Nenggan	50	8.571 S, 115.149 E	Forest, rice	Medium/low	Daily, small	Stream, sawah
Alas Kedaton	350	8.529 S, 115.156 E	Forest, rice	High	Daily, large	Temple/stream
Mesahan	NDC	8.181 S, 115.260 E	NDC	NDC	NDC	NDC
Lemukih	NDC	8.189 S, 115.187 E	NDC	NDC	NDC	NDC
Batumadeg	NDC	8.377 S, 115.601 E	NDC	NDC	NDC	NDC
Bukit Gumang	100	8.505 S, 115.590 E	Dry, tamarind	High	Daily, small	Vendors, offerings
Mekori	60	8.392 S, 115.035 E	Lush forest	Medium	Ceremonial	Vendors, channels
Pejuritan	30	8.306 S, 114.990 E	Forest ravine	Low	Ceremonial	Offerings, sawah
Lempuyang	70	8.237 S, 115.383 E	Cloud forest	Medium	Daily, small	Offerings

Table 13.2 Comparison of activity budgets for *Macaca fascicularis* at one highly provisioned site on Bali and five other sites across Southeast Asia. Adapted from Fooden (1995), van Noordwijk (1985), and Fuentes et al. (2005)

	Feed	Travel	Other
Bali (Padangtegal)	23%	12%	58%
Sumatra	15%	26%	59%
Mauritius	32%	4%	64%
Malaysia	35%	20%	45%
Kalimantan	13%	45%	42%
Bangladesh	39%	9%	53%

Organized provisioning has resulted in an increase in the occurrence of macaque obesity, macaque terrestriality and habituation to humans, and human-macaque interactions. This change in provisioning results in transitions in macaque social dynamics within these populations and contributes to broad scale changes in macaque behaviors and population dynamics. For example, large, highly-provisioned populations have had documented decreases in male emigration, increases in temple-licking, and macaque initiated interactions with humans (Fuentes, in press). In addition, provisioned groups on Bali tend to travel less and spend more time feeding and resting as a result of decreased pressure to locate and obtain food (Table 13.2).

Macaques are not the only primate populations to be affected by the monkey forests, which also impact human communities in the surrounding areas. Many surrounding communities benefit economically from tourism associated with monkey forests. In addition to the direct benefit from voluntary donations or entry fees, the community benefits economically from the influx of money spent in restaurants, shops, hotels, and taxi-stands. While urban development is ongoing on Bali, development surrounding the largest temple structures has resulted in the conversion of hundreds of hectares of rice fields into housing, shops and hotels, and roads over the last decade. While an economic benefit can stem from the presence of a monkey forest in a community, the associated wealth distribution is not equally spread throughout the local community, exiting monkey forests, or the island. Most monkey forests draw no tourists at all, due to either their small size, elusive or transient macaque populations, or remoteness. And, it is these communities that often suffer the greatest amount of crop raiding and "negative" human-macaque interactions resulting in increases in injuries, including bites and scratches. Owing to the economic disparity between monkey forest populations of large and small size, communities throughout the island not traditionally seen as tourist destinations are actively seeking to expand temple infrastructure in an effort to lure additional resources into their community. If successful, these expansions at monkey forests will undoubtedly impact the population dynamics of the macaques island-wide by increasing population sizes, possibly altering behavioral patterns, and likely increasing the intensity of human-macaque interactions.

While traditional Balinese Hindu doctrine plays a role in influencing the view of many Balinese toward nature, including macaques, the relationship between the Balinese and the macaque is not a simple one of protection of the sacred

(Lane et al. in review a; Fuentes et al. 2005). Contextualized by complex parameters, including location, behavior, and human-derived benefit or cost, most Balinese view macaques with an accepted tolerance (Loudon et al. 2006; Lane et al., in review, a). Many individual Balinese feel that while macaques are "sacred" due to their prominent place in Balinese Hindu mythology, individual macaques can be considered a nuisance. For example, populations of macaques are generally tolerated, even "respected," but individual macaques, especially those found crop raiding, are dealt with severely (Lane et al. in review a). When the macaques are on the grounds or in the vicinity of a temple, they are respected and are even treated as sacred or "suci" by many (Fuentes et al. 2005; Loudon et al. 2006; Lane et al. in review a). Moreover, while considered by many to be unpalatable, individuals in certain regions of Bali acknowledge hunting and eating macaques for food, further weakening the argument for ubiquitous sacredness (Lane et al. in review a).

Bali macaques are exposed to contrasting levels and types of human interaction, ranging from feeding or provisioning directly to hunting and eating and from keeping macaques as pets to physical conflict between humans and macaques. Moreover, Balinese visitation to monkey forests ranges from daily to extremely intermittently, with some populations visited only during bi-annual ceremonies. However, tourists often visit only specific macaque populations, visit throughout the year, in massive numbers, and without regard to the ceremonial schedule of the local community. The gradient of human interaction, then, is from extremely limited to extremely prevalent, and it is through this gradient that macaques must navigate.

Population Dynamic Structure: Dispersal

Two ubiquitous and characteristic behavioral aspects of macaques are male-biased dispersal and extreme female philopatry. It has been suggested that many females may never leave their natal groups (Melnick 1990; Napier and Napier 1967) while males begin dispersing when they reach sexual maturity around 7 years of age. Distances traveled, frequencies, and duration of dispersals are still poorly understood for macaques. While it is assumed that most dispersing males attempt entry into one of the nearest neighbor populations, it is well understood that males are capable of long-distance dispersal and males may disperse multiple times within their lifetimes (Southern 2002).

Female macaques remain within the natal population, in groups consisting of largely related females. Within the group, matrifocal lineages determine the dominance of individual females, and the mother's rank is passed down to her daughters, with the youngest female outranking the older sisters. Large populations of Bali macaques at monkey forests can be comprised of as many as four distinct groups, sharing and competing for space, rank, and food resources. This has led to a potential increase in inter-group aggression and may result in modifications to dispersal patterns, either through a decrease in sub-adult male dispersal or through a decrease in emigration/immigration success (Fuentes et al., in review).

Some populations, especially in high tourist areas such as Padangtegal, in Cenrtal Bali, have low rates of male dispersal most likely due to the ease of resource acquisition from provisioning. For example, from 1998 to 2002, 14 males made observed dispersal attempts from this population. This average of 2.8 males per year is lower than expected given the large number of subadult males per annum in the population (>25). In this same population, only one successful immigration was observed during this time period. Acceptance into a new group is often preceded by mutual observation from a distance and occasional closer interactions that may last for period of up to several months. Preliminary evidence suggests that macaque dispersal is characterized by shorter distances, often to neighboring groups within similar habitats to the natal group (Lane et al. unpublished data). Utilizing riparian forest corridors, such dispersal reduces the risks associated with long-distance dispersal. The relative heterogeneity of the landscape may play a role in determining the distance of dispersal (Kennedy et al., in press). High levels of landscape homogeneity may facilitate dispersal as it increases the dispersing male's familiarity with how to successfully traverse the landscape.

Phylogeny and Population Structure

Extreme female philopatry and male-biased dispersal significantly impact the population structure of the macaques (Melnick and Hoelzer 1992; Tosi et al. 2000, 2003). Despite lacking a fully resolved mechanism for this behavior, the result of these behavioral patterns on the genetic structure of macaques has been well elucidated throughout the macaques' range. Melnick and Hoelzer (1992) report that across the entire macaque range, nuclear genes disperse freely, barring any major geophysical barrier, while mitochondrial genome variation is relatively homogenous within populations but highly divergent between regional populations. More recently, comparisons of regions of the Y chromosome and mitochondrial genome have demonstrated significant differences even among assignment of species to macaque species groups, further arguing that the strongest explanatory mechanism for the trends found was the pattern of female philopatry and male dispersal exhibited by all macaque species (Tosi et al. 2000, 2003).

Preliminary sequence evidence parallels the social dynamics of Bali macaques (Lane et al. unpublished data). The two gene genealogies presented in Fig. 13.2 represent the genetic variation across individual macaques sampled at eleven monkey forest populations on Bali. While the mitochondrial phylogeny displays clustering of samples by geography, the Y-chromosomal phylogeny displays no distinguishable pattern. This supports the behaviorally documented strong female philopatry and male dispersal in macaques.

We evaluated the genetic structure of several large populations of macaque across Eastern Java, Bali, and Lombok using microsatellites (Wandia 2007). While F_{st} values of populations on Java and Lombok are relatively low (0.039 and 0.023, respectively), indicating high levels of dispersal within these islands, the F_{st} value

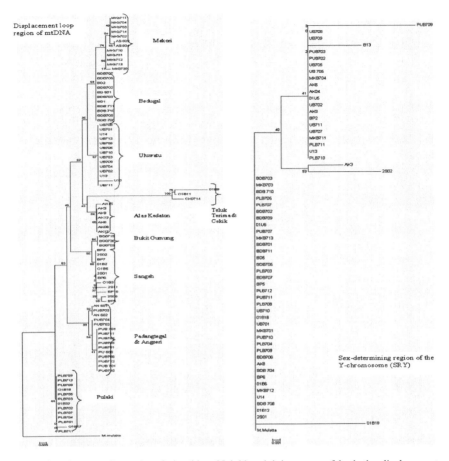

Fig. 13.2 Diagram of genetic relationships. Neighbor joining trees of both the displacement loop (dloop) of the mitochondrial genome (*right*) and the sex-determining region (SRY) of the Y chromosome. Bootstrap consensus values are presented. Note the geographical clustering of mtDNA and the complete lack of clustering in the Y chromosome

across the six populations examined on Bali is significantly greater (0.123), supporting the hypothesis of some philopatry even among males (Wandia 2007). In spite of the larger F_{st} values for Bali, island-wide dispersal, while potentially limited, can and does occur. However, dispersal between islands in this system, while historically possible, is much less likely to be successful. In addition, Lombok's overall lower population numbers would, over time, result in higher levels of inbreeding, while the numerous macaques and macaque populations across Bali would support higher levels of population differentiation.

Within the Bali macaque sites, Nei's standard genetic distances were measured (Nei 1972) and ranged from 0.129 (Alas Kedaton to Bukit Gumang) to 0.384 (Pulaki to Uluwatu; see Table 13.3). The genetic distance between Pulaki, located on the extreme northwestern reaches of the island, and Uluwatu, located at the southernmost

Table 13.3 Nei's standard genetic distances (D_s) and geographic distance (km) between local monkey forest populations on Bali. Adapted from Wandia (2007)

	Pulaki		A. Kedaton		Uluwatu		Sangeh		Ubud		B. Gumang	
	D_s	km	D_s	km	D_s	km	D_s	km	D_s	km	D_s	km
Pulaki	0.00	0.00										
A. Kedaton	0.214	71.93	0.00	0.00								
Uluwatu	0.384	126	0.293	57.49	0.00	0.00						
Sangeh	0.226	63.91	0.221	13.03	0.353	69.09	0.00	0.00				
Ubud	0.268	74.43	0.132	18.4	0.223	66.22	0.226	11.78	0.00	0.00		
B. Gumang	0.357	114.8	0.129	78.82	0.335	110.6	0.267	69.81	0.147	60.46	0.00	0.00

point of the island, likely represents the geophysical distance between these populations (see Fig. 13.2 for population locations). Indeed, comparing genetic distances to the geographic distances between a majority of Bali sites demonstrates a significant, high level of correlation (Mantel test, 0.6284 Pearson correlation coef., $p < 0.01$). Specifically, the interactions demonstrating the greatest level of correlation between genetic and geographic distances involved interactions between either the westernmost population (Pulaki) or the easternmost population (Bukit Gumang). The genetic distances between large macaque populations thriving in the island's core (Alas Kedaton, Sangeh, and Ubud) are too large to be explained by geophysical distance alone and more likely are a result of the heavy anthropogenic modifications to the landscape that have occurred throughout the last centuries. The heterogeneity of the landscape in this region, including many riparian zones, forest patches, and villages, has likely contributed to a fragmentation in the landscape that has limited male dispersal.

Currently, we are undertaking genetic analysis examining both neutral loci and loci under selection from pathogen pressure to fill in the gaps in our current overview of the genetic structure of the macaque populations on Bali. Preliminary evidence supports decreases in heterozygosity in the largest populations of macaques and an increase in rare alleles in the more geographically isolated populations (Lane et al. unpublished data). Coupled with a more refined study of the landscape, this expanded analysis will allow us to uncover the patterns of corridors and barriers to macaque dispersal throughout the island as well as more fully understand the impact of long term anthropogenic landscape alterations on macaque population structure.

Population Management and Conservation Genetics

While many primates are threatened or endangered and suffering ill-effects of habitat degradation and population size reductions, the macaques of Bali are thriving. The success of the long-tailed macaque, in Bali and in other anthropogenically-influenced landscapes, lends itself nicely to predictive modeling, the results of which can be applied to other, less successful systems. One approach to understanding the impact of a complex anthropogenic landscape on the movement and population structure of macaques has been to investigate that impact via the incorporation of

GIS data into an Agent-based model (ABM). While ABMs have been on the ecological scene well over 20 years (Grimm and Railsback 2005), they have only recently begun to be incorporated more readily into broader biological contexts. The utilization of this model, in this and future biological environments, will prove extremely useful in efforts to make predictions of genetic barriers and corridors between monkey forest populations. This tool will provide especially relevant information for conservation management efforts across the island. Through the inclusion of humans as a component of the landscape, we will be able to garner a more powerful understanding of how human attitudes, resource consumptions, and densities facilitate or inhibit macaque movement across the island. With the inclusion of macaque genetic distances as a GIS data layer, we can begin to tease apart the impacts of population dynamics on emigration from the historical landscape change as well as begin to understand how current anthropogenic change can impact the future population structure of the Bali macaque.

To test our predictions of the development of landscape driven corridors and barriers to dispersal, we used an ABM of Bali which incorporated ground confirmed GIS layers as the modeling environment and tracked macaque movement via their ability to transmit a gastrointestinal infection to other macaques across the island. Three macaque populations were chosen as sites of initial infection based on their geophysical positions, and we tested the impact of landscape in inhibiting or enhancing infection as well as identified the specific landscape features most influential in transmission. We found that the landscape significantly impacts macaque dispersal, and thus, the ability of a pathogen to spread between populations. Moreover, we determined that the incorporation of landscape features, per se, do not always inhibit macaque movement and infection. Rather, it is the homogeneity or heterogeneity of the landscape that contributes to the inhibition or enhancement of dispersal opportunities (Kennedy et al., in press). We determined that an anthropogenically fragmented landscape can minimize parasite transmission by serving as a boundary to macaque movement, contrary to what has been previously accepted by disease ecologists and conservationists. Compounded by population density increases (in macaques and/or humans), habitat encroachment, and limited access to resources, the significance of the landscape on macaque dispersal and parasite transmission opportunities should not be underestimated in the development of long-term management strategies for the Balinese macaque.

Humans and Macaques on Bali: Ethnoprimatology and Shared Ecologies

The coexistence of humans and macaques in the Bali landscape has influenced both human and macaque dynamics. Respect for the environment, including monkey forests and macaques results in protection and, in some cases, provisioning of the macaque groups that live in these monkey forests. This has resulted in increased population sizes as well as changes to macaque behavior and ranging patterns.

All things considered, macaques have historically benefited from their relationship with humans although this trend may not continue indefinitely. Humans, in turn, have both benefited and suffered in their relationship with macaques, via the improvement to local economies with increased revenue and job opportunities as well as the increase in human-macaque interactions associated with monkey forests.

Human-wildlife contact, via direct or indirect interaction, has been suggested to be one of the most powerful influences for reported human feelings of respect toward wildlife (Wilson 1984). Moreover, significant health benefits have been reported from human interaction with pets (CDC 2008). In Bali, and throughout Indonesia, it is not uncommon for wild, juvenile macaques to be pets in the family home. Again, while the health benefits of pets are well documented, the risks associated with primates as pets may outweigh the benefits. For example, Jones-Engel and colleagues (2005) examined the amount of gastrointestinal parasite sharing among villagers and their pet macaques and found that there is substantial exchange occurring. The directionality (human to macaque) of this transmission, however, potentially puts wild macaque populations at risk as it is extremely common for pet macaques to be released into the wild once adulthood is reached. For the communities surrounding monkey forests, direct and indirect interactions with the long-tailed macaque is largely unavoidable regardless of pet ownership. In addition to macaques living near temples, macaques frequently travel beyond the bounds of the forest patch. Interactions can have a multifaceted impact on Balinese communities and potentially result in the sustained acceptance of context-associated sacredness (Lane et al. in review a; Loudon et al. 2006).

Direct interactions with wildlife can also result in a multitude of negative outcomes. These interactions can range from fearful or stressful exchanges, resulting in a hormonal cascade including increases or dramatic fluctuations in cortisols, and their associated long term negative effects, to bites and scratches, and their associated pathogen transmission risks. While uncommon, more serious altercations between macaques and humans have occurred, resulting in serious injuries requiring sustained medical attention. Two groups are most at risk from these interactions: tourists and native Balinese that frequent monkey forests. The first category is more likely to sustain bites, scratches, and potentially more serious injuries due to the lack of familiarity with wild primates. The latter group is more likely to be exposed to the variety of pathogens harbored by wild macaques. Thus, it is in this group that the transmission potential from macaque to humans is greatest, potentially resulting in a novel infectious disease emergence. As recently as 2005, simian foamy virus was documented to have infected a few humans working in monkey forests on Bali (Jones-Engel et al. 2005).

The risk for macaques of infection with a human pathogen is often more significant than the risk for humans of infection with a wildlife pathogen (Jones-Engel et al. 2005; Hudson et al. 2001). However, this focus on emerging infectious diseases does not fully address the risk of infectious diseases throughout the island, especially among macaques. The long-term shared evolutionary history as well as the generalized physiologies, morphologies, and ecologies of humans and macaques create unique pathogen sharing opportunities. More specifically, unlike the generalized risk of accidental infection of a wildlife parasite into a human population or individual, the striking similarities between long-tailed macaques and humans allow for even moderately host-specific parasites to infect both humans and macaques.

While human-macaque interactions discussed earlier contribute to these bi-directional pathogen transmission opportunities, environmental factors can also play a direct role in increasing the exposure risk of both human and macaque. Specific guilds of parasites, such as gastrointestinal or vector-borne parasites, are often the most heavily influenced by environmental fluctuations and perturbations. Further, specific parasites, even within a single guild, are significantly influenced by the complex, anthropogenic landscape of Bali (Table 13.4). For example, while common helminth species, including ascarids and hookworms, are associated with macaque populations living in bamboo forests, common protozoa, such as *Entamoeba* spp. and *Giardia lamblia*, are more frequently associated with dry forest habitats.

Ongoing research into the spatial patterns of the gastrointestinal parasites of Bali macaques has uncovered some provocative patterns (Lane et al. in review b). When the impact of environmental factors, including water availability, habitat type, and food availability, on the prevalence of 19 individual parasites was examined, most parasites were most significantly associated with an environmental factor not traditionally associated with high parasite prevalence. For example, while water

Table 13.4 Significant impact of environmental factors on specific parasites, as determined by ANOVA. Note that while most parasites do not respond according to traditional parasitology paradigms, such as increases in prevalence with high water availability, in every instance, at least one parasite does. All categories reported represent categories associated with significant increases in prevalence. Adapted from Lane et al. (in review b)

Parasite	Habitat type	Water availability	Food availability
Ascaris	Bamboo	Moderate	Ceremonial
Ancylostoma	Bamboo	Moderate	Ceremonial
Taenia	Dry forest	Low	Low
Stronglyoides	Rice	–	Low
Enterobius	–	–	–
Trichuris	–	–	–
Acanthocephala	Rice	–	–
Paragonimus	Dry forest	Low	Low
Alaria	–	Low	Low
Entamoeba	Dry forest; scrub	Low	Ceremonial, low, high
Giardia	Dry forest; bamboo	–	
Isospora	Dry forest	–	Moderate
Endolimax	Dry forest; wet forest	–	Low, Moderate
Iodamoeba	–	–	–
Cryptosporidia	Bamboo	High	Ceremonial
Balantidium	–	–	–
Blastocystis	Dry forest	Moderate	Moderate
Trichomonas	Bamboo	Moderate	Ceremonial
Retortomonas	–	–	–

availability is considered a critical component to the successful transmission of a water-borne parasite such as *G. lamblia*, its prevalence among macaque populations was not significantly influenced by the availability of water and, more interestingly, *G. lamblia* burdens significantly increased in macaque populations existing in both dry forest and bamboo forest habitats (see Table 13.4). Moreover, at the population level, water availability played no significant role in determining mean parasite intensity or richness. This research not only suggests complex drivers of parasitism in this system, it hints at the broader role of macaques as agents of gene flow and parasite transmission in this pathogenic landscape.

Conclusion

Balinese macaques thrive amidst dense and complex anthropogenic landscapes. Despite this and the abundance of existing knowledge on behavior, diet, and evolutionary history of these macaques, fine scale studies are required to explore the influence of this landscape and this level of human overlap in order to understand how the human landscape limits or enhances macaque population dynamics. Ongoing research is focused on how the complex anthropogenic landscape influences macaque population structure and how that, in turn, influences the spatial patterns of pathogens on Bali. Further work explores the dynamics of obesity in specific macaque populations, a problem likely related to large-scale provisioning of poor quality food. The Balinese environment has been engineered over millennia, and all along humans and macaques have interacted on this ever-changing medium. This historical and contemporary interplay between the humans and macaques makes this system one of the most informative, interesting, and engaging places to study such interactions.

References

CDC (2008) http://www.cdc.gov/HEALTHYPETS/health_benefits.htm

Fa JE, Lindburg DG (1996) Evolution and ecology of macaque societies. Cambridge University Press, Cambridge

Fooden J (1995) Systematic review of Southeast Asian long-tailed macaques, *Macaca fascicularis*. Fieldiana, Zoology series no. 81. Field Museum of Chicago Publication

Fuentes A, Southern M, Suaryana KG (2005) Monkey forests and human landscapes: is extensive sympatry sustainable for *Homo sapiens* and *Macaca fascicularis* in Bali? In: Patterson J (ed) Commensalism and conflict: the primate-human interface. American Society of Primatology Publications

Fuentes A (2006) Human culture and monkey behavior assessing the contexts of potential pathogen transmission between macaques and humans. Am J Primatol 68:880–896

Fuentes A, Rompis ALT, Arta Putra IGA, Watiniasih NL, Suratha IN, Soma IG, Wandia IN, Harya Putra IDK, Stephenson R, Selamet W (in press) Macaque behavior at the human-monkey interface: the activity and demography of semi-free ranging Macaca fascicular is at Padangtegal, Bali, Indonesia. In Gumert M, Jones-Engel L and Fuentes A. "Managing commensalism in long tailed macaques: The human-macaque interface." Cambridge University Press.

Grimm V, Railsback SF (2005) Individual-based modeling and ecology. Princeton University Press, Princeton, NJ

Hudson PJ, Rizzoli AP, Grenfell BT, Hesterbeek JAP, Dobson AP (2001) The ecology of wildlife diseases. Oxford University Press, Oxford

Jones-Engel L, Engel GA, Schillaci MA (2005) An ethnoprimatological assessment of disease transmission among humans and wild and pet macaques on the Indonesian island of Sulawesi. In: Patterson J (ed) Primate commensalism: the primate-human interface. American Society of Primatology Publications. Ecol Environ Anthro

Kennedy RC, Lane KE, Arifin SMN, Fuentes A, Hollocher H, Madey G (in press) A GIS aware agent-based model of pathogen transmission. International Journal of Intelligent Control and Systems.

Lane KE, Arta Putra IGA, Wandia IN, Rompis ALT, Hollocher H, Fuentes A (in review a) Balinese perceptions of behavioral risks relating to pathogen transmission in human communities surrounding long-tailed macaque populations. Ecol Environ Anthro

Lane KE, Holley C, Hollocher H, Fuentes A (in review b) Environmental variation and parasite intensity in Balinese long-tailed macaques (*Macaca fascicularis*). Amer J Primat

Lansing SJ (2007) Priests and programmers: technologies and power in the engineered landscape of Bali. Princeton University Press, Princeton, NJ

Loudon J, Howell M, Fuentes A (2006) The importance of integrative anthropology: a preliminary investigation employing primatological and cultural anthropological data collection methods in assessing human-monkey co-existence in Bali, Indonesia. Ecol Environ Anthropol 2(1):2–13

Melnick DJ (1990) Molecules, evolution and time. Trends Ecol Evol 5:172–173

Melnick DJ, Hoelzer GA (1992) Differences in male and female macaque dispersal lead to contrasting distributions of nuclear and mitochondrial DNA variation. Int J Primatol 13(4):379–393

Napier JR, Napier PH (1967) A handbook of living primates. Academic, New York

Nei M (1972) Genetic distance between populations. Am Nat 106:283–292

Southern MW (2002) An assessment of potential habitat corridors and landscape ecology for long-tailed macaques (*Macaca fascicularis*) on Bali, Indonesia. Master's thesis, Central Washington University, WA

Thierry B (2007) The macaques: a double-layered social organization. In: Campbell CJ, Fuentes A, MacKinnon KC, Panger M, Bearder SK (eds) Primates in perspective. Oxford University Press, New York

Tosi AJ, Morales JC, Melnick DJ (2000) Comparison of Y chromosome and mtDNA phylogenies leads to unique inferences of macaque evolutionary history. Mol Phylogenet Evol 17(2):133–144

Tosi AJ, Morales JC, Melnick DJ (2002) Y-chromosome and mitochondrial markers in *Macaca fascicularis* indicate introgression with Indochinese *M. mulatta* and a biogeographic barrier in the Isthnus of Kra. Int J Primatol 23:161–178

Tosi AJ, Morales JC, Melnick DJ (2003) Paternal, maternal, and biparental molecular markers provide unique windows onto the evolutionary history of macaque monkeys. Evolution 57(6):1419–1435

van Noordwijk MA (1985) The socioecology of Sumatran long-tailed macaques (*Macaca fascicularis*) II. The behavior of individual. Proefschrift Drukkerij Elinkwijk, Utrecht

Wandia IN (2007) Strukur dan Keragaman Genetik Populasi Lokal Monyet Ekor Panjang (*Macaca fascicularis*) di Jawa Timur, Bali dan Lombok. Disertasi, Sekolah Pascasarjana, Institut Pertanian Bogor

Wheatley BP (1999) The sacred monkeys of Bali. Wavelend Press, Inc., New York

Whitten T, Soeriaatmadia RE, Afiff SA (1997) The ecology of Java and Bali. Periplus Press, New York

Wilson EO (1984) Biophilia. Harvard University Press, Boston

Chapter 14
The Not-So-Sacred Monkeys of Bali: A Radiographic Study of Human-Primate Commensalism

Michael A. Schilaci, Gregory A. Engel, Agustin Fuentes, Aida Rompis, Arta Putra, I. Nengah Wandia, James A. Bailey, B.G. Brogdon, and Lisa Jones-Engel

Introduction

Humans and nonhuman primates have coexisted and interacted for millennia in Asia. Interspecies interaction is particularly intensive at religious sites that are commonly referred to by Westerners as "monkey temples" or "monkey forests". These monkey temples are found throughout South and Southeast Asia, and some have evolved into significant tourist destinations, often contributing substantially to the economic base of the communities in which they are located. In Bali, the location and structural layout of temples are guided by the Balinese Hindu philosophies of *Nawa Sanga* (the ritual grid organizing space) and *Tri Hita Karana* (the three ideals for achieving balance between humans, gods, and the natural world). Balinese culture emphasizes harmony between humankind and nature. Adherence to these traditional values has contributed to the preservation of forests and other natural landscape features associated with the temple areas and protection of nonhuman denizens found at these sites (Fuentes et al. 2005).

Far from being threatened, Bali's primate populations in these settings are often protected, provisioned, and treated with great tolerance. As a result, the macaques often thrive, and populations may increase substantially over time (Fuentes et al. 2005; Wheatley 1999), leading to conflict as humans and macaques compete for limited resources, such as space and food. Crop-raiding by monkeys can become a nuisance to local farmers adjacent to monkey temples. Aggressive primate-human interactions can lead to scratch and/or bite injuries for humans, increasing the likelihood of pathogen transmission between species (Engel et al. 2006; Fuentes 2006; Fuentes and Gamerl 2005). Here we present a study illustrating just one of the myriad and complex ways in which humans and primates interact in these settings.

Located east of the island of Java, Indonesia, Bali is a small (~5,633 km²) island with relatively large human (~3.2 million) and monkey populations. Unlike Java, or

M.A. Schilaci (✉)
Department of Social Sciences, University of Toronto Scarborough,
1265 Military Trail, Toronto, ON M1C 1A4, Canada
e-mail: schillaci@utsc.utoronto.ca

S. Gursky-Doyen and J. Supriatna (eds.), *Indonesian Primates*,
Developments in Primatology: Progress and Prospects,
DOI 10.1007/978-1-4419-1560-3_14, © Springer Science+Business Media, LLC 2010

other islands of Indonesia, which are largely Muslim, Bali's residents are predominantly Hindu (>90%). As an important part of Balinese culture and society, Hindu temples are located throughout the island. Many of these temples are inhabited by troops of long-tailed macaque monkeys (*Macaca fascicularis*), which are protected and often provisioned.

There are at least 63 such sites on the island (Fuentes et al. 2005). Each site has fifteen to over three hundred monkeys with densities ranging from one to over twenty individuals per square kilometer, while human densities average over 500 individuals per square kilometer across the island. Over 68% of these sites are associated with a temple or shrine. These religious complexes can be as small as a simple shrine consisting of a few stones and an altar to elaborate temple complexes that are heavily used by Balinese, and, in some cases, foreigners (Fuentes et al. 2005). Many of these macaque groups receive some provisioning.

Bali's Sacred Monkeys

The most comprehensive, and perhaps well-known, study of Bali's macaques appeared in Bruce Wheatley's (1999) monograph titled, "The Sacred Monkeys of Bali". In that monograph, the author describes his study of macaque behavior and commensalism with the Balinese residents. Wheatley also discusses the Balinese perception of monkeys and its cultural and historical basis. Monkeys *often* (see below) enjoy a prominent, and even sacred, standing in Balinese culture and religion. Their prominence emanates in large part from the role played by monkeys in the Ramayana (Fig. 14.1), a Hindu epic poem that is particularly popular amongst Balinese Hindus.

The Balinese attitude toward monkeys is complex. Monkeys associated with Hindu temples are often tolerated and even treated with kindness in Balinese society. According to Wheatley, their standing in Balinese society is perhaps best described as liminal figures (1999, p. 35), occupying a border area between the animal/demonic world and the world of humans, or, alternatively, between the world of humans and the world of gods. As such the monkey has the power to move between worlds. Thus, a monkey's role or standing at any given point in time is likely defined by context. A monkey engaged in crop-raiding, for example, might be considered animal/demonic, while a monkey residing within the confines of the temple would be regarded as having economic and perhaps even religious significance. An appreciation of the monkey's liminal role in Balinese culture, and of the importance of context, therefore, are essential for understanding human-primate commensalism[1] in Bali (Fuentes et al. 2005; Loudon et al. 2006).

The study presented here was undertaken in the context of a multi-year, multidisciplinary collaborative research on human-monkey commensalism that

[1] As pointed out by a reviewer, "commensalism" implies that one organism benefits while the other neither benefits, nor is harmed. Mutualism implies that both organisms benefit, while parasitism implies that one organism benefits at the expense of the other. The relationship between monkeys and humans in Bali is perhaps better described as mutualism or parasitism, depending on context. We use the term commensalism, however, to be consistent with common usage in ethnoprimatology.

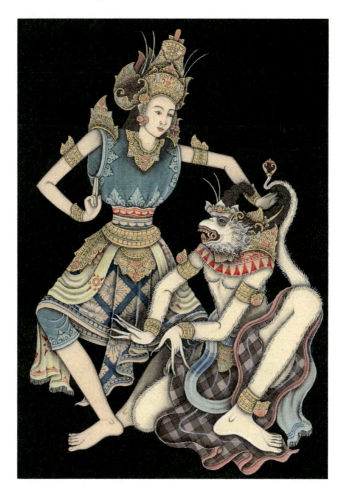

Fig. 14.1 Balinese painting of Hanuman and Sita. The epic story of the Ramayana tells the story of Rama, the seventh incarnation of Vishnu and an ancient king in India, whose wife Sita is abducted by a king of demons Rawana. Hanuman, the incarnation of Shiva, takes the form of a monkey and is sometimes referred to as the "monkey god", is befriended by Rama, and eventually finds the kidnapped Sita. Hanuman helps Rama by leading an army of monkeys, ultimately defeating Rawana. Painting in possesion of M. Schillaci.

examined cross-species transmission of infectious agents, and interspecies behavioral interactions, and their implications for public health and primate conservation.

Methods

Lateral and/or anterior-posterior thoracic radiographs of 91 long-tailed macaques (*M. fascicularis*) were taken in the field at seven different monkey temples in Bali, Indonesia as part of a study examining the incidence of *Mycobacterium tuberculosis*

Table 14.1 Temple information and results from radiographic analysis

Site	Description of site	Estimated size of monkey population	Intensity of tourism	Positive X-rays N (%)	Negative X-rays
Sangeh	Rural/farming	180	+++	7 (36.8%)	12
Teluk Terima	Rural	120	+	0	9
Pulaki	Rural	Approx.120	+	0	21
Uluwatu	Urban	200	+++	0	14
Bedugal	Rural/roadside	Approx. 200	++	0	5
Alas Kedaton	Rural	400	+++	1 (4.3%)	22
Total		1,220		8 (8.8%)	83

in monkey populations. The sample comprised both males ($n=69$) and females ($n=22$) from three different age groups: adults (>5 years old, $n=56$), subadults (3–5 years old, $n=21$), and juveniles (0–3 years old, $n=14$). Ages were determined based on dental eruption patterns. The ecological contexts of the seven sites are noted in Table 14.1.

Results

Evaluation of the exposed X-ray films revealed radio-opaque foreign bodies of uniform size and shape in 8 of the 91 (8.79%) macaques (Table 14.1). These foreign bodies were found in 6 of 69 (8.7%) males and 2 of 22 (9.1%) females. All affected animals were either adults ($n=6$) or subadults ($n=2$). Seven of the eight affected animals were X-rayed at a single temple site, Sangeh. Multiple foreign bodies were visible in two macaques (Fig. 14.2).

The appearance of these foreign bodies is consistent with that of pellets, such as those fired from low-velocity air rifles. Using the classification system presented by Bailey (2007), all of the pellets visible on film appear to be .177 caliber domed diabolo style pellets. Because most monkeys were radiographed in only one plane, determining the depth and precise position of the air gun pellets was not possible (Stockmann et al. 2007). Although we were not looking specifically for pellet wounds, our routine exam of skin and pelage condition did not reveal any evidence of external wounds. Orientation of the air gun pellets, as determined by the head and skirt of the pellets, was variable. Slight deformation of the pellet head was apparent on the radiographs of two animals (Fig. 14.3). Pellet fragmentation was not observed.

Discussion and Conclusions

Our analysis revealed foreign bodies, most likely domed-head diabolo air rifle pellets, in 8 of 91 macaques that were radiographed. This type of pellet is typically used with .177 caliber air rifles with muzzle velocities typically between 350 and 800 fps.

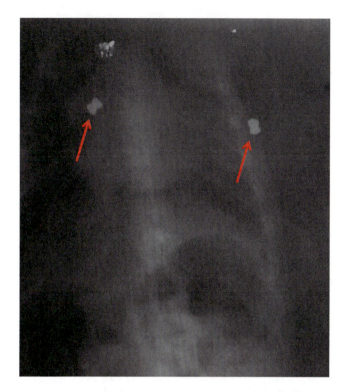

Fig. 14.2 Ventral radiograph of a long-tailed macaque showing two lead pellets.

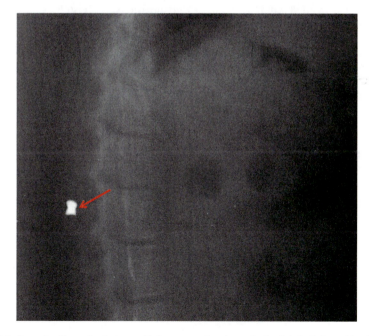

Fig. 14.3 Lateral radiograph of a long-tailed macaque showing a single slightly deformed lead pellet.

Pellets shot from a rifle with a muzzle velocity within this range are capable of inflicting serious tissue injury and death in humans (Laraque 2004; Scribano et al. 1997; Radhakrishnan et al. 1996) and presumably primates. The effective range of most air rifles with muzzle velocities between 350 and 800 fps is approximately between 55 and 65 feet. Pellets typically retain much of their kinetic energy at this range. Considerable variation due to rifle and pellet make, however, is common. The minimum velocity required to perforate human skin is approximately 331 fps with a .177 caliber pellet. The kinetic energy of the pellet diminishes as the distance it travels increases, thus limiting its penetrating ability.

It is important to point out that our evaluation has several important limitations. For example, our sample cannot be considered random or unbiased because it comprises animals we were able to dart. These animals may, therefore, be more accustomed to, and less afraid of, humans. If so, our results may overestimate the prevalence of monkey sniping in Bali. Three of the animals with radiographic evidence of pellets were shot more than once, suggesting they, like Hanuman, may be bold and unafraid. On the other hand, we only radiographed the upper thorax of these monkeys. Any pellets present in the arms, legs or lower abdomen would not have been detected by our study. Also, it is possible that some macaques, having been shot, die, leave the troop, or do not retain the projectile. The latter two factors would both result in underestimates of the prevalence of pellet injuries. It is interesting to note that there is one previous study that suggests monkeys are being shot in Bali. In a study by Loudon et al. (2006) interviews with local village residents at 11 monkey temple sites in Bali demonstrated that humans had occasionally hunted or shot at monkeys.

This is not the first report of primates with air rifle pellets visible on radiographs. In previous research, we report the presence of a pellet in a pet macaque from Sulawesi, Indonesia, as an incidental finding (Schillaci et al. 2001). Several reports in the lay media allude to vervet monkeys in South Africa being wounded or killed by air rifles (Mbanjwa 2004; Ross 2003). Another media account, also from South Africa, reported the prosecution of a man for shooting a vervet monkey that was raiding his garden (News24.com, 04/06/2004, available at: http://www.news24.com/Content/SouthAfrica/News/1059/59419317bb3a404dbedaf9a7667749a1/04-06-2004-01-05/Man_fined_for_shooting_monkey, accessed 22 Apr 2008). That news report also stated that a local SPCA member had found 8 wounded vervet monkeys during the three previous months, all had been shot by pellet guns.

In Indonesia, including Bali, where there are laws prohibiting possession of firearms, air rifles are commonly used for hunting birds and small mammals. The monkeys in our study were likely not hunted, but rather were shot by farmers to deter crop-raiding or even perhaps for sport. In contrast to the other six sites, Sangeh, whose monkey population exhibited the highest prevalence of pellets, is located adjacent to agricultural fields. The use of pellet guns against crop-raiding monkeys points to potential tensions between humans and monkeys. We know of no examples of individuals shooting monkeys on temple grounds. Wheatley (1999, p. 55) describes wounds in several adult male monkeys caused by farmers defending their crops. Although it seems unlikely that farmers armed with air rifles constitute a significant threat to the existence of macaques on the island of Bali, strategies for limiting crop-raiding by monkeys are needed. Such strategies should include guidance regarding how farmers deter monkeys.

Sacred?

Our radiographic findings of air rifle pellets in more than 8% of the monkeys we radiographed would seem, initially, to suggest that at least some Balinese do not regard the island's monkeys as sacred. We would suggest that the monkeys' status is context specific. Monkeys on temple grounds have both a religious and an economic value and are thus protected. However, when monkeys leave the temple and raid a farmer's field they become an economic liability. Context and religious belief, therefore, define the value of the monkeys, and hence how they are treated. Arguably, it is the monkeys' role as tourist attractions, akin to "cash cows" rather than "sacred cows," that give them such high standing in Bali. In other words, they provide a financial incentive for tolerance.

With human populations encroaching increasingly on primate habitats and vice-versa, conflict is becoming common place. This conflict often takes place in an urban context. Some primates, such as macaques, are rapidly adapting to urban environments, bringing monkeys into daily contact – and conflict – with city residents. In Asia, primate commensalism and conflict are often inextricably linked to religion and other cultural elements. Understanding the role primates play in local culture and religion, therefore, is essential for understanding the underlying basis of commensalisms and conflict. Primate conservation in Asia would benefit from the holistic study of primate commensalism and conflict in urban contexts by ethnoprimatologists and applied primatologists (Fuentes 2006). Such research is needed to inform the development of culturally-relevant and sustainable conservation policies for primates and other mammals.

Acknowledgments   We thank the communities and temple committees of Sangeh, Alas Kedaton, Pulaki, Teluk Terima, and Uluwatu as well as Lembaga Ilmu Pengetahuan Indonesia. This project was funded in part by a Research Allocations Committee grant from the University of New Mexico (to G. Engel) and Chicago Zoological Society (to L. Jones-Engel). We are grateful to J. Heidrich for supplies and technical assistance and the University of New Mexico and Central Washington State Bali Field School students for assistance.

References

Bailey JA (2007) .177 caliber pellet classification system and identification key. J Forensic Sci 52:1314–1318
Engel GA, Jones-Engel L, Schillaci M, Suaryana KG, Putra A, Fuentes A (2006) Human culture and monkey behavior: assessing the contexts of potential pathogen transmission between macaques and humans. Am J Primatol 68:880–896

Fuentes A (2006) Human culture and monkey behavior: assessing the contexts of potential pathogen transmission between macaques and humans. American Journal of Primatology 68:880–896

Fuentes A, Gamerl S (2005) Disproportionate participation by age/sex classes in aggressive interactions between longtailed macaques (*Macaca fascicularis*) and human tourists at Padangtegal Monkey Forest, Bali, Indonesia. Am J Primatol 66:197–204

Fuentes A, Southern M, Suaryana K (2005) Monkey forests and human landscapes: is extensive sympatry sustainable for *Homo sapiens* and *Macaca fascicularis* on Bali. In: Patterson J (ed) Commensalism and conflict: the primate human interface. American Society of Primatology Publications, Norman OK, pp 168–195

Laraque D (2004) Injury risk of nonpowder guns. Pediatrics 114:1357–1361

Loudon JE, Howells ME, Fuentes A (2006) The importance of integrative anthropology: a preliminary investigation employing primatological and cultural anthropological data collection methods in assessing human-monkey co-existence in Bali, Indonesia. Ecological and Environmental Anthropology 2(1):2–13

Mbanjwa X (2004) Durban's monkeys are in the firing line. The Sunday Tribune (Sept 19th edn). http://www.iol.co.za/index.php?set_id=1&click_id=31&art_id=vn20040919113144308C353408. Accessed 23 Apr 2008

Radhakrishnan J, Fernandez L, Geissler G (1996) Air rifles-lethal weapons. J Pediatric Surg 31:1407–1408

Ross K (2003) Famed monkeys slaughtered in brutal attacks. The Mercury (Sept 2nd edn). http://www.themercury.co.za/index.php?fArticleId=219240. Accessed 23 Apr 2008

Schillaci MA, Jones-Engel L, Heidrich JE, Miller GP, Froehlich JW (2001) Field methodology for lateral cranial radiography of nonhuman primates. Am J Phys Anthropol 116:278–284

Scribano PV, Nance M, Reilly P, Sing RF, Sebst SM (1997) Pediatric nonpowder firearm injuries: outcomes in an urban pediatric setting. Pediatrics 100:1–3

Stockmann P, Vairaktaris E, Fenner M, Tudor C, Neukam FW, Nkenke E (2007) Conventional radiographs: are they still the standard in localization of projectiles? Oral Surg Oral Med Oral Pathol Oral Radiol Endodontol 104:71–75

Wheatley BP (1999) The sacred monkeys of Bali. Waveland Press, Illinois

Chapter 15
Male–Male Affiliation in Sulawesi Tonkean Macaques

Erin P. Riley

Introduction

Research during the early years of field primatology was primarily centered on the more conspicuous individuals (i.e., males) and behaviors (i.e., aggression), and in particular, males engaged in aggression (Bygott 1974; Hausfater 1975; Popp and DeVore 1979). Since that time, subsequent field research has increasingly revealed the importance of affiliation within primate social groups (Strum 1982, 2001; Smuts 1985; Strier 1994; Gould 1994; Silk 2002), leading some to argue for a renewed attention to the potential role it played in the evolution of primate sociality (Sussman et al. 2005). The primary focus of most of this work has been on the importance of affiliative and cooperative relationships between females and between males and females. Male–male relationships, however, remained largely viewed through the lenses of aggression and dominance (Hill and van Hooff 1994). This is because primate socioecological theory predicts that males and females compete for different resources (i.e., access to mates and food, respectively), and affiliative and cooperative behavior is expected to be high among females and low among males (Trivers 1972; Wrangham 1980). There is, however, increasing evidence that the nature of male–male relationships may be more diverse than previously thought (van Hooff and van Schaik 1994). For example, in 1994, an entire volume of the journal *Behaviour* was devoted to the topic of male–male bonding. Six years later, an edited volume titled *Primate Males* (Kappeler 2000) provided further evidence of the complexity of primate males, particularly with regard to male–male interactions and the role males play in shaping social organization. A number of these papers explore the key variables, both proximate and ultimate, that explain the occurrence of affiliation among males. Male philopatry and kinship have been identified as two of the most important variables (van Hooff and van Schaik 1994). The common chimpanzee represents a good example of male bond-

E.P. Riley (✉)

Department of Anthropology, San Diego State University, San Diego, CA, 92182-6040, USA
e-mail: epriley@mail.sdsu.edu

S. Gursky-Doyen and J. Supriatna (eds.), *Indonesian Primates*,
Developments in Primatology: Progress and Prospects,
DOI 10.1007/978-1-4419-1560-3_15, © Springer Science+Business Media, LLC 2010

ing in a male philopatric species; male chimps have been observed to form strong male–male alliances and engage in high levels of mutual grooming (Nishida and Hiraiwa–Hasegawa 1987). At the same time, although less common, male bonding has been observed in species in which males disperse, thereby suggesting that kinship need not be a prerequisite for male–male affiliation (Silk 1994; Hill and van Hooff 1994).

The genus *Macaca* is a good example of a primarily female philopatric group of species, whereby most males disperse from their natal groups at sexual maturity. Strong affiliative relationships among females (with strongest amongst closest kin) and antagonistic relationships among males were therefore considered, at least initially, to be typical of macaques (Caldecott 1986; Thierry et al. 1994). Empirical evidence, however, shows that while male–male relationships are antagonistic among rhesus, Japanese, and pigtailed macaques, other species, such as the bonnet, Assamese, Tibetan, and Barbary macaques show high levels of affiliative behavior between males (Hill 1994; Silk 1994; Preuschoft and Paul 2000; Cooper and Bernstein 2000; Berman et al. 2007).

These deviations from the macaque model have led researchers to explore other factors, besides kinship, that may influence male relations. For example, Hill (1994) emphasized the importance of demographic variables, such as group size and sex ratio, in shaping patterns of social interactions within a group. Based on a review of 23 studies on eight macaque species, Hill (1994) found that male–male affiliative behavior was rare in groups that had many more adult females than males, but was more frequent in groups where the sex ratio was closer to even. Adult sex ratio is considered to be an important factor shaping intra-sexual relations because a more even ratio between the sexes may lead to a shortage of preferred social partners for males, leading them to form relationships with other males as an alternative. Hill (1994) also observed an association between the rates of male–male affiliation and group size, whereby affiliation was more frequent when group size was smaller. One of the limitations of his analysis, however, is that a majority of the studies reviewed involved macaque groups that were food provisioned. Food provisioning fundamentally alters the abundance and distribution of food, thereby affecting not only population demography (e.g., increased group sizes and density) and activity budgets, but also patterns of social interaction within primate social groups (Asquith 1989; Fa 1991; O'Leary and Fa, 1993; Wheatley and Harya Putra 1994).

As a species group, the Sulawesi macaques differ remarkably from the better known macaque species (e.g., rhesus, long-tailed, and Japanese macaques) in that they exhibit a relaxed dominance style characterized by greater symmetry in agonism and higher rates of reconciliation in dyadic interactions (Bernstein and Baker 1988; Thierry 1985; Thierry et al. 1994; Matsumura 1998). Given their tolerant nature, one might expect lower rates of agonism and higher levels of affiliation among males compared to other macaque species. Adult sex ratio may be another feature in which the Sulawesi macaques differ; a nearly even adult sex ratio has been observed in multiple populations of at least two of the Sulawesi macaque species (Watanabe and Brotoisworo 1982; Okamoto and Matsumura, 2001; Pombo et al. 2004), but not

Table 15.1 Group size and adult sex ratio for wild populations of Sulawesi macaques

Species	Study	Group size	Sex ratio[a]
Macaca nigra	O'Brien and Kinnaird (1997)		
	RAM group	97	4
	Dua group	57	5
	Malonda group	50	2.33
M. nigrescens	Kohlhaas and Southwick (1996)	9–17.7	1.7[b]
M. ochreata	Hillyar (2001)	21	2
M. maura	Watanabe and Brotoisworo (1982)		
	Group A	38–40	1.33
	Group B	20	7
	Group C	19	0.5
	Okamoto and Matsumura (2001)		
	April 1999 Group B	43	1.75
	August 1999 Group B split		
	B1	27	0.88
	B2	16	1.75
M. tonkeana	Pombo et al. (2004)		
	Group A	25	1.2
	Group B	14	1.33

[a]Calculated as the number of females per male (AF/AM).
[b]The authors note that this ratio may be higher, given that many of the unclassified animals in the group likely were females.

in all of them (Kohlhaas and Southwick 1996; O'Brien and Kinnaird, 1997; Hillyar 2001; Table 15.1).

The Sulawesi Tonkean macaque (*M. tonkeana*) is one of the best-known of the seven taxa that comprise the species group (Fooden 1969; Groves 2001). Although the ecology and conservation of Tonkean macaques have been the subjects of recent field investigations (Pombo et al. 2004; Riley 2005a, 2007, 2008), most of the available information on Tonkean macaque social behavior comes from captive research. This body of work (Thierry 1984, 1985; Thierry et al. 1990) has found no evidence of intense aggression (i.e., biting) between males; no evidence of either aggressive or affiliative clasping behavior among adult males; and, has found that grooming among males is infrequent compared to the frequency of adult male–adult female grooming. Thierry (1984) noted, however, that the small number of adult males in the study groups does not permit firm conclusions.

The purpose of this paper is to provide an initial assessment of the nature of male–male relationships among free-ranging, non-provisioned Sulawesi Tonkean macaques (*Macaca tonkeana*). I explore the role demography, specifically sex ratio and group size, plays in shaping male–male relationships by using behavioral data collected on two Tonkean macaque groups that showed a nearly even adult sex ratio, but that differed in group size to test the following hypotheses: (1) Given an even adult sex ratio and the tolerant dominance style of the species, the proportion of observation time males affiliate with other males will be greater than the proportion of available adult males in the social group; (2) Alternatively, given that females are typically preferred social partners for males, the proportion of observation time males affiliate with other males will be less than the proportion of observation time adult males affiliate with adult females; and (3) The proportion of observation time males affiliate with other males will be greater in the smaller group than in larger group. I discuss the results in relation to what is known from other macaque species and consider possible future research directions.

Study Site and Species

The research was conducted from January 2003 to April 2004 in Lore Lindu National Park (LLNP), located at 01°15' to 01°30' S; 119°50' to 120°20'E, in Central Sulawesi, Indonesia (Fig. 15.1). LLNP, comprising a total area of 217,982 ha, was established in 1993 from two existing reserves and is designated as a UNESCO Man and the Biosphere Reserve. LLNP provides habitat for a number of Sulawesi's endemic mammals, including the Tonkean macaque. The subjects of this study were two wild groups of Tonkean macaques that were the focus of a broader ecological study (Riley 2005a). Details regarding group size, composition, adult sex ratio, and social activity patterns are provided in Table 15.2.

Fig. 15.1 Lore Lindu National Park, Sulawesi, Indonesia

Table 15.2 Demographic features and social activity patterns of two study groups

Feature	Group 1 (Anca)	Group 2 (CH)
Group size (range)	6–9	26–28+
Group composition[a]	3AM, 2AF, 1SJ, 2I	9AM, 9AF, 5LJ, 3SJ, 1+U
Adult sex ratio[b]	0.66	1.0
Activity patterns[c]		
Social	11%	11%
Grooming	8.8%	10%

[a]AM = Adult male, AF = Adult female, LJ = Large juvenile, SJ = Small juvenile, I = Infant, U = Unknown. Source: Riley, 2005b.
[b]Calculated as the number of females per male (AF/AM).
[c]Calculated as percentage of scans engaged in socializing overall (i.e., sexual, grooming, play, agonism) and grooming, specifically. Source: Riley, 2005a, 2007.

Methods

Data Collection

In this study, I define affiliation as the proportion of observation time in proximity and grooming. Although proximity is not typically considered an active form of affiliation, I consider it an important measure of affiliation because the maintenance of proximity is by no means passive (Sussman et al. 2005). Data on affiliation were collected as part of a study on Tonkean macaque activity patterns (Riley 2007), for which scan sampling was employed (Altmann 1974). Specifically, I conducted group scans at 30-min intervals on two social groups, which were observed from 0600 to 1800h on 3–4 days per month for 15 months. For each individual located during a period of 10 min, I instantaneously recorded their age/sex class, behavior (i.e., resting, foraging, moving, feeding, allogrooming, autogrooming, agonistic, or sexual behavior), and the age/sex class of their nearest neighbor. Age/sex classes were defined as: adult male, adult female, large juvenile, small juvenile, and infant. The age-sex class of all individuals grooming the subject (or being groomed by the subject) was recorded. The nearest neighbor was defined as the individual who was in the closest proximity to the subject being observed at that moment. The nearest neighbor dataset included a total of 339 data points (monthly mean = 22.6, SD = 13.42) for the Anca group and 166 data points for the CH group (monthly mean = 11.06, SD = 5.93). The grooming dataset included a total of 90 data points for the Anca group (monthly mean = 6.43, SD = 3.84) and 57 data points for the CH group (monthly mean = 4.07, SD = 2.79). The mean number of adult males (range) and mean number of adults (range) observed per scan sample for each group were as follows: Anca: mean adult males = 1.97 (0–3); mean adults = 3.43 (0–5); CH: mean adult males = 2.46 (0–8); mean adults = 4.46 (0–15).

Data Analysis

Affiliation was measured as the proportion of scans spent (a) in proximity to a nearest neighbor and (b) grooming another individual. These proportions were calculated by dividing both the number of adult male–adult male (AM–AM) and adult male–adult female (AM–AF) affiliations by the total number of adult affiliations per scan sample. These proportions were then averaged to render monthly mean proportions that were used in statistical tests. Arcsine transformations were performed on these proportional data to meet assumptions of normality and equal variances (Sokal and Rohlf 1981). To test the hypothesis that the proportion of AM–AM affiliations is greater than the proportion of available adult males (i.e., what would be expected by chance alone), I used a one-sample test for the nearest neighbor and grooming data. The expected (test) value was the proportion of available adult male social partners in the group. This value was calculated as the number of other adult males available to an adult male subject divided by the total number of adults in the group. A one sample t-test was also used to determine whether there were significant differences between the proportion of AM–AM and AM–AF interactions in each group by comparing the mean difference of these proportions to the test value of 0. Because multiple statistical tests were run on each dataset (i.e., grooming and nearest neighbor), the Bonferroni correction was used to determine an alpha level of 0.025. An independent samples t-test was used to evaluate the hypothesis of differences in AM–AM affiliative behavior between the two social groups. The effect of mating was not included in the data analysis because *ad libitum* recording of the timing of female sexual swellings indicated a non-seasonal pattern of reproduction; a finding in line with observations of non-seasonal breeding in other Sulawesi macaques (e.g., *Macaca maura*; Okamoto et al. 2000).

Results

Affiliation: Nearest Neighbor

For the Anca group, the mean proportion of AM–AM in nearest neighbor proximity (.40) was not significantly greater than the proportion based on the adult sex ratio (.50) (Fig. 15.2; $t=-1.797$, df$=14$, $p=0.047$). There was also no significant difference between the mean proportion of AM–AM and AM–AF nearest neighbor proximity (.60) ($t=1.958$, df$=14$, $p=0.070$). For the CH group, the mean proportion of AM–AM in nearest neighbor proximity (.29) was significantly less than the proportion of AM–AF interactions (.71) ($t=3.440$, df$=14$, $p=0.002$) (as predicted) and significantly less than the proportion of available adult male partners (.47) ($t=-2.932$, df$=14$, $p=0.0055$) (the opposite of the predicted outcome).

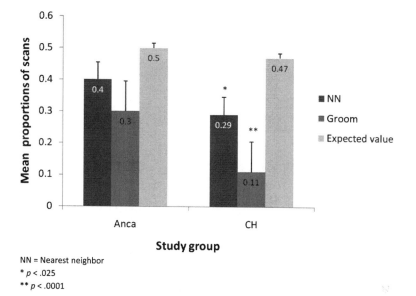

Fig. 15.2 Mean proportion of scans of adult males as nearest neighbors and as grooming partners per group compared to the expected value (i.e., proportion of available adult male social partners). Error bars represent ±1 standard error of the mean.

Affiliation: Grooming

For the Anca group, the mean proportion of AM–AM grooming interactions (.30) was not significantly different than the proportion based on the adult sex ratio (.50) (Fig. 15.2; $t=-1.405$, df= 13, $p=0.0915$) but was significantly less than AM–AF grooming interactions (.70) ($t=2.273$, df= 13, $p=0.0205$). For the CH group, the mean proportion of AM–AM grooming interactions was significantly less (.11) than the proportion of available adult males (.47) ($t=-6.534$, df= 11, $p<0.0001$) and significantly less than the proportion of AM–AF grooming interactions (.89) ($t=6.609$, df= 11, $p<0.0001$).

Affiliation: Between-Group Comparisons

The mean proportion of AM–AM in nearest neighbor proximity in the Anca (smaller) group (.40) was not significantly greater than the observed proportion in the CH (larger) group (.29) (Fig. 15.3; $t=1.355$, df=28, $p=0.093$). The mean proportion of AM–AM grooming interactions in the ANCA group (.30) was greater than that of the CH group (.11), but this difference just barely approached significance ($t=1.679$, df=24, $p=0.053$).

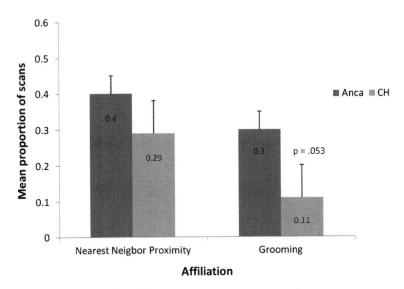

Fig. 15.3 Between-group comparison of mean proportion of AM–AM nearest neighbor and grooming scans (CH = larger group). Error bars represent ±1 standard error of the mean.

Agonism Rates

In the Anca group, there were a total of seven agonistic interactions observed (across 7 scans) out of a total of 747 scans performed on the group. Out of those seven, two were between adult males. In the CH group, there were a total of 11 incidences of agonism observed (across 11 scans) out of a total of 473 scans performed on the group; two of those were between adult males.

Discussion

Researchers are increasingly documenting that male–male interactions among primates extend beyond competition and conflict to include affiliation and bonding (Hill and van Hooff 1994; Kappeler 2000; Cooper and Bernstein 2000; Cooper et al. 2004). The objective of the present study was to investigate the nature of male–male relationships among wild Tonkean macaques. Given nearly even adult sex ratios of the study groups and the species' tolerant style, I predicted that the proportion of AM–AM proximity and grooming would be greater than what would be expected, given the high number of available adult male social partners. Alternatively, I predicted that despite a nearly even sex ratio males would affiliate with females more often than with other males. The first prediction was not supported by the data from either group, thereby suggesting that Tonkean macaque males are not any more likely to affiliate with other adult males just because more are available. Instead, the results suggest that adult females remain their preferred social partners, particularly in grooming. This finding is in line with what is known for other macaque species, including both tolerant species (e.g., bonnet macaques; Kurup, 1988) and despotic species (e.g., Assamese macaques; Cooper and Bernstein 2000; rhesus macaques; Cooper and Bernstein 2008).

An important question to ask regarding the relationship between demography and male–male relationships is whether male–male affiliation and social tolerance allow for more even adult sex ratios (Ogawa 1995), or whether nearly even sex ratios lead to social tolerance among adult males (Preuschoft and Paul 2000; Berman et al. 2006). A comparison of the results on the overall nature of male–male interactions in Tonkean macaques from this study with those from other Sulawesi macaque studies lends support for the latter hypothesis. In this study, rates of AM–AM agonism were extremely low in both groups, and grooming has been noted to comprise more than 80% of their social activities (Riley 2005a, 2007). Moreover, it is noteworthy that in one of the groups 30% of all adult grooming interactions was between males. In contrast, Reed et al. (1997) found little AM–AM affiliation and noted that adult male interactions were characteristically aggressive in a group of wild, Sulawesi crested black macaques (*Macaca nigra*) where the sex ratio was highly skewed (5 females: 1 male). Similarly, Thierry et al. (1990) reported that grooming interactions between adult males were rare in a group of semi-free ranging Tonkean macaques where the adult sex ratio was also highly skewed (3.5 females: 1 male).

Hill (1994) proposed that group size might be another important demographic factor that influences male–male relationships. In this study, one of the social groups was very small in size (see Table 15.1), and therefore, adult males could often be in close proximity to one another across a range of ecological contexts, including during feeding. For example, all members of the group were frequently observed eating in the same Aren palm tree: a preferred food tree that produces a small amount of fruit at any given time (Riley 2007). This might explain why the patterning of nearest neighbors did not differ significantly from a random distribution in this group. Nonetheless, a possible group-size effect was only partially supported by the data; while there was no significant difference between the groups in the proportion of observation time that adult males were in proximity to one another, AM–AM grooming was more frequent in the smaller (Anca) group. These results should be interpreted with caution for two reasons. First, data were only available for a sample of one per group size condition. Second, because a small proportion of all the adult males in the larger (CH) group was observed per scan, it is difficult to determine whether the results reflect a valid test of the group size effect or uneven sampling of the two groups.

The study reported herein represents the first step toward an understanding of the nature of AM–AM relationships in wild, Tonkean macaques and the factors that shape those relationships. It is important, however, to acknowledge the limitations of the study, including a limited data set, a small number of social groups, and a lack of variability in adult sex ratio. The next step requires research focused specifically on male–male dyadic interactions, including both affiliative and agonistic; observations on multiple social groups with varying group sizes; and, further documentation of adult sex ratios in Tonkean macaques, as well as in other Sulawesi macaque species (Bynum 1999). Future work will also require attention to the possible ultimate factors that explain a fundamental underlying question: why are there so many adult males in Tonkean macaque groups? Finally, extensive field observations on the social organization and behavior of the other, lesser-known Sulawesi macaques (e.g., *Macaca ochreata, M. nigrescens*) are needed to better understand the patterning of male social tolerance and bonding in the Sulawesi macaque phyletic group, and in the genus *Macaca* as a whole.

Acknowledgments I thank the Indonesian Institute of Sciences (LIPI) and the Indonesian Ministry of Forestry (PHKA) for permission to conduct the research, and Noviar Andayani and Amir Hamzah for their sponsorship. Financial support was provided by the National Science Foundation, Wenner Gren Foundation, Wildlife Conservation Society, and American Society of Primatologists. I offer many thanks to Matt Cooper, Chia Tan, and the editors of this volume whose comments on earlier drafts greatly improved this manuscript. I also thank Li An for his statistical guidance, and my students, Laura Graves and Jeff Peterson, for their help with the data analyses. I am forever grateful to the following individuals whose assistance in the field made this work possible: Manto, James, Papa Denis, Pak Asdi, Pias, Tinus, and Papa Tri.

References

Altmann J (1974) Observational study of behavior: sampling methods. Behavior 49:227–267

Asquith PJ (1989) Provisioning and the study of free-ranging primates: history, effects, and prospects. Yearb Phys Anthropol 32:129–158

Berman CM, Ionica C, Dorner M, Li J (2006) Postconfllict affiliation between former opponents in *Macaca thibetana* on Mt. Huangshan, China. Int J Primatol 27:827–854

Berman CM, Ionica C, Li J (2007) Supportive and tolerant relationships among male Tibetan macaques at Huangshan, China. Behaviour 144:631–661

Bernstein IS, Baker SC (1988) Activity patterns in a captive group of Celebes Black Apes (*Macaca nigra*). Folia Primatol 51:61–75

Bygott JD (1974) Agonistic behavior and dominance in wild Chimpanzees. Cambridge University, Cambridge

Bynum EL (1999) Biogeography and evolution of Sulawesi macaques. Trop Biodiversity 6(1–2):19–36

Caldecott JO (1986) Mating patterns, societies, and ecogeography of macaques. Anim Behav 34:208–220

Cooper MA, Aureli F, Singh M (2004) Between-group encounters among bonnet macaques (*Macaca radiata*). Behav Ecol Sociobiol 56:217–227

Cooper MA, Bernstein IS (2000) Social grooming in Assamese macaques (*Macaca assamensis*). Am J Primatol 50:77–85

Cooper MA, Bernstein IS (2008) Evaluating dominance style in Assamese and rhesus macaques. Int J Primatol 29:225–243

Fa JE (1991) Provisioning of Barbary Macaques on the rock of Gibraltar. In: Box HO (ed) Primate responses to environmental change. Chapman and Hall, London

Fooden J (1969) Taxonomy and evolution of the Monkeys of Celebes. Karger, Basel

Gould L (1994) Patterns of affiliative behavior in adult male ringtailed lemurs (Lemur catta) at Beza-Mahafaly Reserve, Madagascar. Ph.D. thesis. St Louis, Washington University

Groves C (2001) Primate taxonomy. Smithsonian Institution Press, Washington, DC

Hausfater G (1975) Dominance and reproduction in Baboons. Contributions to Primatology. vol 7. Basel: Karger

Hill DA (1994) Affiliative behavior between adult males of the genus *Macaca*. Behaviour 130(3–4):293–308

Hill DA, van Hooff JARAM (1994) Affiliative relationships between males in groups of nonhuman primate: a summary. Behaviour 130(3–4):143–149

Hillyar J (2001) Time budgeting of the Buton macaque. Macaca ochreata brunnescens, Operation Wallacea

Kappeler PM (2000) Primate males. Cambridge University Press, Cambridge

Kohlhaas A, Southwick CH (1996) *Macaca nigrescens*: grouping patterns and group composition on a Sulawesi Macaque. In: Fa JE, Lindburg DG (eds) Evolution and ecology of Macaque societies. Cambridge University Press, Cambridge, pp 132–145

Kurup GU (1988) The grooming pattern in bonnet macaques (*Macaca radiata*). Ann N Y Acad Sci 525:414–416

Matsumura S (1998) Relaxed dominance relations among female Moor Macaques (*Macaca maurus*) in their natural habitat, South Sulawesi, Indonesia. Folia Primatol 69(2):346–356

Nishida T, Hiraiwa-Hasegawa M (1987) Chimpanzees and bonobos: cooperative relationships among males. In: Smuts BB, Cheney DL, Seyfarth RM, Wrangham RW, Struhsaker TT (eds) Primate societies. University of Chicago Press, Chicago

O'Brien TG, Kinnaird MF (1997) Behavior, diet, and movements of the Sulawesi crested black macaque. Int J Primatol 18(3):321–351

Ogawa H (1995) Bridging behavior and other affiliative interactions among male Tibetan macaques (*Macaca thibetana*). Int J Primatol 16(5):707–729

Okamoto K, Matsumura S (2001) Group fission in moor macaques (*Macaca maurus*). Int J Primatol 22(3):481–493

Okamoto K, Matsumura S, Watanabe K (2000) Life history and demography of wild Moor Macaques (*Macaca maurus*): summary of ten years of observations. Am J Primatol 52:1–11

O'Leary H, Fa JE (1993) Effects of tourists on Barbary Macaques at Gibraltar. Folia Primatol 61:77–91

Pombo AR, Waltert M, Mansjoer SS, Mardiasuti A, Muhlenberg M (2004) Home range, diet and behavior of the Tonkean macaque (*Macaca tonkeana*) in Lore Lindu National Park, Sulawesi. In: Gerold G, Fremerey M, Guhardja E (eds) Land use, nature conservation the stability of Rainforest margins in Southeast Asia. Springer, Berlin, pp 313–325

Popp J, DeVore I (1979) Aggressive competition and social dominance theory. In: Hamburg D, McCown ER (eds) The great apes: perspectives on human evolution, vol 5. Holt, Rinehart, and Winston, New York, pp 317–340

Preuschoft S, Paul A (2000) Dominance, egalitarianism, and stalemate: an experimental approach to male-male competition in Barbary macaques. In: Kappeler PM (ed) Primate males. Cambridge University Press, Cambridge, pp 205–217

Reed C, O'Brien TG, Kinnaird MF (1997) Male social behavior and dominance in the Sulawesi crested black macaque (*Macaca nigra*). Int J Primatol 18(2):247–260

Riley EP (2005a) Ethnoprimatology of Macaca tonkeana: the interface of primate ecology, human ecology, and conservation in Lore Lindu National Park, Sulawesi, Indonesia. Ph.D. Dissertation, University of Georgia

Riley EP (2005b) The loud call of the Sulawesi Tonkean macaque, *Macaca tonkeana*. Trop Biodiversity 8(3):199–209

Riley EP (2007) Flexibility in diet and activity patterns of *Macaca tonkeana* in response to anthropogenic habitat alteration. Int J Primatol 28(1):107–133

Riley EP (2008) Ranging patterns and habitat use of Sulawesi Tonkean macaques (*Macaca tonkeana*) in a human-modified habitat. Am J Primatol 70(7):670–679

Silk JB (1994) Social relationships of male bonnet macaques: male bonding in a matrilineal society. Behaviour 130(3–4):271–291

Silk JB (2002) Introduction. What are friends for? The adaptive value of social bonds in primate groups. Behaviour 139:173–176

Smuts BB (1985) Sex and friendship in Baboons. Aldine de Gruyter, New York

Sokal RR, Rohlf FJ (1981) Biometry, 2nd edn. W.H. Freeman & Co., San Francisco

Strier KB (1994) Brotherhoods among atelins: Kinship, affiliation, and competition. Behaviour 130(3–4):151–167

Strum SC (1982) Agonistic dominance in male baboons: an alternative view. Int J Primatol 3:175–202

Strum SC (2001) Almost Human. University of Chicago, Chicago

Sussman RW, Garber PA, Cheverud JM (2005) The importance of cooperation and affiliation in the evolution of primate sociality. Am J Phys Anthropol 128:84–97

Thierry B (1984) Clasping behavior in *Macaca tonkeana*. Behaviour 89:1–28

Thierry B (1985) Patterns of agonistic Interactions in three species of macaque (*Macaca mulatta, M. fascicularis, M. tonkeana*). Aggress Behav 11:223–233

Thierry B, Anderson JR, Demaria C, Desportes C, Petit O (1994) Tonkean Macaque behavior from the perspective of the evolution of Sulawesi Macaques. In: Roeder JJ, Thierry B, Anderson JR, Herrenschmidt N (eds) Current primatology. Universite Louis Pasteur, Strasbourg, pp 103–117

Thierry B, Gauthier C, Peignot P (1990) Social grooming in Tonkean macaques (*Macaca tonkeana*). Int J Primatol 11(4):357–375

Trivers R (1972) Parental investment and sexual selection. In: Campbell BG (ed) Sexual selection and the descent of man, 1871–1971. Aldine de Gruyter, Chicago, pp 136–179

van Hooff JARAM, van Schaik CP (1994) Male bonds: affiliative relationships among nonhuman primate males. Behaviour 130(3–4):309–337

Watanabe K, Brotoisworo E (1982) Field observation of Sulawesi macaques. Kyoto University Overseas Research Report on Asian Nonhuman Primates 2:3–9

Wheatley BP, Harya Putra DK (1994) Biting the hand that feeds you: Monkeys and tourists in the Balinese monkey forests. Trop Biodiversity 2(2):317–327

Wrangham RW (1980) An ecological model of female-bonded primate groups. Behaviour 75:262–299

Chapter 16
Ecology and Conservation of the Hose's Langur Group (Colobinae: *Presbytis hosei, P. canicrus, P. sabana*): A Review

Vincent Nijman

Introduction

The grey-backed langurs *Presbytis hosei* sensu lato are little-known colobines from northern Borneo. Comprising one, two or possibly three distinct species, the attention these taxa have received from conservationists or from the primatological community is limited. Some people may be familiar with the species from Leo Berenstain's "The Wind Monkey and other stories" published in 1994, in which *P. (h.) canicrus* is the wind monkey referred to in the title. This same taxon gained some fame when it was included as Miller's Grizzled Surili on the 2004–2006 "Top 25 Most Endangered Primates" (Brandon-Jones 2005). While Brandon-Jones (2005) indicated that the species was known only from the north-east Indonesian part of Borneo, the sorry state of the forest in Kutai National Park, the only protected area of its recorded range, led him to suggest that *P. (h.) canicrus* was probably Critically Endangered or even Extinct. Reflecting the lack of attention to the species, he did indicate that no surveys had been undertaken. Apart from Indonesia – *P. (h). hosei, P. (h.) canicrus* and possibly *P. (h.) sabana*- grey-backed langurs occur in the Malaysian State of Sarawak and the Brunei Sultanate – *P. (h.) hosei-* and in the Malaysian State of Sabah – *P. (h). hosei* and *P. (h.) sabana*.

Here I aim (1) to provide a comprehensive overview of the ecology and habitat use of the different taxa comprising the Hose's langur group; (2) to give an overview of the densities at which the different taxa occur in pristine forest areas and in selectively logged forest; and (3) to review the threats Hose's langurs face, with a strong focus on hunting. Throughout this review, there is a strong emphasis on the situation in Indonesia, noting that some aspects of the species biology are better studied in the Malaysian part of their range.

V. Nijman (✉)

Department of Anthropology and Geography, School of Social Sciences and Law,
Oxford Brookes University, OX3 0BP Oxford, UK
e-mail: vnijman@brookes.ac.uk

S. Gursky-Doyen and J. Supriatna (eds.), *Indonesian Primates*,
Developments in Primatology: Progress and Prospects,
DOI 10.1007/978-1-4419-1560-3_16, © Springer Science+Business Media, LLC 2010

Methods

Study Species

There is considerable taxonomic confusion concerning the status of the grey-backed *Presbytis* taxa in the Sundaic Region. Their tripartite distribution, with populations in western Java (*comata/fredericae*), northern Sumatra (*thomasi*) and northern Borneo (*hosei/canicrus/everetti/sabana*), has been a protracted issue of debate. Pocock (1935) considered these taxa as constituting four different species, one each on Sumatra and Java and two on Borneo (*P. sabana* and *P. hosei*). The latter was based on the fact that on Borneo, some populations (*P. hosei*) show adult sexual dimorphism in crest shape and extent of white on the brow, while others are monomorphic (*P. sabana*). Chasen (1940) subsequently considered them to be races of a single species, *P. comata*. The three distribution ranges following the periphery of Sundaland were regarded as areas of convergent evolution by Medway (1970), and in his more cautious interpretation, the three taxa (*comata*, *thomasi*, and *hosei*) were considered to be separate species. This view has been supported by most subsequent workers (Groves 1993, 2001, Napier 1985; Bennett and Davies 1994). While there are diagnostic differences in vocalisations (Geissmann et al. 2008) and craniometry (Meijaard and Groves 2004), Brandon-Jones (1984, 1993, 1996, 1997), focussing on pelage coloration alone ("...for reasons of consistency, geographic variation in vocalisation must remain subordinate in taxonomic status to geographic variation in pelage colour..." Brandon-Jones, 1996: 72) regarded them as relicts of a single population, differentiated at the subspecific level. For Borneo, he recognised four taxa, *P. c. hosei*, *P. c. everetti*, *P. c. canicrus* and *P. c. sabana*. *Presbytis c. hosei* was considered restricted in its distribution to a small area in northern Sarawak's Baram District, whereas *P. c. everetti* occupies the remainder of the western part of North Borneo, with *P. c. sabana* occurring in the north and *P. c. canicrus* in the southeast.

Following Meijaard and Groves (2004), and accepting the need for a more thorough taxonomic analysis, I here treat these taxa as three distinct species, with *everetti* considered a synonym of *hosei*. In terms of pelage coloration and the absence of sexual dimorphism, *P. sabana* and *P. canicrus* (chromatically monomorphic) appear to be more similar to each other than either is to *P. hosei* (chromatically dimorphic). A similar, yet reverse, pattern is apparent in body size although differences are not statistically significant. Both *P. sabana* and *P. canicrus* are sexually dimorphic with females being bigger than males (*P. sabana*: tail length females 77.7 ± 3.3 cm, males 72.5 ± 10.2 cm; *P. canicrus*: tail length females 76.2 ± 3.5 cm, males 72.0 ± 2.3 cm; total length 122.8 ± 5.7 cm, males 118.1 ± 3.2 cm) but in *P. hosei* is sexually monomorphic.

The exact distribution ranges of the three species are not clear (Fig. 16.1). In the remainder of this paper, when referring to Hose's langurs, this is taken as the three species in general, i.e. *P. hosei* sensu lato, otherwise the three species will be indicated separately by their scientific names. Note that in some of the older

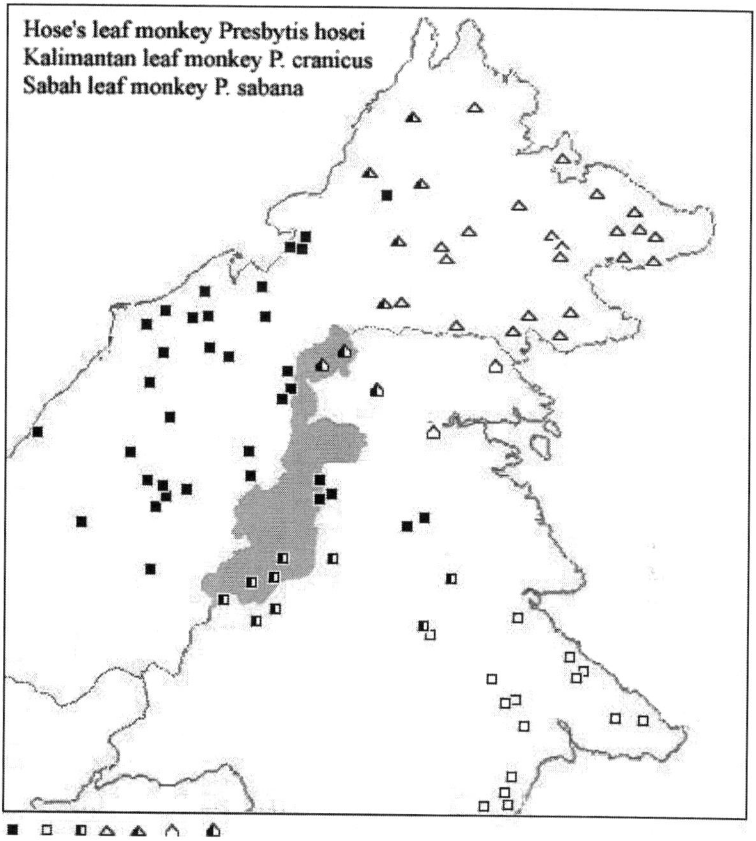

Fig. 16.1 Distribution of Hose's langurs (*Presbytis hosei* ■, *P. canicrus* □, *P. sabana* △) in North Borneo. Intermediate symbols indicate uncertainty to whether which of the species are present in that part of Borneo (from Nijman 2003)

publications, the name *Presbytis aygula* was used for both the Bornean and Javan grey-backed langurs and that this has led to confusion in comparative studies (see e.g. Newton and Dunbar 1994 and Kamilar and Paciulli 2008 who mistakenly include data on *P. canicrus* in their entries for Javan *P. comata*).

Data Acquisition

I studied *P. hosei* in the northern part of East Kalimantan, Indonesia, including Kayan Mentarang National Park and environs (Sept–Nov 1996, Jun–Jul 2003), and *P. cancirus* during surveys in other parts of East Kalimantan including Kutai

National Park and the Sangkulirang Peninsula in Sept and Dec 1996, Nov–Dec 1999, May 2000, Jun–Jul 2003, and Feb 2005.

The study in Kayan Mentarang in 1996 focussed on densities and habitat preferences, both in primary forest and secondary forests of differing age. At this time, hunting of *P. hosei* within the study area was probably absent or it occurred at a very low level. In 2003, the situation had worsened in that there had been an increase in hunting in the preceding years (Nijman 2005). Most of the ecological data reported here refers to the data collected in 1996.

With reference to the ecology of *P. canicrus* and *P. sabana,* I mainly rely on two intensive studies. Rodman (1973, 1978, 1980, 1988) studied the synecology of primates in Kutai National Park, East Kalimantan, Indonesia, including *P. canicrus*, for a period of 17 months between 1970 and 1975. At that time, the forest in the park was largely in pristine conditions and hunting of langurs was probably completely absent. For *P. sabana* data are largely derived from Mitchell (1994) who over a period of 18 months studied two groups of *P. sabana* in the Silabukan Forest Reserve and the Tabin Wildlife Reserve in Sabah, Malaysia, to asses the effects of selective logging on their ecology. A number of studies have been conducted on the effects of (selective) logging on vertebrates within the range of Hose's langurs, including *P. cancirus* (Wilson and Wilson 1975; Wilson and Johns 1982; Howell 2003), *P. sabana* (Johns 1992a, b; Grieser-Johns, 1997), and *P. hosei* (Bennett and Dahaban 1995; Hedges and Dwiyahreni 1995); where relevant, these data are included.

Additional data were collected from the study of museum specimens, and associated information on the specimen labels (collector's measurements, locality data etc.). I considered specimens in the zoological collections in Amsterdam, Leiden, London, Oxford, Paris, Bogor and Singapore.

Assessment of Remaining Habitat

I used the Geographic Information System software ArcGIS 9.2 (ESRI 2006) for distribution modelling. Forest cover data was obtained from the Southeast Asian Mammal Databank (Boitani et al. 2006) in the form of the REM dataset, a model depicting multiple vegetation layers derived from the Global Land Cover Classification map from the year 2000. Following the methods described in Meijaard and Nijman (2003), I extracted the appropriate habitat types and altitudinal limits (<1,500 m asl) from the environmental data layers to clip the extent of occurrence map of each of the three species to the areas of appropriate remnant forest cover. This allows for calculating the current area of occurrence; past area of occurrence is based on the assumption that in the past, prior to high levels of human deforestation and human-induced forest degradation, Hose's langurs were able to live throughout the geographical area.

Results and Discussion

Ecology and Habitat Use

Hose's langurs are arboreal colobines and this is reflected in their use of the forest strata. While it is well-documented that the langurs come down to the ground to drink from "sungans" (salt seepages or salt springs), especially after a period of a few rainless days, most of the time they spent in the trees. Three studies investigated the utilization of vertical strata in the forest for *P. hosei* (Nijman 1997), *P. sabana* (Mitchell 1994) and *P. canicrus* (Rodman 1978). All three studies indicate that the majority of the activities are in the middle layer of the forest between 10 and 30 m from the ground. In Kayan Mentarang, *P. hosei* was mostly observed in the stratum below 20 m and in Kutai *P. canicrus* was mostly observed in the stratum above 20 m (Table 16.1). All three species spent small amounts of time at the upper canopy, above 30 m, or at the lowest levels of the forest below 10 m.

Hose's langurs are primarily folivorous with a substantial amount of seeds eaten (Rodman 1978; Mitchell 1994; Leighton and Leighton 1983). The diet of *P. canicrus* in primary forest consists mainly of leaves and leaf shoots (66% of the feeding observations on first contact), followed by fruit (28%) with the remainder comprising either flowers, buds or insects (6%). Likewise, *P. sabana* in primary forest feeds mainly on leaves (78%), fruits (19% – with 17% including seeds) and flowers (3%). In logged forest, the proportion of leaves and flowers went down (60% and <1% respectively), and the proportion of fruit and seeds went up (40% – with 21% including seeds). The chemical composition of their food items shows that *P. sabana* and *P. canicrus* (Tables 16.2 and 16.3) rely on a low quality diet when compared with other folivorous primates. The crude protein/acid detergent fiber (CP/ADF) ratio for leaves, 0.38, is at the lower range of what has been reported for other colobines (*Nasalis* 0.29, *Rhinopithecus* 0.37, *Trachypithecus* 0.42, other *Presbytis* 0.58, *Colobus* 0.58, *Procolobus* 0.67: ratios calculated from data presented in Waterman and Kool 1994 and Nijboer and Clauss 2006) and comparable with values recorded for hypometabolic lemurs (Mutschler 1999). It appears that fruits are not selected for their protein content, but fruits and arils are an important source for appreciable quantities of fat.

Table 16.1 Utilization of vertical strata. the percentage of heights above the ground at first contact is presented for two species. All observations are included regardless of activity of animals when sighted. Tree data and *P. hosei* from Kayan Mentarang National Park (Nijman 1997) and data from *P. canicrus* from Kutai National Park (Rodman 1978)

Height (m)	Trees, dbh > 10 cm, $n = 100$	*P. hosei*, $n = 34$	*P. canicrus*, $n = 269$
0–10	27	17	6
10–20	55	50	29
20–30	12	25	62
>30	6	7	3

Table 16.2 Chemical composition of food plants eaten by Hose's langurs; all nutrients are on a dry matter basis (source Nijboer et al. 1997; Waterman and Kool 1994)

Item	Species	%-diet	CP	CT	NDF	ADF	Lignin	Fat
Leaves	P. sabana	78	16±4	–	55±20	42±19	24±13	–
Seeds	P. sabana	17	11±5	–	–	–	–	17±15
	P. canicrus	–	10	2.4	–	22	–	–
Fruit flesh	P. canicrus	6	7	2.8	–	–	–	32
Arils	P. canicrus	6	10	2.9	–	22	–	29
Flowers	P. sabana	3	10±3	–	52±7	40±8	27±9	–

Key: *CP* crude protein, *CT* condensed tannins, *NDF* neutral detergent fiber, *ADF* acid detergent fiber.

Table 16.3 Mineral composition of food plants eaten by *Presbytis sabana*; all nutrients are on a dry matter basis (source Nijboer et al. 1997)

Item	%-diet	Ash	Ca	Mg	Cu	Fe	Mn	Zn
Leaves	78	8±5	0.4±0.1	0.3±0.1	13±4	70±30	580±536	31±12
Seeds	17	8±12	0.5±0.7	0.1±0.1	10±3	65±50	174±151	29±28

Table 16.4 Relative use of primary riverine and adjacent interior forest by two species *Presbytis hosei* and *P. canicrus* both showing higher densities for the interior forests. Data from *P. hosei* from Kayan Mentarang National Park (Nijman 1997) and data from *P. canicrus* from Kutai National Park (calculated from data in Rodman 1988)

		Density in interior forest		Density in riverine forest	
	Group size	Groups (km⁻²)	Individuals (km⁻²)	Groups (km⁻²)	Individuals (km⁻²)
P. hosei	8.3	2.3	18.9	1.2	9.5
P. canicrus	8.0	2.9	22.3	2.3	18.4

Two studies investigated the use of riverine and interior forest, both under pristine conditions. Rodman (1978) assessed this in Kutai National Park by the relative use of 0.04 km² gridcells and Nijman (1997) assessed this in Kayan Mentarang National Park by means of transect walks. Both study areas included the home ranges of seven groups and the results are comparable. Riverine forest in Kutai was situated along the Sengata River, and was mainly below 40 m asl; the interior forest was adjacent to this up to an altitude of 300 m asl (Rodman 1988). In Kayan Mentarang, the riverine forest was situated along the Nggeng Bio River situated at 350 m asl, and the interior forest was adjacent to this up to an altitude of 550 m asl. While the densities for both species was similar in the interior forest, i.e. 2.3 and 2.9 groups km⁻² for Kayan Mentarang and Kutai, respectively, densities along the Nggeng Bio River were decisively less than along the Sengatta River (Table 16.4).

Distribution and Densities

North Borneo is one of the most species-rich areas in the world in terms of colobine monkeys (Meijaard and Nijman 2003). Apart from the silvered langur *Trachypithecus cristatus* and the proboscis monkey *Nasalis larvatus*, accepting the three-species arrangement of the Hose's langurs, there are five species of *Presbytis* langurs. In any one area, up to five colobines (*T. cristatus*, *N. larvatus* and three *Presbytis* langurs, only one of which is a member of the *P. hosei* group) can live sympatrically. This is more than in any other area in the world (Oates and Davies 1994).

The different Hose's langurs live sympatrically with a number of congeners. *P. sabana* lives sympatrically with *P. rubicunda*; *P canicrus* lives sympatrically with *P. rubicunda* and *P. frontata*; *P. hosei* lives sympatrically with *P. rubicunda* in the northern part of its range, with *P. rubicunda* and *P. chrysomelas* in the westernmost part of its range, and with *P. rubicunda* and *P. frontata* in the remaining part. A number of researchers have noted the checkerboard distribution pattern of especially *P. sabana* and *P. rubicunda* with one of the other species being present in any given forest area, without, however, any clear pattern (Davies and Payne 1982; Oates and Davies 1994). Figure 16.2 illustrates this well: it appears that where and when one of the two is abundant (>2 groups km^{-2}), either the

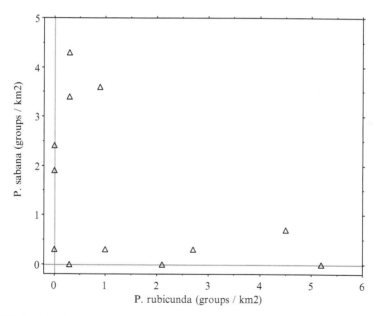

Fig. 16.2 Relationship between densities of *P. sabana* and *P. rubicunda* in Sabah (data from Davies and Payne 1982; for areas where either species was reported present at such a low density that no estimate was made, the density was arbitrarily set at 0.3 groups km^{-2}), showing that in general where the one species is common, the other is less common or absent

other is absent or it occurs at low densities. There are few areas where both species appear to be equally common or equally rare. A similar relationship appears to be present in northern East Kalimantan (either *P. hosei* or *P. rubicunda* common and the other rare, and *P. frontata* absent or rare) and eastern East Kalimantan (mostly *P. canicrus* common and *P. rubicunda* and *P. frontata* rare) (Nijman 1997; Rodman 1978; Wilson and Johns 1982).

In terms of altitudinal distribution, it appears that Hose's langurs can be found over a large elevational range (Fig. 16.3). Combining the data from the three species from primary forest sites, altitude is not a significant predictor for explaining

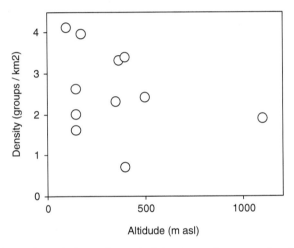

Fig. 16.3 Densities of Hose's langur (sensu lato) in primary forest areas in relation to altitude showing a non-significant relationship (data from Table 16.5)

Table 16.5 Densities of Hose's langurs in primary forest sites where they are deemed common enough to allow density estimates to be made

Species	Area	Altitude	Groups km^{-2}	Reference
P. hosei	Kayan Mentarang, E Kalimantan	300–500	1.2–2.3	Nijman 2004
	Bukit Ibul, Sabah	1,100	1.9	Davies and Payne 1982
	Ulu Temburong, Brunei	366	3.3	Bennett et al. 1987a
P. sabana	Ulu Segama, Sabah	150	2.0	Johns 1992a
	Ulu Segama, Sabah	150	1.6	Anonymous 1989
	Ulu Segama, Sabah	400	0.7	Davies and Payne 1982
	Tabin, Sabah	150–200	3.6–4.3	Davies and Payne 1982
	Ulu Sapulut, Sabah	400	3.4	Davies and Payne 1982
	Semantulang, Sabah	500	2.4	Davies and Payne 1982
P. canicrus	Kutai, E Kalimantan	50–250	2.6	Rodman 1978
	Kutai, E Kalimantan	100	4.1	Wilson and Wilson 1975

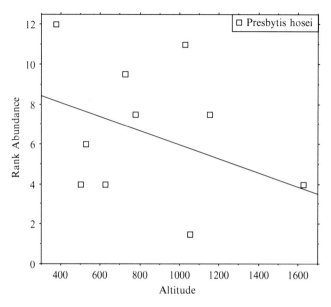

Fig. 16.4 Rank abundance of *P. hosei* in relation to altitude in Kayan Mentarang National Park (data from Wulfraat and Samsu 2000); the relationship is a non-significant one and the trend line is for illustrative purposes only

densities (Table 16.5). In the westernmost part of Sabah, in the Crocker Range and on Mt Kinabalu, *P. hosei* and *P. sabana* be found at altitudes of up to 1600 m asl and occasionally higher especially on the highest mountains (Goodman 1989; Medway 1970). In Kayan Mentarang National Park, *P. hosei* shows no clear relationship between abundance and altitude, apart from that densities appear to be low at altitudes above 1,600 m asl (Fig. 16.4). Davis (1958) noted that in Sarawak's Kelabit highlands at 1,000–1,100 m asl, just across the border of Kayan Mentarang, *P. hosei* was the commonest monkey of the mountain sides. Within the distribution range of *P. canicrus,* there are few high mountains (the exception is Mt Mantan at 2,467 m asl) and this is dully reflected in its smaller altitudinal range.

Average group sizes of Hose's langurs are relatively small, ranging from 7 to 10 individuals. *Presbytis hosei* in Kayan Mentarang lived in groups between 1 and 11 individuals averaging 8.3 individuals in primary forest, and in Temburong, Brunei, group sizes averaged 10 individuals. Rodman (1978) reports an average group size of 8 individuals for *P. canicrus* and Mitchell (1994) studied two groups of *P. sabana* of 7 and 9 individuals respectively. Groups normally contain a single adult male, although all-male groups of 4 individuals and solitary individuals have been observed in Kayan Mentarang (Nijman 2004). In undisturbed forest areas, densities range from less than one group km^{-2} to over 4 groups km^{-2} (Table 16.5), translating to densities of up to 25–30 individuals km^{-2} (Wilson and Wilson 1975; Bennett et al. 1987).

Habitat Loss, Selective Logging and Hunting

Hose's langurs are threatened by a number of largely anthropogenic factors, including habitat loss (either through habitat conversion-oil palm or wood pulp plantations-or fires associated with the ENSO events), selective logging (additionally increasing the risk of forest fires) and hunting (Berenstain et al. 1986; Bennett et al. 1987b; Nijman 2005). The Red List status of *P. hosei* s.l. is Vulnerable, with that of *P. canicrus*, *P. sabana* and *P. hosei* (all listed as subspecies in the IUCN Red List, the latter as *everetti*) all as Endangered (Nijman et al. 2008). For all three species, habitat loss has been considered 50% in the past 20 years and in the next 10 years will certainly be well over the 50% threshold.

Analysis of the year 2000 forest data shows that throughout the range of the three species over half of the forest has been lost (Table 16.6), and of the three species *P. sabana* proportionally has lost most of its forest. While *P. canicrus* has been included in the Top 25 Most Endangered Primates (Brandon-Jones 2005) in fact, based on 2000-forest cover data, of the three species it has proportionally lost least of its habitat. Conservative estimates suggest that both *P. sabana* and *P. canicrus* have some 20,000 km² of habitat remaining, and *P. hosei*, about 40,000 km². Given the patchy distribution of Hose's langurs within their range, it is difficult to extrapolate this to numbers, but a population of >10,000 mature individuals requires an average population density of one group per 10–20 km². Data from Table 16.4 and data presented in the following section suggest that densities are often considerably higher than this.

A number of studies have been conducted on the effect of logging or habitat disturbance in Hose's langurs. Most of these studies are summarised by Meijaard et al. (2005: 76). Johns (1992b) found densities of *P. sabana* to be twice as high in forest plots selectively logged 6 and 12 years prior to the study, than in an adjacent undisturbed plot. Davies and Payne (1982), however, reported *P. sabana* about half as common in a 19-year-old logged forest than in adjacent selectively logged plots. For *P. canicrus,* Howell (2003) reported a progressive decline in densities with increasing time since logging (2–4 years) suggesting a time-lag between the onset of logging and declining primate densities. Wilson and Johns (1982) commented that *P. canicrus* was able to withstand the pressures imposed by logging, but do not provide quantitative data. For *P. hosei*, Nijman (2000) found 30–40% lower

Table 16.6 Estimates of extent of occurrence (in km²) for Hose's langurs. Ranges are based on Fig. 16.1, with conservative estimates derived from the area encompassed by the confirmed localities of each species, and the liberal estimate by inclusion of unconfirmed localities

Species	Conservative estimate		Liberal estimate	
	Original	Remaining (% of original)	Original	Remaining (% of original)
P. hosei	91,120	41,840 (46)	123,010	57,780 (47)
P. sabana	54,850	22,260 (41)	85,410	33,040 (39)
P. canicrus	37,570	19,330 (51)	76,570	41,110 (54)
P hosei s.l.			246,340	111,300 (45)

densities in 45 and 10–20-year-old secondary forest when compared with nearby primary forest. Hedges and Dwiyahreni (1995) counted 10–11 groups of *P. hosei* in c. 6.5 km of transect walks in undisturbed lowland and 9 groups in c. 2.5 km of transect walks in partially disturbed lowland forest, suggesting higher densities in the disturbed forest types. During timed mammal searches, however, this difference was no longer apparent, with in fact a higher number of groups encountered in the undisturbed forest (27 groups per 100 h) than in the disturbed forest (22 groups per 100 h). Combined these data suggest that selective logging in itself may not be detrimental to any of the Hose's langurs, however, as noted by Meijaard et al. (2005) logging operations often lead to an increase in hunting and this clearly has a negative impact on population densities.

Hunting of Hose's langurs can have a dramatic impact on local populations (Nijman 2005). Hunting of primates is especially prevalent in Borneo's interior as many of the coastal areas are inhabited by people who adhere to Islamic principles and do not eat primates. The various interior Dayak and Punan (or Penan) tribes generally do hunt and eat primates. While most primates in Borneo are hunted for their meat, *Presbytis* langurs are additionally targeted for their bezoar stones (visceral excretions used in Traditional Chinese Medicine; locally the stones are known as *batu geligu*: Nijman 2005; Eghenter 2001; Banks 1931; Pfeffer 1958). Bezoar stones are found as a calculus or concretion in the stomachs or intestines of various ruminants (antelopes, deer, goats), porcupines and colobines. Bezoar stones are principally calcium phosphate, but the active ingredient is the crystalline mineral brushite. The name bezoar comes from the Persian *padzahr*, meaning "to expel poison". Once it was speculated that the bezoar stone originated from the Unicorn and would protect its possessor from evil, and the stone would be especially effective in preventing poisoning. From the Middle Ages on down to our own time, the bezoar stone's reputed efficacy has grown to include all manner of diseases and maladies. In the Bornean context, bezoar stones are nowadays mainly traded to supply the Traditional Chinese Medicine market.

Prices of bezoar stones may fluctuate and increase greatly with increasing distance to the area of origin (i.e. the island's more remote parts), but fetched the highest unit prices of all the forest products, i.e. USD 21 g^{-1} for small (<10 g) stones and USD 28 g^{-1} for high quality larger (10–35 g) stones. Hunting for bezoar stones often occurs near "sungans" (salt seepages or salt springs); after a period of a few rainless days, hunters hide nearby and wait for the langurs to decent from the trees to drink at the sungans. Langurs suffering from bezoar stones are often sick and old, and local hunters, particularly Punan, claim that they can identify individuals that carry stones. Many outside collectors lack the necessary experience and specialized knowledge for this, however, and do hunt indiscriminately. Some resort to more drastic methods such as poisoning sungans killing hundreds of langurs in the process (Eghenter 2001; Nijman 2005). Eghenter (2001) reports that in a survey among 38 Dayaks in Apo Kayan, East Kalimantan the respondents admitted to a dual aim of all forest expeditions, namely collecting gaharu (a fragrant resin produced by *Aquilaria* trees in response to fungal attack, which is traded internationally; Soehartono and Newton 2001) and bezoar stones from *P. hosei*.

In another survey reported by Eghenter (2001), covering 43 forest expeditions in a different village in Apo Kayan more than 50% of the informants stated that the purpose of the expeditions had been to find gaharu trees and hunt *P. hosei* when there were guns available.

Eghenter (2001) asserted that, in the past, the collection of bezoar stones was largely in the hands of Punan, who in turn traded the stones with Dayak headmen in exchange for rice and other goods, and that as the number of people involved in collecting activities was probably limited and the pressure exerted discontinuous, hunting proved a sustainable activity. Cowlishaw and Dunbar (2000) present a framework according to which harvesting of primates might be sustainable from an economic perspective. Two factors are important to consider, i.e. the population intrinsic growth rate (r) and the economic discount rate (d).

The population intrinsic growth rate is the maximum possible rate of increase that a species is capable under the best possible conditions. For primates, it has been found empirically that growth rate correlates with body mass (larger species breed more slowly), and when corrected for body mass, it correlates negatively with the amount of forest cover characteristic for a species' typical habitat (Ross 1988). Thus, it can be expected that the intrinsic growth rate for relatively large species in a rainforest environment are amongst the lowest for primates as a whole. Ross (1988: 218) gives an intrinsic growth rate of 0.15 for *Presbytis* langurs.

The economic discount rate is the rate at which the value of an investment declines over time. A primate population represents capital and the harvest of that population represents interest on that capital. If the financial gain through harvesting is less than the potential financial gain through interest on the equivalent sum of money invested elsewhere, then the economic rational course of action is to immediately and totally convert that primate population to cash and reinvest the capital so acquired.

For harvested populations, the critical question therefore becomes the relative magnitudes of the population's intrinsic growth rate (r) and the economic discount rate (d). When d equals r, or when d is larger than r, from an economic perspective, harvesting to extinction becomes the most rationale action. When considering sustainability in hunting levels of Hose's langurs, the frequency of occurrence of bezoar stones within populations need to be taken into account as this is consistently indicated as the main incentive for hunting (although the animal thus acquired does provide a source of meat). Pfeffer (1958) and Banks (1931), referring to *P. hosei* and *P. frontata*, respectively, mentioned that both species were sought after for their bezoar stones, but only a few small stones were found at a time (cf. Nijman 2005). If we assume a relative high frequency of occurrence of stones of 10% (which assumes at least for Hose's langurs that in almost all groups at least one individual suffers from this ailment), harvesting to extinction becomes the economic most rationale action when d equals 1.5%. This rate is far below the current inflation rates in Indonesia and Malaysia, and it goes beyond saying that just bringing your money to the bank will give you a higher return on investment than sustainably managing populations of Hose's langurs.

Conclusions

1. The three colobines, *P. hosei*, *P. canicrus* and *P. sabana* are confined to northern Borneo with the first two or possibly all three species occurring within the Indonesian part of the island. They are largely arboreal, spending most of their time in trees between 10 and 30 m, but will come to the ground level to drink from salt seepages or salt springs.
2. *Presbytis canicrus* and *P. sabana* (and presumably also *P. hosei*) are primarily folivorous, eating a substantial amount of seeds. The chemical composition of their diet suggesting reliance on low-quality food items.
3. The three langurs live sympatrically with one or two other *Presbytis* langurs and the available data suggest that where the one is common, the other is rare or absent, suggesting competitive exclusion. Average group sizes range from 7 to 10 individuals with one adult male, with densities in undisturbed forests ranging from <1 to >4 groups km^{-2}.
4. The main threats to the species are habitat loss and hunting. Each species has lost about half of its habitat mainly because of logging and fire with, conservatively, some 20,000 km^2 of forest remaining for both *P. sabana* and *P. canicrus* and about 40,000 km^2 for *P. hosei*. Hunting, especially for bezoar stones, has a major impact, such that it can lead to local extinctions.

Acknowledgements I wish to thank the Indonesian authorities, through the Directorate General of Forest Protection and Nature Conservation (PHKA), the Provincial Agency for the Conservation of Natural Resources (SBKSDA), and the Indonesian Institute for Sciences (LIPI) for permission to conduct my research. I would like to thank WWF Indonesia, and in particular Christina Eghenter, Tim C. Jessup, Tonny Soehartono, and Stephan Wulfraat for giving me the opportunity to participate in the Kayan Mentarang Project. All WWF staff are thanked for help and support. The curators of the zoological museums, Adri Rol (Amsterdam), Chris Smeenk (Leiden), Paula Jenkins (London), Malgosia Nowak-Kemp (Oxford), Jacques Cuisin (Paris), Boeadi (Bogor) and C.M. Yang (Singapore), are thanked for access to the specimens in their care. Financial support was received from the Society for the Advancement of Research in the Tropics and the Netherlands Foundation for International Nature Protection. Dave Smith kindly helped with the GIS analysis and Giuseppe Donati with interpretation of the nutritional data. I thank the editors, Drs Sharon Gursky and Jatna Supriatna for inviting me to contribute to this volume.

References

Anonymous (1989) New white leaf monkey observed in Southeast Asia. Primate Conserv 10:30–31
Banks E (1931) A popular account of the mammals of Borneo. J Malays Branch R Asiat Soc 9:21–139
Bennett EL, Dahaban Z (1995) Wildlife responses to disturbances in Sarawak and their implications for wildlife management. In: Primack RB, Lovejoy TE (eds) Ecology, conservation and management of Southeast Asian rainforests. Yale University Press, New Haven, pp 66–86
Bennett EL, Davies AG (1994) The ecology of Asian colobines. In: Davies AG, Oates JF (eds) Colobine monkeys: their ecology, behaviour and evolution. Cambridge University Press, Cambridge, pp 129–171

Bennett EL, Caldecott J, Kavanagh M, Sebastian A (1987a) Current status of primates in Sarawak. Primate Conserv 8:184–187

Bennett EL, Caldecott J, Davisson GWH (1987b) A wildlife survey of Ulu Temburong, Brunei. Brunei Mus J 6:121–169

Berenstain L (1994) The wind monkey and other stories. Random House, New York

Berenstain L, Mitani JC, Tenaza RR (1986) Effects of El Niño on habitat and primates in East Kalimantan. Primate Conserv 7:54–55

Boitani L, Catullo I, Marzetti M, Masi M, Rulli M, Savini S (2006) The Southeast Asian Mammal Databank. A tool for conservation and monitoring of mammal diversity in Southeast Asia. Instituto di Ecologia Applicata, Roma

Brandon-Jones D (1984) Colobus and leaf monkeys. In: Macdonald D (ed) The encyclopaedia of mammals, vol 1. Allen and Unwin, London, pp 398–408

Brandon-Jones D (1993) The taxonomic affinities of the Mentawai Islands Sureli, *Presbytis potenziani* (Bonaparte, 1856) (Mammalia: Primata: Cercopithecidae). Raffles Bull Zool 41:331–357

Brandon-Jones D (1996) *Presbytis* species sympatry in Borneo versus allopatry in Sumatra: an interpretation. In: Edwards S (ed) Tropical rainforest research – current issues. Kluwer, Dordrecht, pp 71–76

Brandon-Jones D (1997) The zoogeography of sexual dichromatism in the Bornean grizzled surili, *Presbytis comata* (Desmarest 1822). Sarawak Mus J 50:177–200

Brandon-Jones D (2005) Miller's Grizzled Surili, *Presbytis hosei canicrus* Miller, 1934. In: Mittermeier RA, Valladares-Pádua C, Rylands AB, Eudey AA, Butynski TM, Ganzhorn JU, Kormos R, Aguiar JM, Walker S (eds) Primates in peril: the World's 25 most endangered primates 2004–2006. IUCN/SSC Primate Specialist Group, International Primatological Society and Conservation International, Washington, DC, p 26

Chasen FN (1940) A handbook of Malaysian mammals. A systematic list of the mammals of the Malay Peninsula, Sumatra, Borneo, Java, including the small adjacent islands. Bull Raffles Mus 15:1–209

Cowlishaw G, Dunbar RIM (2000) Primate conservation biology. University of Chicago Press, Chicago

Davies AG, Payne J (1982) A faunal survey of Sabah. World Wildlife Fund, Kuala Lumpur, Malaysia

Davis DD (1958) Mammals of the Kelabit Plateau, Northern Sarawak. Fieldiana Zool 39:119–147

Eghenter C (2001) Towards a casual history of a trade scenario in the interior of East Kalimantan, Indonesia, 1900–1999. Bijdr Taal Land Volkenkd 157:739–769

ESRI (2006) ArcGIS 9.2. Available from http://www.esri.com/software/arcgis/

Geissmann T, Stuenkel A, Vermeer J, Nijman V (2008) The loud calls of Asian colobines (Primates: Cercopithecidae): a phylogenetic approach. Unpublished report, University Zürich-Irchel, Zürich

Goodman SM (1989) Predation by the grey leaf monkey (*Presbytis hosei*) on the contents of a bird's nest at Mt. Kinabalu Park, Sabah. Primates 30(1):127–128

Grieser-Johns A (1997) Timber production and biodiversity conservation in tropical rain forests. Cambridge University Press, Cambridge

Groves CP (1993) Order primates. In: Wilson DE, Reeder DM (eds) Mammal species of the World: a taxonomic and geographic reference. Smithsonian Institution Press, Washington, DC, pp 243–278

Groves CP (2001) Primate taxonomy. Smithsonian Institution Press, Washington, DC

Hedges S, Dwiyahreni AA (1995) Section 2: mammal surveys. In: O'Brien TG, Fimbel RA (eds) Faunal survey in unlogged forest of the INHUTANI II Malinau timber concession, East Kalimantan, Indonesia. Wildlife Conservation Society, New York

Howell D (2003) The effects of human activity on primates and other large mammals in East Kalimantan, Indonesia. MA thesis, Central Washington University, St Louis

Johns AD (1992a) Vertebrate responses to selective logging: implications for the design of logging systems. Proc R Soc Lond B 335:437–442

Johns AD (1992b) Species conservation in managed forest. In: Whitmore TC, Sayer JA (eds) Tropical deforestation and species extinction. Chapman and Hall, London, pp 15–53

Kamilar JM, Paciulli LM (2008) Examining the extinction risk of specialized folivores: a comparative study of colobine monkeys. Am J Primatol 70:816–827

Lord Medway (1970) The monkeys of Sundaland: ecology and systematics of the cercopithecids of a humid equatorial environment. In: Napier JR, Napier PH (eds) Old World monkeys: evolution, systematics and behavior. Academic, New York, pp 513–553

Leighton M, Leighton DR (1983) Vertebrate response to fruiting seasonality within a Bornean rain forest. In: Sutton SL, Whitmore TC, Chadwick AL (eds) Tropical rain forest ecology and management. Blackwell, Oxford, pp 181–196

Meijaard E, Groves CP (2004) The biogeographical evolution and phylogeny of the genus *Presbytis*. Primate Rep 68:71–90

Meijaard E, Nijman V (2003) Primate hotspots on Borneo: predictive value for general biodiversity and the effects of taxonomy. Conserv Biol 17:725–732

Meijaard E, Sheil D, Rosenbaum B, Iskandar D, Augeri D, Setyawati T, Duckworth W, Lammertink WJ, Rachmatika I, Nasi R, Wong A, Soehartono T, Stanley S, O'Brien T (2005) Life after logging: reconciling wildlife conservation and production forestry in Indonesian Borneo. CIFOR, WCS and UNESCO, Bogor, Indonesia

Mitchell AH (1994) Ecology of Hose's langur (*Presbytis hosei*) in logged and unlogged Dipterocarp forest of northeast Borneo. Ph.D dissertation, Yale University, New Haven

Mutschler T (1999) Folivory in a small-bodied lemur: the nutrition of the Alaotran gentle lemur (*Hapalemur griseus alaotrensis*). In: Rakotosamimanana B, Rasamimanana H, Ganzhorn JU, Goodman SM (eds) New directions in lemur studies. Kluwer Academic/Plenum Publishers, New York, pp 221–239

Napier PH (1985) Catalogue of primates in the British Museum (Natural History) and elsewhere in the British Isles. Part III: Familie Cercopithecidae, subfamily Colobinae. British Museum (Natural History), London

Newton PN, Dunbar RIM (1994) Colobine monkey society. In: Davies GA, Oates JF (eds) Colobine Monkeys, their ecology, behaviour and evolution. Cambridge University Press, Cambridge, pp 311–346

Nijboer J, Clauss M (2006) The digestive physiology of colobine primates. In: Nijboer J (ed) Fibre intake and faeces quality in leaf-eating primates. Ridderkerk, The Netherlands, pp 9–28

Nijboer J, Dierenfeld ES, Yeager CP, Bennett EL, Bleisch W, Mitchell AH (1997) Chemical composition of Southeast Asian colobine foods. In: Proceeding of the second conference of the nutrition advisory group. AZA, Forth Worth

Nijman V (1997) Preliminary survey on colobine monkeys and other primates in north-eastern Kalimantan. WWF-Indonesia, Jakarta

Nijman V (2000) Geographical distribution of ebony leaf monkey *Trachypithcus auratus* (Geoffroy Saint Hilaire 1812) (Mammalia: Primates: Cercopthecidae). Contrib Zool 69:157–177

Nijman V (2003) The primates. In: Suhartono T (ed) Species management plan for Kayan Mentarang National Park. WWF-Indonesia, Jakarta

Nijman V (2004) Effects of habitat disturbance and hunting on the densities and biomass of the endemic Hose's leaf monkey *Presbytis hosei* (Thomas 1889) (Mammalia: Primates: Cercopithecidae) in east Borneo. Contrib Zool 73:283–291

Nijman V (2005) Rapid decline of Hose's langur in Kayan Mentarang National Park. Oryx 39(2):223–226

Nijman V, Meijaard E, Hon J (2008) *Presbytis hosei*. In: IUCN 2008. 2008 IUCN Red List of threatened species. www.iucnredlist.org. Accessed 6 Jan 2009

Oates JF, Davies AG (1994) Conclusions: the past, present and future of the colobines. In: Davies GA, Oates JF (eds) Colobine monkeys, their ecology, behaviour and evolution. Cambridge University Press, Cambridge, pp 347–358

Pfeffer P (1958) Situation actuelle de quelques animaux menacés d'Indonesie. La Terre et la Vie 105:128–145

Pocock RI (1935) The monkeys of the genera *Pithecus* (or *Presbytis*) and *Pygathrix* found to the east of the Bay of Bengal. Proc Zool Soc London 1934:895–961

Rodman PS (1973) Synecology of Bornean primates. A test for interspecific interactions in spatial distribution of five species. Am J Phys Anthropol 38:655–660

Rodman PS (1978) Diets, densities and distributions of Bornean primates. In: Montgomery GG (ed) The ecology of arboreal folivores. Smithsonian Institute Press, Washington DC, pp 465–478

Rodman PS (1980) Why monkeys live together. Int Wildl 10:18–23

Rodman PS (1988) Resources and group sizes of primates. In: Slobodchikoff CN (ed) The ecology of social behavior. Academic, San Diego, pp 83–108

Ross C (1988) The intrinsic rate of natural increase and reproductive effort in primates. J Zool Lond 214:199–219

Soehartono T, Newton AC (2001) Conservation and sustainable use of tropical trees in the genus *Aquilaria* II. The impact of gaharu harvesting in Indonesia. Biol Conserv 97:29–41

Waterman PG, Kool KM (1994) Colobine food selection and plant chemistry. In: Davies GA, Oates JF (eds) Colobine monkeys, their ecology, behaviour and evolution. Cambridge University Press, Cambridge, pp 251–284

Wilson WL, Johns AD (1982) Diversity and abundance of selected animal species in undisturbed forest, selectively logged forest and plantations in East Kalimantan, Indonesia. Biol Conserv 24:205–218

Wilson CC, Wilson WL (1975) The influence of selective logging on primates and some other animals in East Kalimantan. Folia Primatol 23:245–274

Wulfraat S, Samsu T (2000) An overview of the biodiversity of Kayan Mentarang National Park. WWF-Indonesia Programme, Balikpapan

Chapter 17
Thomas Langurs: Ecology, Sexual Conflict and Social Dynamics

Serge A. Wich and Elisabeth H.M. Sterck

Introduction

The Thomas langur (*Presbytis thomasi*) is a colobine species endemic to northern Sumatra, Indonesia. Despite their limited distribution, this species may provide insights into the socio-ecology of folivorous primates. Predictions of the socio-ecological model (van Schaik 1989) suggest that colobine primates feed from nonmonopolizable food sources. Females are therefore expected to experience mainly within-group scramble competition. When this type of competition prevails, the female dominance hierarchy will not be despotic, and female coalitions against other female group members will be rare. Moreover, females may disperse between groups. Many folivorous primates, however, do not fit this predicted pattern (Sterck 1999; Chapman and Pavelka 2005). Some species are presumed to lack scramble competition because group sizes are relatively small and grouping does not seem to entail costs, also known as the folivore paradox (Steenbeek and van Schaik 2001). It has been suggested that not food competition, but male sexual strategies may limit their group size (Crockett and Janson 2000). Different connections between food competition, sexual strategies, and social behavior may exist in folivorous primates, and the Thomas langur may represent one possible connection.

The Thomas langur is a relatively well-studied colobine monkey. Their food sources, food competition, and the effect of group size on behavior have been determined. In addition, male sexual strategies have been investigated, and the dynamics of their social system are well documented. This allows an exploration of the fit and deviations in the predictions of the socio-ecological model and the importance of male sexual strategies for female behavior. Similar social dynamics are found in a number of other Asian colobines (Sterck 1998), a one-male red colobus population (*Colobus badius*: Marsh 1979) and mountain gorillas (*Gorilla gorilla beringei*: Harcourt et al. 1976; Watts 1989; Robbins and Sawyer 2007).

S.A. Wich (✉)
Great Ape Trust of Iowa, 4200 SE 44th Ave, Des Moines, IA, 50320, USA
e-mail: swich@greatapetrust.org

S. Gursky-Doyen and J. Supriatna (eds.), *Indonesian Primates*,
Developments in Primatology: Progress and Prospects,
DOI 10.1007/978-1-4419-1560-3_17, © Springer Science+Business Media, LLC 2010

Therefore, understanding Thomas langur's social dynamics may suggest an explanation for the behavior of folivorous species with a similar social organization and may generate hypotheses concerning folivorous species with a different social organization. In this chapter, we review the distribution, ecology and behavior of the Thomas langur and explore how their social dynamics relate to ecology and sexual strategies.

Thomas Langurs and the Study Areas

The Thomas langur is one of the many langur species that occur in Southeast Asia. While the langur taxonomy is under discussion (Brandon-Jones et al. 2004) and Thomas langur skeletal and pelage features can be hard to distinguish from the neighboring *P. melalophos*, the male loud call is substantially different between the species (Wilson and Wilson 1975). Therefore, they are usually recognized as a distinct species. Thomas langurs, like all colobine monkeys, are able to digest unripe fruits and leaves, while soft sweet fruits form a problem (Bauchop 1971), and this is reflected in their diet (Sterck 1995). The species shows no sexual dimorphism in body size and adults weigh 7 to 8 kg (Sterck 1995). However, similar to many other primates (Plavcan and van Schaik 1992), males have substantially larger canines than females (Sterck unpublished data).

Thomas langur behavior has been studied in detail at two sites, the Ketambe Research Station, Aceh, and at Bohorok, North Sumatra, both located in the Gunung Leuser Ecosystem. The study at the Ketambe Research Station has been the longest and most detailed study of this species and was conducted from 1987 to 2001. The Ketambe habitat consists of alluvial lowland forest. A study of shorter duration (1981–1984) has been conducted in the forest-plantation mosaic of Bohorok (Gurmaya 1986).

At Ketambe, a total of 20 mixed-sex groups were studied, along with 16 all-male-bands (amb's) and solitary males (Sterck et al. 2005; Wich et al. 2007). Most data were collected during full day follows of langur groups during which behavior of individuals was recorded. The methods employ focal observations, wherein during 15 min periods, behavior was recorded each minute and observations rotated between individuals. Full day follows of males were also made (see for details Steenbeek and van Schaik 2001; Sterck 1995). At Bohorok, three groups, two mixed-sex groups and one amb, were studied (Gurmaya 1986). In this chapter, most data come from the Ketambe study, but where possible, we compare these results with those from Gurmaya (1986).

Distribution and Conservation Status

Thomas langurs are found in a wide variety of habitats (Fig. 17.1). The highest recorded density is from a swamp forest (172 kg/km^2), with somewhat lower densities recorded in lowland forest (131 kg/km^2) and lowland alluvial forest

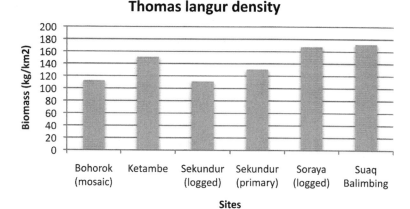

Fig. 17.1 Suaq balimbing (peat swamp forest), Soraya (selectively logged lowland dipterocarp forests), Ketambe (alluvial lowland forest), Sekundur (primary: lowland dipterocarp forest), Bohorok (mosaic of rubber plantations and primary hilly dipterocarp forest) and Sekundur (logged: logged dipterocarp lowland forest). Data from Wich and van Schaik unpublished, Sterck 1995, Priatna 1997. Sites are arranged alphabetically

(151 kg/km^2). Thomas langurs can also live in disturbed forests, such as in a mosaic of rubber plantations and primary hilly dipterocarp forest. Densities in selectively logged areas do not seem to be much lower than in primary rainforests. They are found from lowland to the montane (1,500–2,400 m) and the lower parts of the subalpine zones (2,400–3,400 m) in the Leuser Ecosystem (Wich, unpublished data) and density decreases with elevation. In all these habitats, the Thomas langur is arboreal although they occasionally come to the ground to feed (Sterck 1995). Probably, their most common habitat is alluvial lowland forest, which also forms the habitat of Ketambe Research Area.

Thomas langurs are endemic to the northern part of the island of Sumatra, Indonesia. Island-wide surveys to determine the extent of the *Presbytis melalophos* group report that Thomas langurs are mainly found in inland forests northwest of the Wampu and north of the Alas River, but they also found one population south of the Alas River (Aimi and Bakar 1992; 1996). Since Thomas langurs can survive in mosaics of forests and plantations, it is difficult to determine the precise extent of their range. However, on the basis of data from the literature, extensive survey work by one of the authors (SAW) and recent information by Utami Atmoko (pers comm.), a rough distribution map was made (Fig. 17.2). It should be noted that, although the langurs are found up to high altitudes, some high mountain areas are probably not part of their distribution.

Hunting Thomas langurs for the pet trade occurs as the species can be available on the animal market in Medan, the capital city of Sumatra Utara, albeit in low numbers when compared with long-tailed macaques (*Macaca fascicularis*, Shepherd et al. 2004). Probably, more important for their densities is the extensive logging and conversion of forests to agricultural plantations of northwestern Sumatran

Fig. 17.2 Thomas langur distribution map. © Perry van Duijnhoven

forests (van Schaik et al. 2001; Wich et al. 2008a). Subsequently, the IUCN now categorizes the Thomas langur as Vulnerable (Supriatna and Mittermeier 2008).

Ecology

Thomas langurs inhabit the evergreen tropical rainforests of northern Sumatra. The annual rainfall in the region is high. Suaq Balimbing, in the coastal peat swamps on the west coast of Aceh, receives 3,362 mm (Wich and van Schaik 2000),

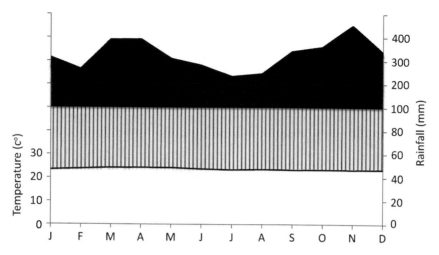

Fig. 17.3 Standard climate diagram (following Walter and Leih 1967) for Ketambe during 1996-2007. The black area indicates rainfall above 100 mm, which illustrates very wet periods. The area below 100 mm, where precipitation is higher than the temperature is hatched to indicate that it is a humid period

Ketambe that lies in between the two Barisan mountain ranges has an annual rainfall of 3,288 mm (Wich and van Schaik 2000), the lowland dipterocarp forest of Sekundur on the east of the Barisan range receives 3,367 mm per year (Wich unpublished data) and the Bohorok mosaic of plantations and secondary hilly dipterocarp forest south of Sekundur receives 4,575 mm per year (Gurmaya 1986).

The forests in northern Sumatra do not show a pronounced dry season in much of their range (Whitten et al. 1987). As an illustration, the mean monthly rainfall in Ketambe is higher than 200 mm per month in all months (Fig. 17.3) and has two rainfall peaks in April–May and September–November (van Schaik 1986). The temperature variation, as is common in the tropics, is very limited (Fig. 17.3). As a result of the nonlimiting rainfall and the fertile volcanic soils, forest productivity is high (Marshall et al. 2009; van Schaik 1986). Production of flowers and fruits, however, is not constant and shows periods with high and low availability within and between years (Fig. 17.4). In addition, in some years, community-wide forest production of flowers and fruits is much higher than other years (Ashton et al. 1988; van Schaik 1986; Wich and van Schaik 2000) and this Southeast Asian phenomenon is called masting. Also, the production of young leaves is seasonal and is more pronounced during the months from November to March and in August (van Schaik 1986, Fig. 17.5), while mature leaves are always available. As all these food items are consumed by Thomas langurs, this indicates that their food availability fluctuates within and between years.

Thomas langurs have a broad diet that differs between sites. At Ketambe, they feed from 218 species of trees and lianas, consisting of 191 fruit species, 28 flower species and 69 leaf species (Sterck 1995). This differs from the number

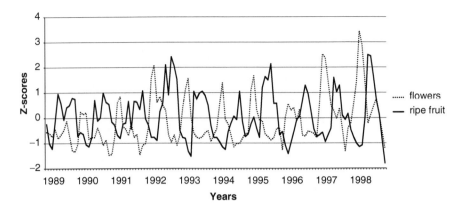

Fig. 17.4 Flower and fruit production for Ketambe for 1989-1998 expressed as standardized z-scores. Details of the methods can be found in Wich and van Schaik 2000

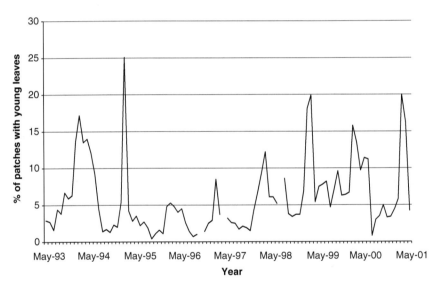

Fig. 17.5 Leaf production in Ketambe. Only trees with more than 750 leaves were included as having leaves since many trees almost continuously have some young leaves and including those would reduce variability in leaf production for trees with large numbers of young leaves that are important for the langurs

of plant species that constitute the Thomas langur diet at the forest-plantation mosaic at Bohorok. At this site, the diet consists of 26 plant species, which were mainly agricultural species such as banana (*Musa* sp.) and rubber (*Hevea brasiliensis*) (Gurmaya 1986). Although at Ketambe there is a larger number of fruit than leaf species in the diet, the number of food patches visited per day is similar for

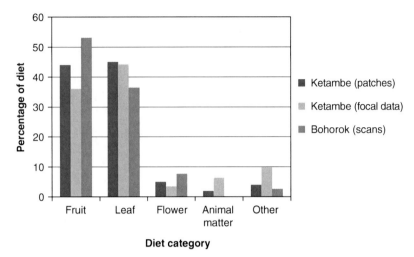

Fig. 17.6 Diet data based on patches visited and focal data from Ketambe from four groups (Sterck 1995) and from scan data from three groups in Bohorok (Gurmaya 1986). Animal matter was not scored separately for Bohorok and were included in the category 'other'

fruits and leaves (Fig. 17.6). The percentage of time spent eating fruits is slightly lower than the time spent eating leaves. At Bohorok, however, the langurs spend more time feeding on fruit than on leaves. This may result from less variation in the availability of fruits in agricultural areas than forests. In addition to fruit and leaves, also flower and animal matter are consumed. At Ketambe, the langurs also feed on items such as snails from small streams, algae during the dry season when pools with algae form in the Ketambe and Alas River, and dirt from termite mounts and from between the roots of toppled trees (Sterck 1995; Wich unpubl. data). Since the largest proportions of the Thomas langur diet consist of fruits and leaves, it can be best described as a foli-frugivore. The variation in their diet is illustrated by an example of the items included on one day in their diet (Fig. 17.7).

Perhaps owing to the substantial contribution of leaves to the diet, Thomas langurs spend a large part of the day resting and spend a smaller part of their time feeding (Fig. 17.8). In general, leafy material requires longer gut retention time than fruits (Clauss et al. 2008), which could result in longer resting time. This idea is supported by a comparison with sympatric more frugivorous primate species in Ketambe. Long-tailed macaques (*Macaca fascicularis*), orangutans (*Pongo abelii*), siamangs (*Hylobates syndactylus*) and white-handed gibbons (*Hylobates lar*) all feed considerably less on leaves than Thomas langurs and also rest less (Morrogh-Bernard et al. 2009; Palombit 1997; van Schaik and van Noordwijk 1986). However, a comparison of feeding habits and time budget of primate communities at multiple sites is needed to test this idea. In Thomas langurs, moving and time spent on social activities such as grooming take up an even smaller part of their activity budget. Despite differences in diet, the activity budgets of Ketambe and Bohorok are very similar.

Fig. 17.7 Example of all items in a day of Thomas langur feeding

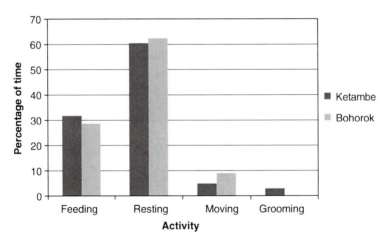

Fig. 17.8 Activity budget data for Ketambe (Sterck 1995) and Bohorok (Gurmaya 1986). Grooming data were not available for Bohorok

The mean home range size (of mid tenure phase groups, see below) at Ketambe measures 27.4 ha (sd = 13.0, $n = 14$ groups, Steenbeek and van Schaik 2001). This is much larger than those reported from Bohorok, where the mean home range is 14.0 ha (sd = 2.4, $n = 2$ groups) (Gurmaya 1986). However, at the Bohorok, number of sampled groups is much smaller and the tenure phase of the male is not known.

Therefore, it is not possible to determine whether this home range smaller size reflects a real difference or results from a sampling bias. The Ketambe Thomas langurs traverse these home ranges with a day journey length (DJL) that is has a mean of 1,067.7 m (sd = 149.7, $n = 14$ groups, Steenbeek and van Schaik 2001), whereas the DJL in Bohorok was much shorter (683.5 m, sd = 61.5, $n = 2$ groups).

Thomas Langur Demography

The life history characteristics of the Thomas langur were calculated for the Ketambe population. From 1987–2001, a total of 164 individuals were individually recognized on the basis of tail characteristics, facial marks, scars, shape of crest and overall physical appearance. Individuals were classified into three categories: infants (0–19.3 months), juveniles (females 19.3–65 months; males 19.3–72 months) and adults. Adults in females were defined according to their mean age of first reproduction and for males with the descending of their testicles and production of loud calls (Wich et al. 2007).

The primary sex ratio, i.e. the average proportion of males at birth, is 0.44 and the mean inter-birth interval is 22 months, with a longer inter-birth interval (27 months) if only surviving infants are included and shorter interval (18 months) when only nonsurviving infants are included (Wich et al. 2007). The birth rate for females did not decrease with age, with on average 0.44 infants per year (Wich et al. 2007). The birth rate showed a peak at six year of age, indicating that the birth of the first infant often occurred at this age. Mortality was high for both males (0.48) and females (0.43) during the first year and thereafter declines into adulthood (Wich et al. 2007). Although our study was too short to produce solid estimates of life span, the oldest female in the population was 20 years old and still alive when the study ended and the oldest male that was observed was 13 years old and subsequently disappeared. Survival curves indicate that once a female reaches adulthood she has a 50% chance to become 21, whereas for males, this is an age of 11 years (Wich et al. 2007). Although detailed comparisons will need to be made in the future, the Thomas langur life history pattern seems to fall well within the range reported for other southeast Asian langurs (Borries et al. 2001; Borries and Koenig 2008).

Food Competition and Social Behavior

Models of primate socio-ecology predict that food sources of folivores will be relatively abundant and nonmonopolizable (Sterck et al. 1997; Isbell and van Vuren 1996). Therefore, they are expected to experience little within-group and between-group contest competition, and mainly experience scramble competition for food sources. This competitive regime will be reflected in female social relationships (van Schaik 1989; Sterck et al. 1997), proposed to be characterized by nondespotic female dominance hierarchies and an absence of female coalitions within and between groups. Also, females were predicted to disperse. Many folivorous primates, however, do not seem to fit the predictions (Chapman and Pavelka 2005). Here, we review the food competition Thomas langurs experience and their social behavior.

Thomas langurs often feed on food sources that are in relatively large and contain many food items (Sterck and Steenbeek 1997). However, they are more often aggressive inside than outside food patches, indicating that they contest with group members for food. It was estimated that about two third of the used food patches elicited these aggressive reactions (Sterck and Steenbeek 1997), while in about one third of the patches food was abundant. These aggressive interactions yield in some groups a clear dominance hierarchy, but in others groups, periods and outside food patches the dyadic relationships are bi-directional and the hierarchy is unclear. Altogether, these results suggest that dominance hierarchies are not clearly linear and therefore are not very despotic (Sterck and Steenbeek 1997). Moreover, female coalitions against female group members were rare and females seem to obtain their rank without the help of female relatives, suggesting that ranks are obtained individually.

When the combined displacements were used to determine a dominance hierarchy, dominant females seem to move less and have more neighbors when feeding than low-ranking ones (Sterck 1995), indicating an easier time budget and less avoidance of others by high-ranking females. However, this is not translated in a higher birth rate or higher survival of high-ranking female's offspring (Sterck 1995). These data indicate that while Thomas langur females behaviorally experience within-group contest competition, this has a minor effect on their time budget and no effect on their reproductive success. The nondespotic and individual dominance relationships fit the female social relationships predicted when experiencing low within-group contest competition.

Abundant food sources suggest that in part of the food sources, within-group scramble competition will be absent. However, about two-third of the food sources were small enough to elicit contest behavior and may not provide sufficient food for all group members, thereby eliciting within-group scramble competition. Indeed, ranging behavior indicates that Thomas langurs experience within group scramble competition (Steenbeek and van Schaik 2001). The size of a group changes over time with male tenure (see below) and home range size follows this pattern: small when a group is newly formed, then large for a stable period during the mid tenure phase, and shrinking at end tenure (Steenbeek 1999a; Sterck 1997). For mid tenure groups, home range sizes are larger and day journey length of are longer for large than for small groups (Steenbeek and van Schaik 2001). Another potential factor that may influence home range size is habitat quality, but for Thomas langurs, habitat quality does not correlate with home range size (Steenbeek and van Schaik 2001). In addition, changes in home ranges size are not correlated with changes in diet. Moreover, group size does not correlate with time budget measures. Similarly, the number of offspring born does not change with group size, and immature survival even tends to increase with group size (Steenbeek and van Schaik 2001).

This indicates that there is a potential for within-group scramble competition, but it is not important for female reproductive output. Alternatively, the costs of within-group scramble may be compensated by between-group contest competition.

Female aggression during between-group encounters is the expected behavioral expression of between-group competition. Female Thomas langurs, however, do not behave aggressively to females of other groups (Sterck 1997). Alternatively, male between-group aggression is evident and may make female aggression unnecessary. However, behavioral observations (Steenbeek 1999b) and playback experiments (Wich et al. 2002a) indicate that male's reaction to the broadcast of another male's vocalization is not stronger in food sources than at other locations. Altogether, this suggests that between-group contest competition is not strong and cannot compensate for within-group scramble competition.

This sketches a picture of food competition in Thomas langurs. While reactions in food sources indicate that only one third of the patches contains abundant food and does not necessitate any competition, they experience both with within-group scramble and contest competition, while between-group contest seems absent. However, competition has only a minor effect on behavior and does not affect female reproductive success. This suggests that group sizes are smaller than would be predicted when within-group competition limits group size. This suggestion is reinforced by the large number of abundant food sources that indicate that group size could be larger. Thus, the proposition of the folivore paradox (Steenbeek and van Schaik 2001), i.e. that folivore group sizes are smaller than expected when food competition limits group size, fits Thomas langurs.

Social Dynamics

Thomas langurs live social lives although occasionally solitary adult males are also encountered (Sterck 1997). Group sizes at Ketambe ranged from 2 to 16 individuals and typically contain one adult male (Wich et al. 2007), only occasionally groups contain two adult males (Steenbeek et al. 2000). At Bohorok, group sizes were similar for mixed-sex groups that contained one adult male (Gurmaya 1986). Bohorok groups containing two males, constituting 13% of the population, were larger (9–21 individuals, Gurmaya 1986).

Group membership, however, is remarkable dynamic since both females and males disperse (Sterck 1997; Steenbeek et al. 2000). The sexes, however, have a different dispersal pattern. Females that have not had any offspring yet (nulliparous females) as well as females that previously had offspring (parous females) disperse (Fig. 17.9). They enter a new group immediately after they left their previous group, a process called transfer, indicating that dispersal mainly concerns a change of group membership (Isbell and van Vuren 1996; Sterck 1998). They usually disperse when their youngest offspring is independent. Older (female) offspring can accompany them, while barely independent (male) offspring is often left in the previous group (Sterck 1997). Immature males usually live some period in an amb, either because

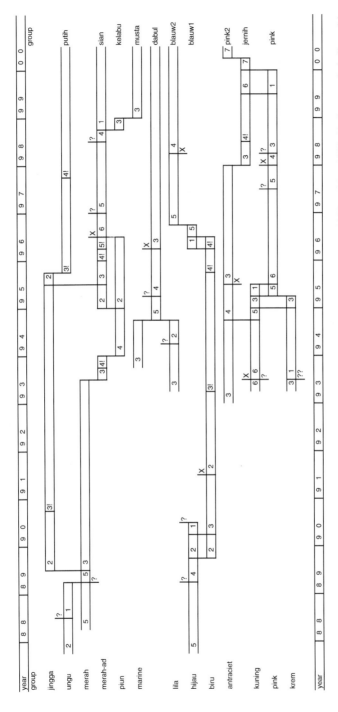

Fig. 17.9 Overview of the flexible grouping pattern in Thomas langurs at Ketambe from November 1987 to May 2000. Each group is indicated by a horizontal bar. An open beginning or ending indicates that the group was present when study started or ended. The name of the group is provided before or after the bar and a new group is formed when females associate with a new male. The numbers indicate the number of adult females in the group and an (!) indicates a female that matures (i.e. 65 mo after her birth date). Vertical lines that connect two groups indicate female transfer. A vertical line that does not connect to a group and that is accompanied by an (?) indicates an immigration from or possible emigration to an unknown destination; and that is accompanied by an (X) indicates a death. Events are rounded off to three month periods

all adult females left their group and they are left with the resident male or because they disperse to an existing amb (Steenbeek et al. 2000; Sterck 1997). Young adult males can range alone and try to associate with adult females by attacking mixed-sex groups, which can result in females associating with the attacking male (Sterck 1995, 1997; Steenbeek 1999a, b).

These different dispersal patterns of males and females result in two types of grouping dynamics. The first type, female split-merger (Sterck 1998), is found when all females voluntarily transfer from a mixed-sex group. All females do not leave at the same moment, typically two adult females join a solitary male, and the remaining adult females follow their example and join the same male within a year (Steenbeek 1999b; Sterck 1997). As a result, the old mixed-sex group turns into an amb consisting of the resident male and remaining offspring. A second grouping pattern is called male take-over, similar to the classic male-take over as described by Hrdy (1974, 1977) for Hanuman langurs (*Semnopithecus entellus*), where a new male replaces the resident male and all females remain in the group (Steenbeek 1999a, b). In Thomas langurs, female split-merger seems more common than male-take over. During our 12.5 years of observations, 12 new mixed-sex groups were formed. Six of these were formed through the process of female split-merger, four by an aggressive male take over, and two through the association of a male with a groups that had lost its resident male through unknown causes (Steenbeek and van Schaik 2001; Sterck et al. 2005). Fights between the resident male and his prospective successor can be fierce and in one case, a male has been observed to succumb to his wounds after being ousted from his group (Wich, unpublished data). A similar incident of male death after aggression has been reported in Bohorok (Gurmaya 1986).

Because of this dynamic group membership, three different phases can be distinguished in the lifespan of mixed-sex groups: the early, middle, and late tenure phase (Steenbeek et al. 2000). The early phase of a male's tenure starts when he associates with at least one adult female and lasts until the first offspring is born. The last 12 months of a male's residency in a mixed-sex group is defined as the late tenure phase since during this period, females can start to transfer to other groups (Steenbeek 1999b; Steenbeek et al. 2000). The stable period between the birth of the first offspring and this last year of a male's tenure is called the middle tenure phase. Total male tenure length varies from 5 days to 72 months, with a median of 60 months (Sterck et al. 2005). Males typically become the resident male in only one mixed-sex group (Sterck 1997; Sterck et al. 2005). They are around seven years old when they settle as the resident male in a group and are around 13 year at the end of their late tenure phase (Wich et al. 2003a).

Male and Female Sexual Strategies

The dynamic grouping pattern may result from the sexual strategies that males and females employ. Thomas langur females invest heavily in offspring during pregnancy and lactation, while direct male care is not needed. In such a situation, males are expected to compete for access to the limited number of females, while females will prefer males that provide benefits. The group composition of Thomas langurs, consisting of one adult male and several females with offspring, and the absence of mating with extra-group males indicate that becoming resident in a group is crucial for male reproductive success. Therefore, male-male competition will mainly concern competition over membership of a mixed-sex group. Male–male pre-mating competition is also suggested by the sexual dimorphy in canine size (Plavcan and van Schaik 1992). Moreover, primate males can coerce females (Smuts and Smuts 1993) and male infanticide is an important coercive tool (van Schaik 2000). Subsequently, we describe how these male and female sexual strategies are expressed in Thomas langurs and how they affect social behavior.

Infanticide occurs when an individual kills a conspecific infant. In many primate species, infanticide is committed by males (Hausfater and Hrdy 1984; van Schaik and Janson 2000). Often, male infanticide takes place after an adult male becomes the new dominant male in a group, either because he ousted the former resident male (through a male take-over in an one-male group) or because he obtains the most dominant position (in a group with multiple males). Also, male Thomas langurs can commit infanticide (Steenbeek 2000; Sterck 1997). Infanticide, however, is committed before an adult male becomes resident in a mixed-sex group (Sterck 1997). Male infanticide may be a crucial factor in Thomas langur grouping dynamics.

Transfer from one group to another can have costs and benefits for females and transferring parous females run the risk that a new male may kill their offspring. To avoid such costs, females should transfer without an infant. Indeed, Thomas langur females transferred significantly more often without than with an infant (Sterck et al. 2005). Their reproductive stage also affects the order in which female Thomas langurs transfer from their group, and females with the oldest offspring transferred first. Although they avoid the costs of infanticide after associating with a new male, transferring parous females have a longer inter-birth interval than females that do not transfer. Such reproductive costs were not found for nulliparous females, since the age at which the first offspring was born did not differ between transferring or resident females. Thus, for parous females, there is a cost to transferring, and the benefits should compensate for the time lost for reproduction. Several hypotheses propose benefits for female transfer (Sterck 1997; Sterck et al. 2005).

The first hypothesis suggests that nulliparous females disperse to avoid inbreeding (Clutton-Brock 1989). This predicts that these females should disperse when their father is still the resident male when they mature. In line with this prediction, nulliparous Thomas langur females transferred almost exclusively when their father was the resident male of their group, while those in a group with an unrelated resident male did not disperse. This result indicates that transfer by nulliparous female is best explained by inbreeding avoidance (Sterck 1997; Sterck et al. 2005). However, this cannot explain dispersal of parous females.

The second hypothesis states that females transfer to smaller groups to decrease the costs of food competition (van Schaik 1983). Indeed, females transferred to groups significantly smaller than their original group when total group size was used, but this difference disappeared when only the number of adult females was considered.

Therefore, the smaller group size was mainly the result of the virtual absence of infants and juveniles in the new groups (Sterck 1997). Group size, however, is not limited by food competition size in Thomas langurs groups (Steenbeek and van Schaik 2001; Sterck and Steenbeek 1997; see also above). Consequently, it is unlikely that female dispersal is determined by the costs of food competition.

Third, it has been hypothesized that females transfer to reduce the risk of predation because groups below a certain group size may suffer higher predation rates (van Schaik 1983). This hypothesis predicts that group size after transfer should be larger than before. However, as described earlier, this was not found (Sterck 1997; Sterck et al. 2005). Therefore, reduction of the predation risk is not likely to direct female transfer decisions.

Fourth, dispersing females may express female mate choice. This hypothesis has been suggested for several other primate species with dispersal of parous females, such as gorillas (*Gorilla gorilla beringei*: Harcourt et al. 1976; Watts 1989) and red colobus monkeys (*Colobus badius*: Marsh 1979). The expectation is that females transfer to a male that provides superior protection of her future offspring and, therefore, reduces the risk of infanticide (van Schaik 1996; Treves and Chapman 1996). Indeed, the last cohort of infants born before Thomas langur females transfer experience a significantly higher mortality rate than the first cohort born after transfer to a new male (Sterck et al. 2005). This suggests that older, long resident males do not provide adequate protection against infanticide and predation. Moreover, females transfer towards relatively young males and not to experienced males who have been the resident male in another group (Sterck et al. 2005). Females may prefer these young adult males above the older resident male because these are better at protecting future offspring than older males. This preference for a strong and protective male will apply to both parous and nulliparous females.

Thus, in Thomas langurs, inbreeding avoidance most likely explains dispersal of young nulliparous females, whereas dispersal of reproductive females most likely concerns female mate choice for the male with the best protective abilities. The male strategy to commit infanticide before he has formed a mixed-sex group fits female choice patterns: by committing infanticide he may simultaneously show that his competitive abilities are better than those of the resident male and induce in females through infant killing the reproductive stage where females reconsider group membership.

Male Loud Calls and Male Sexual Strategies

The vocal repertoire of Thomas langurs was first described by Gurmaya (1986). He distinguished 13 kinds of vocalizations, ranging from soft tonal infant vocalizations to the loud long-distance male loud call emitted by adult males. The loud call has been the focus of further investigation, since long-distance vocalizations can function in mate attraction and male-male competition (reviewed in Delgado 2006). This call is given in the early morning before animals leave the tree in which they slept, but also during between-group encounters and as an alarm call against predators (Wich et al. 2003b). The acoustic characteristics of loud calls differ between individual males (Steenbeek and Assink 1998; Wich et al. 2003b), but also between the different contexts (Wich et al. 2003b), male tenure phases (Wich et al. 2003a) and populations (Wich et al. 2008b).

It is a long-standing issue whether or not primates have voluntary control over vocal production. Recently, studies are reporting some form of control over sound and vocal production (e.g. Cartmill and Byrne 2007; Cheney and Seyfarth 1985, 1990; Hopkins et al. 2007; Papworth et al. 2008; Poss et al. 2006; Wich et al. 2008c). We investigated whether Thomas langur males may have control over production of loud calls and alarm calls. Predator model studies that employed a fake tiger sheet were conducted at Ketambe and Bohorok (Wich and Sterck 2003; Wich and de Vries 2006). These studies showed that males appear to have some form of control over their vocal production, since males living in mixed-sex groups produced more loud calls when exposed to a tiger sheet than solitary males. As there may be motivational differences between males in mixed-sex groups and those that are solitary that influence loud call production, a second tiger model experiment was conducted with males in mixed-sex groups. In this experiment, it was found that males continue to give calls until all independent individuals in a group had given alarm calls (Wich and de Vries 2006). This seems to indicate that males keep track of which individuals in their group have given alarm calls and control the continuation of their own calling behavior.

Like other langur species (van Schaik et al. 1992), Thomas langurs are considered territorial. When two groups encounter each other, the two resident males make loud calls and chase one another, while females usually ignore each other. It therefore seems that Thomas langur males actively defend their home ranges. It has been hypothesized that males may react aggressively to other males to defend their females (Trivers 1972) or to defend food sources that indirectly provide reproductive access to females (Emlen and Oring 1977). The third hypothesis suggests that males may engage in encounters to attract females from other groups by committing infanticide (van Schaik 1996; Steenbeek 1999a, b; Sterck 1997), while resident males will react to such attempts by defending their females and infant offspring. These hypotheses were tested in the Thomas langurs using observational and experimental data (Steenbeek 1999a, b; Sterck 1997; Wich et al. 2002a; Wich and Sterck 2007).

The observational studies by Steenbeek (1999b) showed that male aggression during between-group encounters did not depend on whether the encounter took place inside or outside a food resource, suggesting that aggression during between-group encounters reflected mate defense rather than food resource defense. During between-group encounters, male loud calls are exchanged between the males of the two groups (Steenbeek et al. 1999), creating the opportunity to evaluate these hypotheses with experimental playback studies (Wich et al. 2002a, 2004).

During the first set of experiments, neighboring male's loud calls were broadcast from the centre or edge of the test male's home range. If males were to defend females, no difference was expected (Wilson et al. 2001), but if food resources were important, a stronger reaction in the centre would be expected. The test male's reaction was more vigorous to centre than to edge playbacks. Similar to the results of the observational study, in the edge experiments, there was no difference in male reaction depending on whether the male and the group were on a tree with food or not, which may indicate that on the edge, the less vigorous reaction is not influenced by food resources (Wich et al. 2002a). Natural observations on the Thomas langurs indicate that infanticidal attempts occur more often in the centre of a home range than the edge (Wich et al. 2002a). Therefore, not defense of food sources, but a reaction to a potential infantical male is the most likely explanation for the male's reaction during the centre-edge experiments.

Further support for the protection against infanticide hypothesis was provided by a second playback study, in which playbacks of neighboring and stranger males were broadcast (Wich et al. 2002b). Since unfamiliar males were more likely than neighboring males to conduct infanticidal attempts (Steenbeek 1999b; Sterck 1997), it was expected that males would react more vigorously toward playbacks from strange males than neighboring males. This effect was found (Wich et al. 2002b). Moreover, females reacted more cautiously in the stranger condition and moved away from the speaker. Also, this experiment supported the protection against infanticide hypothesis.

Observations at Ketambe also show that infanticidal attempts were more often conducted by all-male band (amb) males (Wich et al. 2004). Such males aim to associate with females and it is hypothesized that infanticide is a strategy to show females in mixed-sex groups that their resident male's quality is declining (Steenbeek 1999a; Sterck 1997). Loud calls of younger males are of shorter duration than calls of older males (Wich et al. 2003a). To test whether males and females can differentiate between these calls, we conducted an experiment in which calls from young (amb) males and mixed-sex group males were played back (Wich et al. 2004). As expected, males reacted more vigorously and females reacted more cautiously toward young males' loud calls than to the loud calls of mature males from mixed-sex groups (Wich et al. 2004). Together, these results provide strong support for the hypothesis that males protect infants against infanticide.

Assessing Male Quality

Changes in male physiology and behavior with male tenure phase may reflect changes in male quality. Alternatively, some of the amb males' behavioral differences may also be related to the fact that they do not have to defend offspring. It is expected that end tenure males are of relatively low quality since females leave these groups and infant survival is low (Sterck et al. 2005). They transfer to groups at start tenure phase, suggesting that the males residing in such groups are relatively strong. They do not leave during mid tenure phase and infant survival is, at least initially, high, similarly indicating that the resident male is relatively strong. Males in all male band males, however, are probably of low quality since they are either not yet mature or do not (yet) seem able to attract females.

Male fecal testosterone levels were significantly lower in the amb than during the other male tenure phases (Wich et al. 2003a) and may underlie differences between male behavior in amb and in mixed-sex groups. Although it is not known how testosterone levels influence male loud calls, the largest change between loud call characteristics was found between young males in amb's and males in early tenure phase (Wich et al. 2003a). In addition, the tendency to start or answer a loud call bout was lower for amb males than for resident males in all three tenure phases (Steenbeek et al. 1999) and they gave fewer loud calls during the morning chorus (Steenbeek et al. 1999; Sterck et al. 2005). Moreover, amb males responded less often to loud calls from extra-group males than residents males of all tenure phases (Steenbeek et al. 1999) and approached extra-group calling males less often than early and middle tenure males (Steenbeek et al. 1999). These results support the idea that amb males are of relatively low quality.

Behavior of a resident male may depend on his tenure phase. Resident males of different tenure phases, however, did not differ in the number of loud calls produced or answers to loud calls from others (Steenbeek et al. 1999; Sterck et al. 2005). Although resident males associated most of the time with the females of their group, they occasionally left their group and went on their own to a neighboring mixed-sex group. During these encounters, the male silently approached the other group and attempted to attack the individuals of another group during these silent encounters. Infanticidal attempts occurred, especially during these silent encounters (Steenbeek 1999b). Males in the end tenure phase less often initiated silent encounters than early and middle tenure phase males (Steenbeek 1999b; Sterck et al. 2005). Moreover, end tenure males approached extra-group calling males less often than early and middle tenure males (Steenbeek et al. 1999). Also, the effectiveness in preventing aggression from extra-group males towards females and/ or infants is lower in the end tenure phase males than early and middle tenure phase males. This indicates that end tenure phase males have a relatively low quality (Steenbeek 1999a, b).

In contrast, male herding occurred more often during the early tenure phase than the middle tenure phase, whereas no difference was found between middle and late tenure (Steenbeek 1999b). This indicates that males put most effort to keep females away from extra-group males during early tenure phase. Herding was probably not needed once females had an infant and themselves avoided extra-group males.

Altogether, amb and resident males in different tenure phases seem to differ in their relatively quality and, as proposed, early and mid tenure phase males seem to be of higher quality then amb and end tenure phase males.

Cues of Male Quality

Assessing male quality is a crucial factor for both females and males. It is important for a female to assess the quality of the resident male relative to extra-group males and especially detecting a decline in the quality of the resident male is expected to be crucial. Females may use different cues to determine a male's strength. First, when an outside male succeeds in killing an infant, this can form a simultaneous indication that the current resident male is no longer a good protector and the outside male may be able to protect. Indeed, females are known to transfer to a male after he probably killed her offspring (Sterck 1997). Second, females can also use a male's behavior toward other males as an indicator of male strength. The changes in a male's behavior with tenure phase may provide ideas of the cues females may use. In addition, changes in the structural characteristics of the male loud calls can provide females with an indication that males are in their end tenure phase (Wich et al. 2003a). In a statistical analysis, 66% of late tenure loud calls were correctly assigned to the late tenure phase (Wich et al. 2003a), indicating that loud call characteristics could at least be part of the cues that females use to assess whether a male's quality is diminishing. Similarly, they may use loud calls to determine which males are young and in a phase of their career where their strength is increasing (Wich et al. 2003a), since young males (amb and begin tenure) have loud calls that are shorter or have fewer elements in a loud call than middle and end tenure males (Wich et al. 2003a). Altogether, females have access to several behavioral sources that indicate a male's strength.

Two hypotheses concerning proximate mechanisms have been explored in the context of male–male aggression in Thomas langurs (Wich and Sterck 2007). The first hypothesis emphasizes that familiarity between the contestants influences encounter intensity (Getty 1989; Ydenberg et al. 1988). Getty (1989) proposes that animals fight to learn about each other, i.e. they learn what is to be gained from or lost to the other contestant, depending on the males' relative strength. The second hypothesis emphasizes that the potential threat of the opponent explains variation in the intensity of the encounter (e.g. Harcourt 1978; Temeles 1990). The opponent's threat may depend of what a male can lose during an encounter. These losses can vary from temporary access to a food source, to extra-group copulations, attracting females away from the opponent's group and, in the most extreme case, a take-over of the opponent's group (Steenbeek 1999a; Temeles 1990, 1994).

To examine the influence of these proximate mechanisms, observations of Thomas langur groups that varied in familiarity or in threat level were compared. Males that were more familiar with each other, but had a similar threat level as determinate by a similar male tenure phase, reacted less strongly to each other than to unfamiliar males (Wich and Sterck 2007). In addition, males that differed in threat level had more intense encounters when familiarity was held constant (Wich and Sterck 2007). Thus, both the familiarity and the threat level difference of males determine part of the variation observed in their aggression during interactions. Familiar males may have been less risky since their behavioral tendencies were known. Moreover, males of a different tenure phase differed in the risk they posed, suggesting that these males differed in their tendencies. Thus, multiple cues are available to assess a male's quality and whether he poses a threat. In addition, we showed that both male and female behavior depends on the threat posed by an extra-group male.

Conclusions

The Thomas langur is folivore primate species that has the potential for within-group scramble and contest competition. However, a substantial part of their food sources contains an abundant number of food items, suggesting that group size may become larger than actually found. Moreover, a larger group size and a lower female dominance rank do not diminish female reproductive success. On the contrary, females in larger groups seem to have more surviving offspring. Therefore, food competition does not seem to limit group size and groups can potentially become larger.

Other factors may limit group size. One possibility is that resting time cannot be diminished further (Korstjens and Dunbar 2007). However, resting time was not affected by group size or female dominance rank. In addition, time budgets at Ketambe and Bohorok were very similar, despite different ecological conditions. This does not suggest that a lack of resting limits group size.

Alternatively, male sexual strategies may limit group size. Thomas langur males commit infanticide before they associate with females and attack larger groups more often than small ones. Thus, females may benefit from living in relatively small groups, resulting in the folivore paradox. This suggests that we have to ask ourselves what the minimum, and the maximum, group size is in which animals like Thomas langurs can live. For Thomas langurs, one male and two females seems to be the minimum size since newly started groups typically contained two females. However, what determines the minimum group size remains to be further explored.

The Thomas langur social dynamics seem to result from male and female sexual strategies. Males can commit infanticide, a sexual coercive strategy, and their ability to protect their offspring seems crucial. Female preferences for males with superior protective capacities determine female transfer decisions provided that they have no dependent offspring. Male infanticide before female transfer enforces on females the reproductive stage that makes them willing to transfer, creating the opportunity for the formation of a new mixed-sex group. All these features result in the social dynamics observed. Crucial features are the ability of females to determine differences in male quality, to express female choice through dispersal and an increase of male threat with a larger group size. The folivore paradox may only be found in these conditions. This reasoning implies that habitual dispersal by parous females is a crucial feature of the folivore paradox.

Not all folivore species show habitual dispersal of parous females (Sterck 1999; Sterck and Korstjens 2000; Chapman and Pavelka 2005). Habitual female dispersal is only possible when females do not gain substantial benefits from continuously associating with a particular group of females and do not have to maintain their group size at all times. Theory suggest that continuous females associations will be important when female coalitions are crucial in securing a dominance position or when strong between-group contest requires continuously maximum group strength (van Schaik 1989). Alternatively, females may not be able to live in small groups due to the distribution of food sources or predation risk. The factors that are crucial in promoting permanent female associations at group sizes that result in food competition remain to be established for folivore primates. Therefore, we suggest that the causes and consequences of group cohesion may be a good starting point to further investigate the link between female ecology and social behavior.

Acknowledgments We gratefully acknowledge the co-operation and support of the Indonesian Institute of Science (LIPI, Jakarta), the Indonesian Nature Conservation Service (PHPA) in Jakarta, Medan, and Kutacane (Gunung Leuser National Park Office), Universitas National (UNAS, Jakarta), Universitas Syiah Kuala (UNSYIAH, Banda Aceh), and the Leuser Development Program (LDP, Medan), especially M. Griffiths and Dr. K. Monk. We also thank the LDP staff in Ketambe, especially Abu Hanifah Lubis, for providing an excellent research environment and strong logistical support. We thank a large number of assistants and students for helping with data collection, without their hard work, this long-term study would not have been possible. Financial support was generously provided by the Netherlands Foundation for the Advancement of Tropical Research (WOTRO), the Treub Foundation, the Dobberke Foundation and the Lucie Burgers Foundation for Comparative Behavior Research. Jan van Hooff and Carel van Schaik are also acknowledged for their support throughout the whole study. We thank the European Commission and the Government of Indonesia as the funding agencies for the Leuser Development Program.

References

Aimi M, Bakar A (1992) Taxonomy and distribution of Presbytis melalophos group in Sumatera, Indonesia. Primates 33:191–206

Aimi M, Bakar A (1996) Distribution and deployment of *Presbytis melalophos* group in Sumatera, Indonesia. Primates 37:399–409

Ashton PS, Givnish TJ, Appanah S (1988) Staggered flowering in the Dipterocarpaceae: new insights into floral induction and the evolution of mast fruiting in the aseasonal tropics. Am Nat 132:44–66

Bauchop T (1971) Stomach microbiology of primates. Ann Rev Microbiol 25:429–436

Borries C, Koenig A, Winkler P (2001) Variation of life history traits and mating patterns in female langur monkeys (*Semnopithecus entellus*). Behav Ecol Sociobiol 50:391–402

Borries C, Koenig A (2008) Reproductive and behavioral characteristics of aging in female Asian colobines. In: Atsalis S, Margulis SW, Hof PR (eds) Primate reproductive aging: cross-taxon perspectives on reproduction. Karger, Basel, pp 80–102

Brandon-Jones D, Eudey AA, Geissman T, Groves CP, Melnick DJ, Morales JC, Shekelle M, Stewart CB (2004) Asian primate classification. Int J Primatol 25:97–164

Cartmill EA, Byrne RW (2007) Orangutans modify their gestural signaling according to their audience's comprehension. Curr Biol 17:1345–1348

Chapman CA, Pavelka MSM (2005) Group size in folivorous primates: ecological constraints and the possible influence of social factors. Primates 46:1–9

Cheney D, Seyfarth R (1985) Vervet monkey alarm calls: manipulation through shared information? Behaviour 94:150–166

Cheney DL, Seyfarth RM (1990) How monkeys see the world. Chicago University Press, Chicago

Clauss M, Streich WJ, Nunn CL, Ortmann S, Hohmann G, Schwarm A, Hummel J (2008) The influence of natural diet composition, food intake level, and body size on ingesta passage in primates. Comp Biochem Physiol Part A 150:274–281

Clutton-Brock TH (1989) Female transfer and inbreeding avoidance in social mammals. Nature 337:70–72

Crockett CM, Janson CH (2000) Infanticide in red howlers: female group size, male membership, and a possible link to folivory. In: van Schaik CP, Janson CH (eds) Infanticide by males and its implications. Cambridge University Press, Cambridge, pp 75–98

Delgado RA (2006) Sexual selection in the loud calls of male primates: signal content and function. Int J Primatol 27:5–25

Emlen ST, Oring LW (1977) Ecology, sexual selection, and the evolution of mating systems. Science 197:215–223

Getty T (1989) Are dear enemies in a war of attrition? Anim Behav 37:337–339

Gurmaya KJ (1986) Ecology and behaviour of *Presbytis thomasi* in northern Sumatra. Primates 27:151–172

Harcourt AH (1978) Strategies of emigration and transfer by primates, with particular reference to gorillas. Zeit Tierpsychol 48:401–420

Harcourt AH, Stewart KS, Fossey D (1976) Male emigration and female transfer in wild mountain gorilla. Nature 263:226–227

Hausfater G, Hrdy SB (1984) Infanticide: comparative and evolutionary perspectives. Aldine, New York

Hopkins WD, Taglialatela JP, Leavens DA (2007) Chimpanzees differentially produce novel vocalizations to capture the attention of a human. Anim Behav 73:281–286

Hrdy SB (1974) Male–male competition and infanticide among the langurs (*Presbytis entellus*) of Abu, Rajasthan. Folia Primatol 22:19–58

Hrdy SB (1977) Infanticide as a primate reproductive strategy. Am Sci 65:40–49

Isbell L, van Vuren D (1996) Differential costs of locational and social dispersal and their consequences for female group-living primates. Behaviour 133:1–36

Korstjens AH, Dunbar RIM (2007) Time constraints limit group sizes and distribution in red and black-and-white colobus. Int J Primatol 28:551–575

Marsh CW (1979) Female transference and mate choice among Tana River red colobus. Nature 281:568–569

Marshall AJ, Ancrenaz M, Brearley FQ, Fredriksson GM, Ghaffar N, Heydon M, Husson SJ, Leighton M, McConkey KR, Morrogh-Bernard HC, Proctor J, van Schaik CP, Yaeger C, Wich SA (2009) The effects of forest phenology and floristics on populations of Bornean and Sumatran orangutans. In: Wich SA, Utami Atmoko SS, Mitra Setia T, van Schaik CP (eds) Orangutans: Geographic variation in behavioral ecology and conservation. Oxford University Press, New York, pp 97–118

Morrogh-Bernard HC, Husson SJ, Knott CD, Wich SA, van Schaik CP, van Noordwijk MA, Lackman-Ancrenaz I, Marshall AJ, Kanamori T, Kuze N, Bin Sakong R, Bin Sakong R (2009) Orangutan activity budgets and diet. In: Wich SA, Utami Atmoko SS, Mitra Setia T, van Schaik CP, van Schaik CP (eds) Orangutans: Geographic variation in behavioral ecology and conservation. Oxford University Press, New York, pp 119–134

Palombit RA (1997) Inter- and intraspecific variation in the diets of sympatric siamang (*Hylobates syndactylusi*) and lar gibbons (*Hylobates lar*). Folia Primatol 68:321–337

Papworth S, Boese AS, Barker J, Schel AM, Zuberbuehler K (2008) Male blue monkeys alarm call in response to danger experienced by others. Biology Letter 4:472–475

Plavcan JM, van Schaik CP (1992) Intrasexual competition and canine dimorphism in primates. Am J Phys Anthropol 87:461–477

Poss SR, Kuhar C, Stoinski TS, Hopkins WD (2006) Differential use of attentional and visual communicative signaling by orangutans (*Pongo pygmaeus*) and gorillas (*Gorilla gorilla*) in response to the attentional status of a human. Am J Primatol 68:978–992

Robbins MM, Sawyer SC (2007) Intergroup encounters in mountain gorillas of Bwindi Impenetrable National Park, Uganda. Behaviour 144:1497–1519

Shepherd CR, Sukumaran J, Wich SA (2004) Open season: an analysis of the pet trade in Medan, Sumatra 1997–2001. TRAFFIC Southeast Asia, Petaling Jaya, Malaysia

Smuts BB, Smuts RW (1993) Male aggression and sexual coercion of females in nonhuman primates and other mammals: evidence and theoretical implications. Adv St Behav 22:1–63

Steenbeek R (1999a) Female choice and male coercion in wild Thomas's langurs. Unpublished Ph.D. dissertation, Utrecht University

Steenbeek R (1999b) Tenure related changes in wild Thomas's langurs I: between group interactions. Behaviour 136:595–626

Steenbeek R, Assink PR (1998) Individual differences long-distance calls of male wild Thomas langurs (*Presbytis thomasi*). Folia Primatol 69:77–80

Steenbeek R, Assink P, Wich SA (1999) Tenure related changes in wild Thomas's langurs II: loud calls. Behaviour 136:627–650

Steenbeek R (2000) Infanticide by males and female choice in wild Thomas's langurs. In: van Schaik CP, Janson CH (eds) Infanticide by males and its implications. Cambridge University Press, Cambridge, pp 153–177

Steenbeek R, Sterck EHM, de Vries H, van Hooff JARAM (2000) Costs and benefits of the one-male, age-graded and all-male phase in wild Thomas's langur groups. In: Kappeler PM (ed) Primate Males. Cambridge University Press, Cambridge, pp 130–145

Steenbeek R, van Schaik CP (2001) Competition and group size in Thomas's langurs (*Presbytis thomasi*): the folivore paradox revisited. Behav Ecol Sociobiol 49:100–110

Sterck EHM (1995) Females, foods and fights. Unpublished Ph.D. dissertation, Utrecht University, Utrecht

Sterck EHM (1997) Determinants of female dispersal in Thomas langurs. Am J Primatol 42:179–198

Sterck EHM (1998) Female dispersal, social organization, and infanticide in langurs: are they linked to human disturbance? Am J Primatol 44:235–254

Sterck EHM (1999) Variation in langur social organization in relation to the socioecological model, human habitat alteration, and phylogentic constraints. Primates 40:199–213

Sterck EHM, Korstjens AH (2000) Female diserpsal and infanticide avoidance in primates. In: van Schaik CP, Janson CH (eds) Infanticide by males and its implications. Cambridge University Press, Cambridge, pp 293–321

Sterck EHM, Steenbeek R (1997) Female dominance relationships and food competition in the sympatric Thomas langur and long-tailed macaque. Behaviour 134:749–774

Sterck EHM, Watts DP, van Schaik CP (1997) The evolution of female social relationships in nonhuman primates. Behav Ecol Sociobiol 41:291–309

Sterck EHM, Willems EP, van Hooff J, Wich SA (2005) Female dispersal, inbreeding avoidance and mate choice in Thomas langurs (*Presbytis thomasi*). Behaviour 142:845–868

Supriatna J, Mittermeier RA (2008) *Presbytis thomasi*. In: IUCN 2008. 2008 IUCN Red List of Threatened Species. www.iucnredlist.org. Downloaded on 18 Nov 2008

Temeles EJ (1990) Northern harriers on feeding territories repond more aggressively to neighbours than floaters. Behav Ecol Sociobiol 26:57–63

Temeles EJ (1994) The role of neighbours in territorial systems: when are they 'dear enemies'? Anim Behav 47:339–350

Treves A, Chapman CA (1996) Conspecific threat, predation avoidance, and resource defense: implications for grouping in langurs. Behav Ecol Sociobiol 39:43–53

Trivers RL (1972) Parental investment and sexual selection. In: Campbell B (ed) Sexual selection and the descent of man. Aldine, Chicago, pp 136–179

van Schaik CP (1983) Why are diurnal primates living in groups? Behaviour 87:120–144

van Schaik CP (1986) Phenological changes in a Sumatran rain forest. J. Trop. Ecol. 2:327–347

van Schaik CP (1989) The ecology of social relationships amongst female primates. In: Standen V, Foley RA (eds) Comparative Socioecology. Blackwell, Oxford, pp 195–218

van Schaik CP (1996) Social evolution in primates: the role of ecological factors and male behaviour. Proc. Brit. Acad. 88:9–31

van Schaik CP (2000) Vulnerability to infanticide: patterns among mammals. In: van Schaik CP, Janson CH (eds) Infanticide by Males and Its Implications. Cambridge University Press, Cambridge, pp 61–71

van Schaik CP, van Noordwijk MA (1986) The hidden costs of sociality: Intra-group variation in feeding strategies in Sumatran long-tailed macaques (*Macaca fasicularis*). Behaviour 99:296–315

van Schaik CP, Janson CH (2000) Infanticide by Males and Its Implications. Cambridge University Press, Cambridge

van Schaik CP, Assink PR, Salafsky N (1992) Territorial behavior in Southeast Asian langurs: resource defense or mate defense? Am J Primatol 26:233–242 van Schaik CP, Monk KA, Robertson JMY (2001) Dramatic decline in orang-utan numbers in the Leuser Ecosystem, Northern Sumatra. Oryx 35:14–25

Watts DP (1989) Infanticide in Mountain Gorillas: New Cases and a Reconsideration of the Evidence. Ethology 81:1–18

Whitten AJ, Damanik SJ, Jazanul A (1987) & Nazaruddin H. Gadjah Mada University Press, The Ecology of Sumatra, Yogyakarta, Indonesia

Wich SA, van Schaik CP (2000) The impact of El Nino on mast fruiting in Sumatra and elsewhere in Malesia. J. Trop. Ecol. 16:563–577

Wich SA, Assink PR, Becher F, Sterck EHM (2002a) Playbacks of loud calls to wild Thomas langurs (Primates; *Presbytis thomasi*): The effect of familiarity. Behaviour 139:79–87

Wich SA, Assink PR, Becher F, Sterck EHM (2002b) Playbacks of loud calls to wild Thomas langurs (Primates; Presbytis thomasi): The effect of location. Behaviour 139:65–78

Wich SA, Sterck EHM (2003) Possible audience effect in Thomas Langurs (Primates; Presbytis thomasi): An experimental study on male loud calls in response to a tiger model. Am J Primatol 60:155–159

Wich SA, Koski S, de Vries H, van Schaik CP (2003a) Individual and contextual variation in Thomas langur male loud calls. Ethology 109(1):1–13

Wich SA, van der Post DJ, Heistermann M, Mohle U, van Hooff JARAM, Sterck EHM (2003b) Life-phase related changes in male loud call characteristics and testosterone levels in wild Thomas langurs. Int J Primatol 24:1251–1265

Wich SA, Assink PR, Sterck EHM (2004) Thomas langurs (Presbytis thomasi) discriminate between calls of young solitary versus older group-living males: A factor in avoiding infanticide? Behaviour 141:41–51

Wich SA, de Vries H (2006) Male monkeys remember which group members have given alarm calls. Proc. Roy. Soc. Lond 273:735–740

Wich SA, Steenbeek R, Sterck EHM, Korstjens AH, Willems EP, van Schaik CP (2007) Demography and Life History of Thomas Langurs (Presbytis thomasi). Am J Primatol 69:641–651

Wich SA, Sterck EHM (2007) Familiarity and threat of opponents determine variation in Thomas langur (*Presbytis thomasi*) male behaviour during between-group encounters. Behaviour 144:1583–1598

Wich SA, Meijaard E, Marshall AJ, Husson S, Ancrenaz M, Lacy RC, van Schaik CP, Sugardjito J, Simorangkir T, Traylor-Holzer K, Doughty M, Supriatna J, Dennis R, Gumal M, Knott CD, Singleton I (2008a) Distribution and conservation status of the orang-utan (*Pongo* spp.) on Borneo and Sumatra: how many remain? Oryx 42:329–339

Wich SA, Schel AM, de Vries H (2008b) Geographic variation in Thomas langur (*Presbytis thomasi*) loud calls. Am J Primatol 70:566–574

Wich SA, Swartz KB, Hardus ME, Lameira AR, Stromberg E, Shumaker RW (2008c) A case of spontaneous acquisition of a human sound by an orangutan. Primates. doi:10.1007/s 10329-008-0117-y

Wilson ML, Hauser MD, Wrangham RW (2001) Does participation in intergroup conflict depend on numerical assessment, range location, or rank for wild chimpanzees? *Anim.* Behav. 61:1203–1216

Wilson WL, Wilson CC (1975) Species-specific vocalizations and the determination of phylogenetic affinities of the Presbytis aygula-melalophos group in Sumatra. In: Kondo S, Kawai M, Ehara A (eds) Contemporary Primatology. S. Karger, Basel, pp 459–463

Ydenberg RC, Giraldeau LA, Falls JB (1988) Neighbours, strangers, and the assymetric war of attrition. Anim Behav 36:343–347

Chapter 18
Dominance and Reciprocity in the Grooming Relationships of Female Long-Tailed Macaques (*Macaca fascicularis*) in Indonesia

Michael D. Gumert

Introduction

It has been long known that females form the stable core of macaque societies (Bernstein and Sharpe 1966; Vandenbergh 1967; Drickamer 1976). They have strong relational ties and develop lifelong relationships with other females in their groups, thus they are considered to be female-bonded (Wrangham 1980). They have particularly close relationships with kin that are characterized by high levels of affiliation (Sade 1965; Drickamer 1976; Kurland 1977; Chapais 1983; Gouzoules and Gouzoules 1987; Kapsalis 2004; Silk 2006). This pattern results because females are philopatric and remain in their natal group for life, while males disperse and emigrate from their natal groups shortly after reaching sexual maturity (van Noordwijk and van Schaik 1985; Pusey and Packer 1987). Since female family lineages generally remain in the same location across generations, macaque groups are based on a cross-generational matrilineal social structure of closely related females. Macaque groups are typically multi-male/multi-female, and consist of several female matrilines (i.e., families), their young, and unrelated immigrant adult males that have migrated from neighboring communities and maintain transient relationships with the females until they emigrate again (de Ruiter and Geffen 1998). Due to the matrilineal structure of macaque societies, females must maintain long-term affiliative relationships with other females in their group. Consequently female cercopithecine primates, such as macaques, are equipped with adaptations for developing and maintaining close female–female bonds because females that can develop larger relationship networks tend to have higher fitness (Silk et al. 2003; Silk 2007).

Female macaques are not only nepotistic, but they maintain strict dominance hierarchies in their societies due to the socioecological pressures influencing them to have strong within-group competition (van Schaik 1983; Sterck et al. 1997).

M.D. Gumert (✉)
Division of Psychology, School of Humanities and Social Sciences, Nanyang Technological University, Singapore, 639798, Singapore
e-mail: gumert@ntu.edu.sg

S. Gursky-Doyen and J. Supriatna (eds.), *Indonesian Primates*,
Developments in Primatology: Progress and Prospects,
DOI 10.1007/978-1-4419-1560-3_18, © Springer Science+Business Media, LLC 2010

These hierarchies are stable over long periods of time (Sade 1972a), and each matriline maintains a relative rank status with the other matrilines (Kapsalis 2004). In this structure, females support their close kin in agonistic conflicts and females develop into their adult rank position with the support of their kin (Kawai 1958; Cheney 1977; Datta 1983a, b, c; Chapais 1992; Pereira 1995). Across individuals, there is also a stable linear dominance hierarchy and females compete over dominance rank (Kawai 1958; Walters and Seyfarth 1987). Rank is attained and maintained with the support of closely-bonded kin, but can also be influenced by support from non-kin allies (de Waal 1977; Chapais and Gauthier 2004). In addition, social rank is related to reproductive success in female long-tailed macaques (*Macaca fascicularis*) (van Noordwijk and van Schaik 1987, 1999) and some other species (Silk 2006). As such, maintaining close bonds with both kin and non-kin are critical to a female's social and reproductive success, and therefore females are expected to be adapted to develop and sustain social relationships beneficial to their social status.

The importance of female–female relationships is evident in how females interact with each other. Females groom each other, remain in close proximity, feed together, offer support in conflicts, and engage in reconciliation. The patterning of these social interactions is largely structured by kinship, (Bernstein 1991; Kapsalis 2004; Silk 2006), and dominance hierarchy (de Waal 1989; Thierry 1990). Grooming patterns have been of particular focus in measuring and understanding female relationships, and are a useful indicator of female social bonding (Hemelrijk 2005). Moreover, the direction and balance of grooming performance can be used to empirically quantify the relationship quality of dyads by measuring the time matching of grooming between pairs (Barrett et al. 1999, 2000; Manson et al. 2004). Relationships characterized by bi-directional exchange of grooming can be an indicator of close social bonding. In contrast, if grooming interactions are one-sided, aspects of the relationships may be based on service exchanges (Barrett and Henzi 2001; Barrett et al. 2002; Gumert and Ho 2008). In macaques, females generally have reciprocal grooming relationships indicating close social bonds. Despite this, grooming is also biased up-rank and this may be due to attempts by lower-ranked females to gain tolerance (Henzi and Barrett 1999; Barrett et al. 2002) and/or support (Seyfarth 1977; Schino 2007) from higher-ranked females, as a form of service exchange.

Grooming in Female Relationships

Grooming is the most commonly studied form of affiliation in primates and is considered to be the social "glue" of cercopithecine interpersonal relationships (Curley and Keverne 2005). Grooming has both hygienic and social functions, and plays a central part in the societies of many primates. Grooming is when one macaque engages in "manual brushing and picking at that hairs" on a passive receiving individual (Goosen 1987) (Plate 18.1). It is also generally relaxed and friendly, and is related to several positive physiological influences in both the actor

Plate 18.1 The α-female, Helen, grooms a mid-ranked female, Dawn, atop a cottage at an eco-tourist lodge in Tanjung Puting National Park. The amount of time that female long-tailed macaques engage in grooming varies across individuals and can consist of 5 to 30% of their activity budget

and the receiver (Fabre-Nys et al. 1982; Boccia 1987; Boccia et al. 1989; Keverne et al. 1989; Shutt et al. 2007). Grooming appears to be rewarding to those that give it (de Waal et al. 2008), and tension-relieving to those that receive it (Terry 1970; Boccia 1987; Schino et al. 1988). Fundamentally, grooming serves a basic hygienic function. The pelage of the receiver is cleaned by the groomer through removing louse, debris, and other matter from the hair and skin. It is often directed to areas unreachable by the receiver alone (Hutchins and Barash 1976; Tanaka and Takefushi 1993) and is more likely to occur on areas infested with louse eggs (Zamma 2002).

The social function of grooming has now surpassed the adaptive significance of its hygienic functions (Dunbar 1991), and females that don't perform grooming might be disadvantaged in forming relationships with others. Grooming is a way to direct low costs benefits to a receiver, and potentially manipulate or influence the state of the receiver to return social acts at a later time (Dunbar and Sharman 1984; Reiss 1984). Dyads that perform grooming also cement social bonds and maintain relationships with one another (Matheson and Bernstein 2000). In the anthropoid primates particularly, grooming functions in social exchange to coordinate cooperation between partners, such as when handling another's infant or attempting to successfully mate (Gumert 2007a, b). Moreover, grooming can increase the likelihood that a partner will offer social support during agonistic conflicts (Seyfarth and Cheney 1984; Hemelrijk 1994), and grooming has been considered highly important in the development of alliances critical in maintaining a female's social rank (Seyfarth 1976, 1977, 1980; Schino 2007). Lastly, grooming may also establish tolerance between pairs of distant rank, and thus aid to ameliorate conflicts that can arise due to competition (Henzi and Barrett 1999; Barrett and Henzi 2001).

Seyfarth's Model of Female Grooming

The most widely applied model of social grooming has been Seyfarth's (1976, 1977, 1983) theory of female grooming relationships. In this model, Seyfarth (1977) suggested that females have a limited amount of time to engage in grooming activities, but that from grooming they need to obtain two important benefits, (1) hygiene and (2) the development of alliances (i.e., relationships that lead to aggressive support in social conflicts). Females are rather equal in their ability to clean each other, but a female's ability to support another during conflict will vary, and thus the limited amount of grooming a female can engage in should be based on establishing these alliances. What accounts for a female's ability to provide social support is her status in the dominance hierarchy. Therefore, the model predicts that female grooming relationships will be based on an attraction to groom higher-ranked females in order to develop the most effective social alliances. This will result in subordinate females competing over access to higher-ranked females by providing more grooming than they receive from higher-ranked partners (Seyfarth 1983).

Another aspect of the model, competitive exclusion, will not allow every female to have grooming access to the highest-ranked females, and consequently, females should direct the majority of their grooming toward adjacently-ranked partners (Seyfarth 1977, 1983). Each female will be attempting to groom the highest-ranked female that she can, but the higher-ranked females above her will exclude her access to more distantly-ranked females through competition. Naturally, the female will then have the best competitive ability to groom the female directly above her, because she can exclude all females below her for access to that female, while the females above her seek access to even higher-ranked females. Consequently, top-ranked females will have fewer constraints on their grooming choices than the bottom-ranked females, and thus higher-ranked females should have more access to the top-ranked females. The expected outcome of grooming that is based on rank attraction and competitive exclusion is that the majority of grooming should be toward those of adjacent rank and that there is a linear relationship between rank and grooming where the highest-ranked females receive the most amount of grooming, and the lowest-ranked females receive the least.

Support for Seyfarth's model are mixed and tests of the model have not always held up (de Waal and Luttrell 1986; Henzi and Barrett 1999; Henzi et al. 2003). Despite this, there is substantial evidence to suggest that attraction to groom high-ranked partners does play a role in the societies of many primate species (Seyfarth 1976, 1977, 1980; Schino 2001). The confusion on the subject warrants continued investigation of the role of dominance on grooming in macaques and others species because it is still unclear exactly how and why dominance influences grooming patterns. For example, newer work has suggested that variation in the dominance gradient will alter grooming patterns (Barrett et al. 2002; Stevens et al. 2005) because it alters the level of need by subordinates to gain tolerance (i.e., reduced aggression and competition) from their higher-ranked partners. This type of model contrasts with Seyfarth's static model because the tolerance model predicts that the role of rank on female grooming is variable and dependent on the current context and relational needs between dominants and subordinates. Seyfarth's model is a static model that predicts female grooming patterns will be consistently related to the hierarchy, only changing when the hierarchy is reorganized.

In this study, I explored the grooming relationships of 20 sexually-mature females (i.e., adult and adolescent) from a group of long-tailed macaques across a 14-month time period. I investigated how kinship (i.e., inferred kinship, see Methods) and rank influenced patterns of grooming reciprocation, and tested Seyfarth's model on female grooming in more detail than previously reported for this species (Butovskaya et al. 1995; Wheatley 1999). I tested for grooming reciprocity at the group level, while controlling for the effects of rank and inferred kinship. Furthermore, I investigated the grooming balance within pairs (i.e., time-matching), the difference between up-rank and down-rank grooming, the relationship of rank and grooming, the affect of rank on grooming adjacently ranked partners, and how a female's rank structured her grooming of partners. In this report, I demonstrate grooming reciprocity and illustrate how rank affects grooming patterns in long-tailed macaques. I end with a discussion on how these patterns relate to female bonding and Seyfarth's model of female grooming.

Methods

Location and Study Group

Data was collected on female grooming patterns in a group of long-tailed macaques between July 2003 and August 2004. The study was conducted at a site on the northwestern border of Tanjung Puting National Park (TPNP), Kalimantan, Tengah, Indonesia, a 304,000 ha nature reserve area located at E 112°49′ and S 02°49′. The research was based at an eco-tourist lodge and the study site included the area surrounding the lodge along the Sekonyer River. During the year of this study there was not much tourist traffic through the lodge, but the macaques were influenced by human activity both at the lodge and at a nearby village, Tanjung Harapan. The macaques were provisioned daily from refuse at the lodge and occasionally raided fruit trees and crops around Tanjung Harapan. The eco-lodge was surrounded by riparian swamp forest, ex-slash and burn fields or ladang (i.e., fields of elephant grass, *Imperata* spp.), and recently reforested ladang areas. The macaques utilized all habitat types, but always stayed within 1 km of the river. The macaques' home range was centered around the lodge, and the group typically roosted in trees within the lodge and close to the river, typical of this species' riverine roosting patterns observed in Sumatra and Eastern Kalimantan (Fittinghoff and Lindburg 1980; Wheatley 1980; van Schaik et al. 1996). The group's home range was an area approximately 1.2 km². Trails and bridges were built in this range and used to follow the macaques daily.

The number of individuals in the group varied during the study between 48 and 53 and group composition changes were due to births, deaths, immigration, and emigration. During the study, 18 adult females and 2 adolescent females were studied. Also during the study 5 adult males and 5 adolescent males were observed. Two of the adult males immigrated into the group during the study, the first in July and the second in Nov, 2003. One adolescent male disappeared from the group during March, 2004 (i.e., emigrated or died). I also observed 28 juveniles and infants (19♂:9♀), of which 17 were born into the group during the 14-month period of the study. Four infants (2♂:2♀) and one juvenile (1♂) died during the study. The juvenile and one male infant were the sons of the second ranked-female, Lucy.

Female composition did not change throughout the study. Both of the two adolescent females came into sexual maturity during the study, and their social activities were highly integrated with the fully adult females. As such, I studied the grooming patterns of adult and adolescent females together. Seventeen out of 18 (i.e., 94%) of the adult females carried an infant during parts of the study, and 15 (i.e., 83%) gave birth at some point during the study. The only female not observed to carry an infant during the study was the α-female, Helen. This same female was also observed during a 3-month study in 1999, and at this time she was also the α-female but showed no evidence of nursing any offspring. Moreover, her teats have never shown signs of lactating or stretching during either study period, and local residents have reported never having observed her bearing an infant. Therefore she is likely to be sterile.

Data Collection and Compilation

I followed the group 4–6 days a week with some breaks over the 14-month period. During group follows, I collected focal samples on all individuals (Altmann 1974), but focal sampling was not continuous as other types of samples were collected for other studies. Focal samples were 10 min in duration, although sometimes samples were cut short if the focal subject moved out of sight. When this happened, the lost time was made up by conducting a shorter follow-up focal sample, or by adding a short amount of additional time to later focal samples on the subject. Focal samples were based on a randomized list. During focal samples grooming activity was recorded, and I scored the identity of grooming partners, the direction of the grooming, and the duration of time the grooming occurred. Agonistic interactions were also recorded both in focal samples and ad libitum (Altmann 1974), and this data was used to determine the dominance hierarchy of the females. During the study 163 h 10 min of focal sampling data were collected on adult and adolescent females with an average of 8 h 10 min per female.

For each female, the time she spent grooming each partner or receiving grooming from them was calculated from focal samples. This raw data was converted into a percentage score, which indicated the percent of time out of the female's total focal sampling time that she engaged in giving or receiving grooming with each of her partners. A sociomatrix was constructed that contained the percentage of time that each female groomed her partners, labeled the grooming–actor matrix (Table 18.1). The sociomatrix was then transposed into the grooming–actor transposition matrix. Comparing these two matrices tested the independent amount of grooming each female gave to each other because it compared the amount of grooming given in female X_y's focal samples to the amount given in the independent set of Y_x's focal samples. A third matrix was constructed that contained the amount of grooming each female received from their partners during her focal samples, labeled the grooming–receiver matrix (Table 18.2). The grooming–receiver matrix was compared with the grooming–actor matrix, providing a comparison of the amount of grooming each female gave and received in their focal samples. This data was partially dependent though because some samples included immediately reciprocated grooming, and thus the data that was compared in each matrix were not entirely independent of each other, but was used because it incorporated immediate reciprocity into the analysis. This allowed me to determine if a sample with immediate reciprocation included would show a higher degree of reciprocity than data from two sets of independent samples.

Three hypothesis sociomatrices were constructed. The first contained the inferred kin relationship of each female (Table 18.3), the second contained each individual's rank status (Table 18.4), and the third contained the rank difference between each pair (Table 18.5). Rank was measured by unidirectional agonistic interactions. The most frequently observed agonistic behaviors used in this matrix were silent-bared teeth displays (van Hooff 1967) and displacements, but all decided conflicts were included. Inferred kin was scored as kin or not kin. Kinship was not inferred through measures of grooming or proximity between females in focal samples because these were not independent from the tested grooming variables, which were taken from focal samples. Rather, I used the stability of roosting partners and the focus of attention (i.e., directing behavior) by adult females toward the same juveniles. Roosting partners were observed after dusk, using a halogen spot lamp, and lineages were determined by identifying the overlap of females in affiliating with specific juveniles. For example, Alexandria, Cleopatra, and Cinta were considered a kin group because they were consistently observed to sleep together in roost trees and they typically interacted with the same two juveniles, Sutomo and Copernicus, of which both were also typically in the matriline's roosting huddle. This method established clear clusters, which were inferred to be matrilineal families.

Matrix Analysis

The grooming sociomatrices were compared to each other to test whether the study group showed a pattern of grooming reciprocation between pairs. The Tau Kr and

Table 18.1 Grooming Actor Matrix – the amount of grooming given (percentage of time) to each female subject's partner in the subject's focal samples. The focal subjects are represented in the rows and the partners in the columns. A transposed version of this matrix was also used in the analysis

	Al	Cl	Ct	Dr	Dw	El	Fl	Hl	Ir	Jn	Kt	Lc	Ng	Nl	Pr	Pt	Rd	Rs	Rz	Vt	
Al	0.00	2.68	0.68	0.00	0.00	0.00	0.04	2.08	0.10	0.00	0.00	0.00	0.00	0.00	2.08	0.69	0.01	0.00	1.09	0.00	**9.5**
Cl	1.50	0.00	0.25	0.00	0.00	0.18	0.00	0.38	0.00	0.90	0.00	0.85	0.00	0.00	1.10	0.00	0.00	0.00	0.00	0.00	**5.2**
Ct	1.44	3.39	0.00	0.00	0.00	0.00	0.15	0.00	0.00	0.00	0.54	0.00	0.00	0.00	0.00	0.00	0.00	0.00	0.62	0.00	**6.1**
Dr	0.18	0.00	0.00	0.00	0.00	0.00	0.93	0.59	1.85	0.00	0.00	1.43	0.11	0.00	0.00	0.00	1.36	0.41	0.00	0.61	**7.5**
Dw	0.00	0.00	0.00	0.00	0.00	0.00	3.85	4.62	1.03	0.00	0.00	0.00	0.00	0.12	0.00	0.00	0.00	2.32	0.00	0.69	**12.5**
El	0.00	0.00	0.00	0.00	0.00	0.00	1.14	1.25	0.00	0.00	0.00	1.87	0.00	0.00	0.00	0.00	0.03	0.00	0.00	0.00	**4.4**
Fl	0.72	0.00	0.38	0.00	0.54	0.99	0.00	1.48	1.70	0.11	0.12	2.92	0.00	0.00	0.18	1.10	0.00	1.33	0.55	0.83	**12.9**
Hl	0.61	0.39	0.00	0.00	0.00	1.76	0.00	0.00	0.78	0.02	0.00	3.83	0.00	1.37	1.02	0.00	1.10	0.05	0.14	0.03	**11.1**
Ir	0.00	0.00	0.56	1.27	0.00	0.00	0.00	0.00	0.00	0.00	0.95	0.00	0.00	0.00	0.15	0.00	1.04	0.96	0.00	2.40	**7.3**
Jn	0.75	0.71	0.17	0.26	0.77	1.33	0.00	0.00	0.00	0.00	0.05	0.00	0.00	0.00	0.32	0.00	0.41	0.00	0.25	0.55	**5.6**
Kt	0.00	0.31	0.22	0.00	0.00	0.00	0.58	0.88	0.00	0.00	0.00	0.00	0.00	0.00	0.00	1.62	0.00	0.99	0.00	0.00	**4.6**
Lc	0.00	0.00	0.00	0.00	0.00	0.00	0.97	0.08	0.00	0.00	0.00	0.00	0.00	0.00	0.00	0.00	0.30	0.00	0.00	0.00	**1.4**
Ng	0.24	0.00	1.16	0.15	0.00	0.38	0.00	1.66	0.00	0.00	0.00	1.65	0.00	0.00	1.04	0.33	0.00	0.00	0.00	0.00	**6.6**
Nl	0.04	0.16	0.00	0.00	0.00	0.14	0.00	0.00	0.00	0.82	0.00	0.00	0.00	0.00	0.00	0.00	0.00	0.00	0.00	0.00	**1.2**
Pr	0.00	0.00	0.59	0.00	0.00	0.00	0.88	2.06	0.00	0.26	0.00	0.00	0.00	1.78	0.00	0.70	0.00	0.00	0.00	0.00	**6.3**
Pt	0.64	0.00	0.00	0.00	0.00	0.00	1.55	0.00	0.00	0.00	0.00	0.00	1.40	0.00	0.00	0.00	0.10	0.00	0.00	0.04	**3.7**
Rd	0.00	0.29	0.00	0.00	0.00	0.00	1.58	0.00	0.09	0.98	0.00	2.12	0.00	0.00	0.00	0.00	0.00	0.50	0.49	1.64	**7.7**
Rs	0.00	0.00	0.00	0.79	0.00	0.00	0.00	0.00	0.00	0.00	0.00	0.00	0.00	0.00	0.00	2.00	0.00	0.00	0.10	0.00	**2.9**
Rz	0.29	0.00	0.00	0.00	0.00	0.00	0.00	0.00	0.00	0.17	0.00	0.00	0.00	0.59	0.00	0.00	0.00	0.00	0.00	0.00	**1.1**
Vt	0.00	0.00	0.00	0.00	0.02	0.00	0.00	1.35	0.00	0.00	0.12	0.30	0.21	0.00	0.11	0.00	0.00	0.00	0.00	0.00	**2.1**
	6.4	**7.9**	**4.0**	**2.5**	**1.3**	**4.8**	**11.7**	**16.4**	**5.6**	**3.3**	**1.8**	**15.0**	**1.7**	**3.9**	**6.0**	**6.4**	**4.4**	**6.6**	**3.2**	**6.8**	**119.6**

Table 18.2 Grooming Receiver Matrix – the amount of grooming received by each female subject from her partners during the subject's focal samples

	Al	Cl	Ct	Dr	Dw	El	Fl	Hl	Ir	Jn	Kt	Lc	Ng	Nl	Pr	Pt	Rd	Rs	Rz	Vt	
Al	0.00	2.20	1.44	0.00	0.00	0.00	0.00	3.84	0.00	0.30	0.00	0.00	0.00	0.00	0.00	0.00	0.27	0.24	0.00	0.00	**8.3**
Cl	2.69	0.00	0.91	0.00	0.00	0.00	0.00	1.62	0.00	0.00	0.00	0.66	0.00	0.00	0.00	0.00	0.00	0.00	0.00	0.04	**5.9**
Ct	0.34	0.60	0.00	0.00	0.00	0.00	0.22	0.62	2.06	0.00	0.00	0.00	0.00	0.00	0.00	0.00	0.00	0.00	0.00	0.00	**3.8**
Dr	0.00	0.00	0.00	0.00	0.00	0.00	0.58	2.72	0.00	0.00	0.00	0.15	0.09	0.00	0.00	0.00	0.00	1.81	0.00	0.00	**5.4**
Dw	0.00	0.00	0.00	0.00	0.00	0.00	1.91	0.12	3.01	0.73	0.00	0.00	0.64	0.00	0.00	0.00	2.62	1.80	0.00	1.15	**12.0**
El	0.00	0.00	0.00	0.00	0.00	0.00	0.59	0.46	0.00	0.08	0.00	0.00	0.00	0.05	0.00	0.00	0.00	0.00	0.00	0.00	**1.2**
Fl	0.00	0.00	1.21	0.00	2.56	0.00	0.00	1.06	0.00	3.06	1.59	0.87	1.18	0.00	0.32	0.48	0.78	1.52	0.00	0.00	**15.6**
Hl	0.93	0.00	0.00	0.00	1.18	0.00	2.02	0.00	1.20	1.42	0.00	0.40	0.00	0.00	0.99	0.00	1.33	0.00	0.00	1.01	**9.6**
Ir	0.00	0.00	0.00	0.88	0.12	0.00	0.00	0.10	0.00	0.18	1.11	0.00	0.00	0.00	1.91	0.00	1.22	0.00	0.00	0.08	**5.6**
Jn	0.00	0.00	0.33	0.14	0.00	0.07	0.00	0.47	0.00	0.00	0.00	0.00	0.00	0.88	0.00	0.00	0.23	0.00	0.00	0.00	**2.7**
Kt	0.00	0.00	0.00	0.00	0.00	0.00	1.16	0.57	0.00	0.58	0.00	0.00	0.00	0.00	0.00	0.45	0.00	0.00	0.00	0.93	**4.0**
Lc	0.00	0.00	0.00	0.00	1.07	0.00	2.81	1.55	0.00	0.55	1.69	0.00	0.00	0.00	1.13	0.00	0.00	0.00	0.00	3.68	**12.5**
Ng	1.06	0.00	0.00	1.34	0.00	0.00	0.00	0.62	0.00	0.64	0.00	0.45	0.00	0.00	2.25	0.00	0.00	0.21	0.59	0.00	**7.2**
Nl	0.00	2.43	0.00	0.00	0.00	0.00	0.00	0.00	1.70	0.85	0.00	0.00	0.00	0.00	0.00	0.00	0.00	0.00	0.00	0.00	**5.0**
Pr	2.47	0.00	0.00	0.00	0.00	0.00	0.00	1.73	1.14	0.16	0.00	0.00	0.00	0.00	0.00	1.00	0.00	0.00	0.00	0.00	**6.8**
Pt	1.09	0.00	0.97	0.00	0.00	0.00	1.00	0.00	0.00	0.00	0.00	0.00	0.00	0.27	0.00	0.00	0.00	0.00	0.00	0.44	**4.0**
Rd	0.00	0.01	0.00	0.00	0.00	0.00	0.00	0.34	0.86	0.65	0.00	0.37	0.00	0.00	0.51	0.00	0.00	0.93	0.33	1.32	**5.3**
Rs	0.00	0.00	0.00	0.12	0.20	0.00	0.00	0.00	1.05	0.00	0.15	0.00	0.00	0.37	0.11	0.00	0.00	0.00	0.00	0.00	**1.5**
Rz	0.07	0.00	0.00	0.00	0.00	0.00	0.39	0.00	0.00	0.77	0.00	0.00	0.00	0.00	0.00	0.00	0.00	0.00	0.00	0.02	**1.5**
Vt	0.00	0.00	0.00	1.41	0.00	0.00	0.00	0.68	0.00	0.00	0.00	1.06	0.00	0.30	0.00	0.00	2.46	0.00	0.00	0.00	**5.6**
	8.7	**5.2**	**4.9**	**3.9**	**5.1**	**0.1**	**10.7**	**16.5**	**11.3**	**10.0**	**4.5**	**4.0**	**1.9**	**1.9**	**7.2**	**1.9**	**8.9**	**6.5**	**0.9**	**9.2**	**123.3**

Table 18.3 Female Inferred-Kinship Matrix – the kin relationship of each female as inferred from roosting patterns and shared affiliation with juveniles. 1 represents inferred as kin, and – represents inferred as not being kin

	Al	Cl	Ct	Dr	Dw	El	Fl	Hl	Ir	Jn	Kt	Lc	Ng	Nl	Pr	Pt	Rd	Rs	Rz	Vt	
Al	–	1	1	–	–	–	–	–	–	–	–	–	–	–	–	–	–	–	–	–	2.0
Cl	1	–	1	–	–	–	–	–	–	–	–	–	–	–	–	–	–	–	–	–	2.0
Ct	1	1	–	–	–	–	–	–	–	–	–	–	–	–	–	–	–	–	–	–	2.0
Dr	–	–	–	–	1	–	1	1	1	–	–	1	–	1	–	–	1	1	–	–	8.0
Dw	–	–	–	1	–	–	1	1	1	–	1	1	–	–	–	–	1	1	–	–	8.0
El	–	–	–	–	–	–	–	–	–	1	1	–	1	1	–	–	–	–	–	–	4.0
Fl	–	–	–	1	1	–	–	1	1	–	–	–	–	–	–	–	1	1	–	–	6.0
Hl	–	–	–	1	1	–	1	–	1	–	–	–	–	–	–	–	1	–	–	–	5.0
Ir	–	–	–	1	1	–	1	1	–	–	1	1	–	–	–	–	1	1	–	–	8.0
Jn	–	–	–	–	–	1	–	–	–	–	1	–	1	1	–	–	–	–	–	–	4.0
Kt	–	–	–	–	1	1	–	–	1	1	–	–	1	–	–	–	–	–	–	–	5.0
Lc	–	–	–	1	1	–	–	–	1	–	–	–	–	–	–	–	1	1	–	1	6.0
Ng	–	–	–	–	–	1	–	–	–	1	1	–	–	1	–	–	–	–	–	–	4.0
Nl	–	–	–	1	–	1	–	–	–	1	–	–	1	–	–	–	–	–	–	–	4.0
Pr	–	–	–	–	–	–	–	–	–	–	–	–	–	–	–	1	–	1	–	–	2.0
Pt	–	–	–	–	–	–	–	–	–	–	–	–	–	–	1	–	–	–	1	–	2.0
Rd	–	–	–	1	1	–	1	1	1	–	–	1	–	–	–	–	–	1	–	1	8.0
Rs	–	–	–	1	1	–	1	–	1	–	–	1	–	–	1	–	1	–	–	1	8.0
Rz	–	–	–	–	–	–	–	–	–	–	–	–	–	–	–	1	–	–	–	1	2.0
Vt	–	–	–	–	–	–	–	–	–	–	–	1	–	–	–	–	1	1	1	–	8.0
	2.0	2.0	2.0	8.0	8.0	4.0	8.0	8.0	6.0	4.0	4.0	6.0	4.0	4.0	2.0	2.0	7.0	7.0	2.0	8.0	98.0

Table 18.4 Female Rank Matrix – the rank of each female's partner, with 1 being the lowest-ranked female and 20 being the highest-ranked female

	Al	Cl	Ct	Dr	Dw	El	Fl	Hl	Ir	Jn	Kt	Lc	Ng	Nl	Pr	Pt	Rd	Rs	Rz	Vt
Al	3	5	4	2	12	13	17	20	1	10	15	19	14	9	6	8	18	11	7	16
Cl	3	5	4	2	12	13	17	20	1	10	15	19	14	9	6	8	18	11	7	16
Ct	3	5	4	2	12	13	17	20	1	10	15	19	14	9	6	8	18	11	7	16
Dr	3	5	4	2	12	13	17	20	1	10	15	19	14	9	6	8	18	11	7	16
Dw	3	5	4	2	12	13	17	20	1	10	15	19	14	9	6	8	18	11	7	16
El	3	5	4	2	12	13	17	20	1	10	15	19	14	9	6	8	18	11	7	16
Fl	3	5	4	2	12	13	17	20	1	10	15	19	14	9	6	8	18	11	7	16
Hl	3	5	4	2	12	13	17	20	1	10	15	19	14	9	6	8	18	11	7	16
Ir	3	5	4	2	12	13	17	20	1	10	15	19	14	9	6	8	18	11	7	16
Jn	3	5	4	2	12	13	17	20	1	10	15	19	14	9	6	8	18	11	7	16
Kt	3	5	4	2	12	13	17	20	1	10	15	19	14	9	6	8	18	11	7	16
Lc	3	5	4	2	12	13	17	20	1	10	15	19	14	9	6	8	18	11	7	16
Ng	3	5	4	2	12	13	17	20	1	10	15	19	14	9	6	8	18	11	7	16
Nl	3	5	4	2	12	13	17	20	1	10	15	19	14	9	6	8	18	11	7	16
Pr	3	5	4	2	12	13	17	20	1	10	15	19	14	9	6	8	18	11	7	16
Pt	3	5	4	2	12	13	17	20	1	10	15	19	14	9	6	8	18	11	7	16
Rd	3	5	4	2	12	13	17	20	1	10	15	19	14	9	6	8	18	11	7	16
Rs	3	5	4	2	12	13	17	20	1	10	15	19	14	9	6	8	18	11	7	16
Rz	3	5	4	2	12	13	17	20	1	10	15	19	14	9	6	8	18	11	7	16
Vt	3	5	4	2	12	13	17	20	1	10	15	19	14	9	6	8	18	11	7	14

Table 18.5 Female Rank-Difference Matrix – the linear rank difference between each female

	Al	Cl	Ct	Dr	Dw	El	Fl	Hl	Ir	Jn	Kt	Lc	Ng	Nl	Pr	Pt	Rd	Rs	Rz	Vt
Al	0	2	-1	1	-9	-10	-14	-17	2	-7	-12	-16	-11	-6	-3	-5	-15	-8	-4	-13
Cl	2	0	1	3	-7	-8	-12	-15	4	-5	-10	-14	-9	-4	-1	-3	-13	-6	-2	-11
Ct	1	-1	0	2	-8	-9	-13	-16	3	-6	-11	-15	-10	-5	-2	-4	-14	-7	-3	-12
Dr	-1	-3	-2	0	-10	-11	-15	-18	1	-8	-13	-17	-12	-7	-4	-6	-16	-9	-5	-14
Dw	9	7	8	10	0	-1	-5	-8	11	-2	-3	-7	-2	3	6	4	-6	1	5	-4
El	10	8	9	11	1	0	-4	-7	12	3	-2	-6	-1	4	7	5	-5	2	6	-3
Fl	14	12	13	15	5	4	0	-3	16	7	2	-2	3	8	11	9	-1	6	10	1
Hl	17	15	16	18	8	7	3	0	19	10	5	1	6	11	14	12	2	9	13	4
Ir	-2	-4	-3	-1	-11	-12	-16	-19	0	-9	-14	-18	-13	-8	-5	-7	-17	-10	-6	-15
Jn	7	5	6	8	-2	-3	-7	-10	9	0	-5	-9	-4	1	4	2	-8	-1	3	-6
Kt	12	10	11	13	3	2	-2	-5	14	5	0	-4	1	6	9	7	-3	4	8	-1
Lc	16	14	15	17	7	6	2	-1	18	9	4	0	5	10	13	11	1	8	12	3
Ng	11	9	10	12	2	1	-3	-6	13	4	-1	-5	0	5	8	6	-4	3	7	-2
Nl	6	4	5	7	-3	-4	-8	-11	8	-1	-6	-10	-5	0	3	1	-9	-2	2	-7
Pr	3	1	2	4	-6	-7	-11	-14	5	-4	-9	-13	-8	-3	0	-2	-12	-5	-1	-10
Pt	5	3	4	6	-4	-5	-9	-12	7	-2	-7	-11	-6	-1	2	0	-10	-3	-1	-8
Rd	15	13	14	16	6	5	1	-2	17	8	3	-1	4	9	12	10	0	7	11	2
Rs	8	6	7	9	-1	-2	-6	-9	10	1	-4	-8	-3	2	5	3	-7	0	4	-5
Rz	4	2	3	5	-5	-6	-10	-13	6	-3	-8	-12	-7	-2	1	-1	-11	-4	0	-9
Vt	13	11	12	14	4	3	-1	-4	15	6	1	-3	2	7	10	8	-2	5	9	0

Mantel Z & R matrix correlation tests were used in this study using MatSquar software (Hemelrijk 1990a, b). These types of tests are nonparametric and more useful for sociomatrix analysis in determining symmetry and unidirectionality between two matrices than parametric regression and correlation techniques. This is because data in a sociomatrix are not independent, as observations of the same individual recur within the matrix. This lack of independence in sociomatrix data can inflate the calculation of p if using a parametric test making interpretation less accurate. The Tau Kr and Mantel matrix analyses can handle this interdependency in the data and therefore are less likely to overestimate the relationship. Moreover, the Tau Kr test accounts well for individual variation and a partial correlation version of the test can be used to factor out potential confounding variables (Hemelrijk 1990a, b; de Vries 1993).

Two matrix correlations were run to test if female grooming relationships were reciprocal. First, a grooming–given matrix was tested against a transposed grooming–given matrix. In this analysis, the comparative cell in each matrix was collected from independent focal samples and compared the grooming given in focal samples of female X_y with the grooming given in the focal samples of the second female, Y_x. In the second analysis, the grooming–given matrix was compared to the grooming–received matrix, and thus only compared data from X's focal samples (i.e., X_y compared to X_x). This latter had less independent data because the compared cells came from the same focal sample set (i.e., female X) and thus included directly reciprocated bouts (i.e., grooming given and received within the same focal sample). All the matrices were further correlated with the rank, rank distance, and inferred kinship hypothesis matrices, and partial correlations were run to control for the effects of any hypothesis matrices that were found to significantly correlate with the data. This tested for any spurious relationships indicating reciprocity (Hemelrijk 1990a).

The matrix correlations were run in the following manner. A Tau Kr test was first used and this tested for the relative form of reciprocity (Hemelrijk 1990b), which is that individuals give more grooming to those that give more grooming to them, but not necessarily in the same amounts. Following any significant Tau Kr test, Mantel Z and R tests were performed on the matrices to further test the grooming balance between two dyads. These subsequent tests examined whether the absolute form of reciprocity occurred (Hemelrijk 1990b). Absolute reciprocity is when proportional amounts of grooming are reciprocated between partners, indicating time-matching of grooming. For all tests, 10,000 permutations were used. Moreover, Bonferonni corrections were used to correct α-level in each family of comparisons by dividing $p_{fw} = 0.05$ by the number of comparisons in a family. For the family of two comparison between grooming given and received (i.e., grooming–actor with grooming–actor transposed, and grooming–actor with grooming–receiver), $p = 0.025$. For the three families comparing each of the three grooming matrices with the kinship, rank, and rank distance hypothesis matrices, α-level was adjusted to $p = 0.0167$ within each family (i.e., group of comparisons of one grooming matrix to all three hypothesis matrices).

Testing for Time-Matching

I used regression and general linear mixed modeling (GLMM) using SPSS 16.0 to more closely inspect the time-matching of grooming in each pair of females by testing if the amount of grooming a female received from a partner was related to how much she gave each partner. Moreover, this approach provided a comparison to validate the findings of the matrix correlations. Before regression was used, all percentage data was transformed using an arcsine-square root method. Two regression models were then used to assess the relationship between how much giving and receiving occurred in each dyad. The first model compared the grooming–actor matrix to the grooming–actor transposed matrix, and thus compared paired data from two different sets of focal samples. The second regression compared the data from the grooming–actor matrix to the grooming–receiver matrix, and thus compared pair data from the same focal sample set.

Since regression can lack independence when using sociometric data, a GLMM was performed to better assess the validity of any results found. In this model, inferred kinship was input as a fixed factor and subject identity was input as a random factor. Factoring in subject identity controlled for the lack of independence of the data caused by having repeated subjects. Factoring in kinship tested whether the amount of grooming a female gave was better accounted for by kinship or by the amount of grooming she received, or whether both influenced grooming. I compared the data in the grooming–actor and grooming–actor transposed matrices, as well as the grooming–actor and grooming–receiver matrices. Grooming given was set as the dependent variable and grooming received was set as a covariate in all models. For each comparison two models were run. In the first model, inferred kinship and the focal subject's identity were input as random factors. In the second model, inferred kinship and the partner's identity were input as random factors. I could not construct a model with both focal subject and partners' identity factored in because there were not enough degrees of freedom to calculate an error term. Consequently, I only tested one subject factor at a time.

Comparing Up and Down-Rank Grooming

I performed a comparison to test if females directed more grooming to partners ranked above or below them. I divided each female's data into grooming with partners ranked higher than her and with partners ranked lower than her. I then calculated total grooming given, received, and the difference between grooming given and received (i.e., grooming balance) in up-rank and down-rank groups for each subject. Following this, I compared the two groups of data for each dependent variable using paired t-tests. The top-ranked female and the bottom-ranked female were excluded from the analysis since they could only groom in one direction. After comparison was completed, I ran Spearman rho correlations and cubic regression on the up-rank and down-rank data set to study how the focal subject's rank influenced her grooming patterns to partners of higher or lower rank. Cubic regression allowed me to determine the effect of any outliers and to test if grooming and rank relationships were best described using a linear or nonlinear model.

Testing Dominance Rank's Effect on Grooming

I used Spearman Rho correlations to test if a females' rank was related to how much grooming she gave and received from others. I also tested for a correlation between her rank and the difference between grooming given and received (i.e., grooming balance). The data for grooming given was data from the grooming–actor matrix and for grooming received I used the transposed grooming–actor matrix. In addition to the correlation, each relationship was fit with a cubic regression line of best fit. Cubic regression provided a better test because it had the potential to incorporate the effect of any outliers in the equation and determine if the data were better explained by a nonlinear model. Therefore, cubic regression could better evaluate if any relationship between grooming and rank was linear across subjects or whether the effect of rank on individuals varied. A linear model is more appropriate if rank affects each female to a similar degree, as in Seyfarth's model. In contrast, a cubic model would better account for data where rank influences varied across individuals in the hierarchy. For example, top-ranked female affiliation patterns might be affected by rank to a different degree than mid and low-ranked females.

Testing Rank Effects in Adjacently Ranked Partners

To test whether females provided more grooming to their higher-ranked adjacent partner than their lower-ranked adjacent partner, I paired data from each female. The amount of grooming given to the female ranked directly above a subject was matched with the amount of grooming given to the female ranked adjacently below her. The top-ranked and bottom-ranked females of the hierarchy had to be excluded from the test because they only had partners in one rank direction. A paired *t*-test was then performed on the data to test for any significant differences between the up-rank and down-rank conditions.

Testing the Influence of Rank on each Individual's Grooming

I plotted the relationship between rank and grooming given for each adult female (i.e., not including adolescents), and used the plot to calculate the slope of relationship between grooming given to each partner and each partner's rank. Each slope is a mathematical calculation of the available points, and was not tested for significance.

Rather, to test for whether these slopes differed from chance, I matched the calculated slope with the rank of each female, and ran a Spearman Rho correlation to test if the variation observed in these calculated slopes correlated with variation in female rank. This test was used to determine if females in the upper portion of the hierarchy groomed other higher-ranked females more than they did lower-ranked females. Finally, a regression analysis was performed to illustrate any relationship between a female's rank and the degree (i.e., slope of relationship) to which her grooming was influenced by the rank of her partners.

Results

Grooming Reciprocity

In this group of 20 females, there were 190 pairs and 380 possible actor-receiver dyads that could occur. When comparing the amount of independent grooming females gave to each other in their focal samples (i.e., comparing female A's focal samples to B's focal samples), I found that 208 actor-receiver dyads groomed (54.7%) and 64 of these dyads (16.8%) were reciprocal, while 144 were unidirectional (37.9%). Of the 208 dyads, only 30.8% of them were reciprocal, while 69.2% were unidirectional. Therefore, grooming reciprocity when tested with independent focal samples was not a characteristic of the majority of female grooming relationships. When I compared grooming within a single set of focal samples (i.e., grooming within female X's focal samples only) the amount of reciprocation was slightly higher, but not much different. I found that 180 actor-receiver dyads groomed (47.4%) and of these 82 (21.6%) were reciprocal and 98 (25.8%) were unidirectional. Of the 180 grooming dyads, 45.6% were reciprocal and 54.4% were unidirectional. The larger amount of reciprocal relationships found in the dependent analysis is most likely due to this sample including immediately reciprocated (i.e., dependent) grooming bouts. In both comparisons, only about one-third to one-half of all grooming relationships were reciprocal, and therefore grooming reciprocity does not appear to occur in the majority of female relationships.

Group-Level Grooming Reciprocity

Tests using Tau Kr, Z, & R tests showed that grooming reciprocity was a significant factor in the relationships of female long-tailed macaques during this study. After testing for reciprocity between each female's focal samples (i.e., comparing Table 18.1 and its transposition), I found the test for relative reciprocity to be significant, but the test for absolute reciprocity was not. This showed that females provided more grooming to females that groomed them more, but that reciprocation was not in proportional amounts (i.e., it was not time-matched) (Kr = 278, Tau

Kr$=0.143$, $p=0.0015$; $Z=51.899$, $p=0.0563$). I also tested for grooming reciprocity within each females focal samples by comparing the grooming given (Table 18.1) to the grooming received data (Table 18.2) collected from the same individual's focal sample set. In this analysis, I found absolute reciprocity where females reciprocated similar proportions of grooming (Kr$=596$, Tau Kr$=0.313$, $p=0.0001$; $Z=84.405$, $p=0.0001$; $R=15,190,884$, $p=0.0001$). Overall, relative grooming reciprocity was a significant factor in the relationships of female long-tailed macaques, but there was a greater degree of time-matching within a single female's data set, which included immediate reciprocity, than in the comparison between focal samples of differing females.

The grooming data was also compared with inferred kinship (Table 18.3), dominance rank (Table 18.4), and rank distance (Table 18.5). Kinship was significantly correlated with group-level grooming patterns (grooming actor: Kr$=264$, Tau Kr$=0.171$, $p=0.0011$; grooming actor transposition: Kr$=378$, Tau Kr$=0.245$, $p=0.0001$; grooming receiver: Kr$=345$, Tau Kr$=0.226$, $p=0.0001$), but rank was not (grooming actor: Kr$=222$, Tau Kr$=0.087$, $p=0.0480$; grooming actor transposition: Kr$=65$, Tau Kr$=0.025$, $p=0.3420$; grooming received: Kr$=227$, Tau Kr$=0.090$, $p=0.1146$). Rank distance was also not significantly correlated to grooming patterns (grooming actor: Kr$=-222$, Tau Kr$=-0.087$, $p=0.9485$; grooming actor transposition: Kr$=-65$, Tau Kr$=-0.025$, $p=0.6604$; grooming received: Kr$=-227$, Tau Kr$=-0.090$, $p=0.8824$). Since kinship was significantly related to grooming, it was partialled out from the previous analysis between grooming given and received. Controlling for kinship did not remove the reciprocity found in female grooming relationships (grooming–actor transposition: Tau Kr$=0.106$, $p=0.0146$; grooming actor-receiver: Tau Kr$=0.286$, $p=0.001$). Therefore, both kinship and reciprocity were significant influences on the group-level grooming patterns in this group of female-long-tailed macaques.

Time-Matching in Grooming Relationships

An analysis at the dyadic level across each female's focal samples, indicated that the amount of grooming a female gave to her partners was related to how much she received from them ($r^2=0.023$, $F=8.924$, $df_1=1$, $df_2=378$, $p=0.003$), showing a very weak, but significant level of time-matching. The analysis of reciprocity within each female's focal samples showed a stronger level of time-matching that was also significant ($r^2=0.158$, $F=71.007$, $df_1=1$, $df_2=378$, $p<0.001$). Generalized linear mixed models were used to factor in inferred kinship and subject identity and thus better test the validity of this relationship. For the across focal sample comparison, inferred kinship was significantly related to the grooming a female gave, but the relationship was no longer significant with grooming received. This result was found in both models used. First, focal subject identity was factored in (kin: $F=37.135$, $p<0.001$; grooming received: $F=0.968$, $p=0.326$), and second, partner identity was factored in (kin: $F=35.789$, $p<0.001$; grooming received:

$F = 0.865$, $p = 0.353$). For the within-focal sample comparison, both kin and groom-ing received were found to be significantly related to the amount of grooming a female gave to others, with focal subject identity as a factor (kin: $F = 15.702$, $p < 0.001$; grooming received: $F = 20.100$, $p < 0.001$) and partner subject identity as a factor (kin: $F = 15.246$, $p < 0.001$; grooming received: $F = 19.653$, $p < 0.001$). These GLMMs indicate that the correlation showing grooming reciprocity between each female's focal samples is better explained by kinship, but that the correlation found within females' focal samples, although related to kinship, is also partly related to direct grooming trade.

Rank Direction and Grooming Balance

Each female's relationships were categorized as up or down-rank depending on her partner's rank relative to her rank position. The top and the bottom-ranked female were removed from these analyses because they could only groom in one direction. I found no significant difference for grooming given or grooming received in up and down-rank conditions, but I did find a significant difference in the grooming balance between up and down-rank grooming. In up-rank relationships, females gave more than they received, and in down-rank relationships females received more than they gave (paired t-test: $t = 2.599$, df = 17, $p = 0.019$) (Fig. 18.1). I found no significant

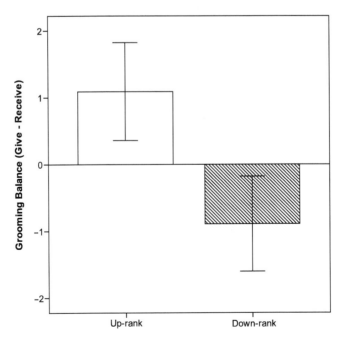

Fig. 18.1 The difference in percentage of time that grooming was given and received (i.e., grooming balance) when grooming higher-ranked and lower-ranked partners

correlation between grooming balance and rank for either up-rank (Spearman's rho: $r_s = -0.408$, $N = 18$, $p = 0.093$) or down-rank ($r_s = 0.094$, $N = 18$, $p = 0.724$). Rather, rank position showed a significant cubic relationship for grooming balance in down-rank relationships ($r^2 = 0.433$, $F = 3.820$, $df_1 = 3$, $df_2 = 15$, $p = 0.032$), indicating that top-ranked females received more than they gave to lower-ranked females than compared to the down-rank relationships of mid and low-ranked females (Fig. 18.2). The cubic relationship was the result of the two highest-ranked females in the analysis, and when removed from the analysis a significant relationship could no longer be found ($r^2 = 0.167$, $F = 0.937$, $df_1 = 3$, $df_2 = 14$, $p = 0.449$). Female grooming relationships were imbalanced in favor of the higher-ranking partner, and this was especially

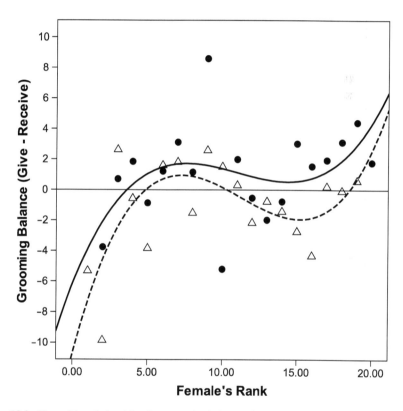

Fig. 18.2 The cubic relationships between the balance of grooming and a female's rank for grooming relationships with higher-ranked (*solid line*) and lower-ranked females (*dotted line*). This relationship was only significant for a female's grooming balance with lower-ranked partners. Higher-ranked females gave less grooming then they received from their lower-ranked partners, and this effect was especially pronounced in top-ranked females. *Closed circles* and the *solid line* represents a female's grooming balance with higher-ranked partners, and *open triangles* and the *dotted line* represents grooming balance with lower-ranked partners. Grooming balance was difference in the percentage of time grooming minus received

pronounced in the elite-ranked (i.e., very top) females. These results support that females are attracted to higher-ranked partners, as suggested in Seyfarth's model.

Dominance and Grooming

The amount of grooming given, received and the difference between giving and receiving were all tested for their correlation with female dominance rank using Spearman Rho correlations. No significant correlations were found. In contrast, when I used cubic regression to assess the possibility of a relationship skewed toward the elite-ranked females I found a significant relationship between a female's rank and the amount of grooming she received. This relationship showed a high degree of skewing toward the two top-ranked females ($r^2 = 0.590$, $F = 7.682$, $df_1 = 3$, $df_2 = 16$, $p < 0.001$) (Fig. 18.3). I also found that if the two top-ranked females were removed from the analysis, the relationship was no longer significant ($r^2 = 0.188$, $F = 1.082$, $df_1 = 3$, $df_2 = 14$, $p = 0.389$). Therefore, the significance of the relationship seemed largely due to the top 10th percentile of the hierarchy receiving more grooming than the lower 90th percentile. Consequently, the amount of grooming received is not simply a linear function of rank, as predicted by the competitive exclusion aspect of

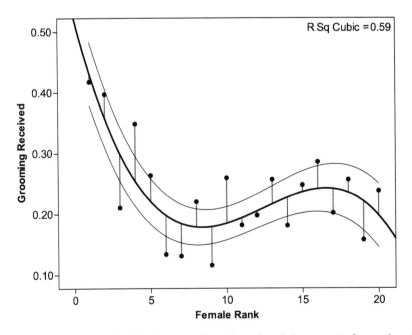

Fig. 18.3 The cubic relationship between a female's rank and the amount of grooming she received. The two outer lines represent the 99% confidence interval. Grooming balance is the arcsine transformation of percentage of time grooming was received

Seyfarth's model. Rather it is a highly skewed relationship toward the elite-ranked females in this group of long-tailed macaques.

Rank Adjacency and Grooming

In Seyfarth's model, it is predicted that females will groom their adjacent partners more because of competitive exclusion and that they should groom the higher-ranked of the two more. I tested if adjacency and rank direction were related to the independent measures of grooming by using a paired t-test to compare each females grooming to their adjacently ranked above and below partner. The test excluded the top and the bottom-ranked female, since they did not have adjacently ranked partners in both directions. The means were in the expected directions for each analysis, but I did not find any significant differences. Moreover, six females groomed their higher-ranked adjacent partner, five groomed the lower-ranked female adjacent to them, and seven did not groom either of their adjacently ranked partners. Rank adjacency did not predict grooming patterns as suggested in Seyfarth's model.

Individual Rank and Grooming

I examined how a female's own rank status was related to her grooming patterns by plotting each female's grooming given with the rank of her partners (Fig. 18.4). For each of these plots, I calculated the slope of the relationship between grooming given and partner rank for each female. A negative slope indicated that the female's grooming data was skewed toward grooming higher-ranked females, and a positive slope indicated the grooming data was skewed toward lower-ranked females. The calculated slopes were as correlated to female rank (Spearman's rho: $r_s = 0.701$, $N = 18$, $p = 0.001$), and regression analysis also showed a significant relationship between a female's rank and the degree to which her grooming was skewed toward grooming higher-ranked females ($r^2 = 0.409$, $F = 11.087$, $df_1 = 1$, $df_2 = 16$, $p = 0.004$) (Fig. 18.5). Higher-ranked females directed more grooming toward other higher-ranked females than lower-ranked females did. This is consistent with Seyfarth's predictions showing that higher-ranked females have out-competed lower-ranked females for access to groom high-ranked females.

Discussion

In this study, grooming reciprocity was a significant factor in the relationships of female long-tailed macaques, but reciprocity did not characterize the majority of relationships. Only between one-third and one-half of all grooming relationships

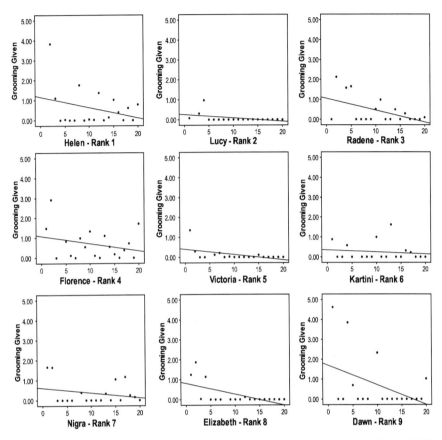

Fig. 18.4 The pattern of giving grooming in relation to a partners' rank for each of the 18 fully adult females observed during this study

showed reciprocity. Most dyads did not groom each other, thus grooming was selective. As expected, grooming was largely predicted by inferred kinship relationships. In contrast, rank was not found to be significantly related to the group-level patterns of grooming, nor were there any clear linear relationships with rank found. Rather, kinship and reciprocity better explained grooming patterns. Despite showing reciprocity at the group level, grooming did not appear to be well-balanced, or time-matched, when analyzed further. Moreover, our dyadic analysis of time matching seemed best accounted for by inferred kinship, as only our sample including immediate reciprocity showed that grooming was reciprocated independently of our measure of kinship. We therefore concluded that grooming patterns were consistent with predictions on socially-bonded female primates. Females held close reciprocal bonds with other females, but relationships were not highly balanced and thus factors other than bonding may be influencing patterns of female affiliation.

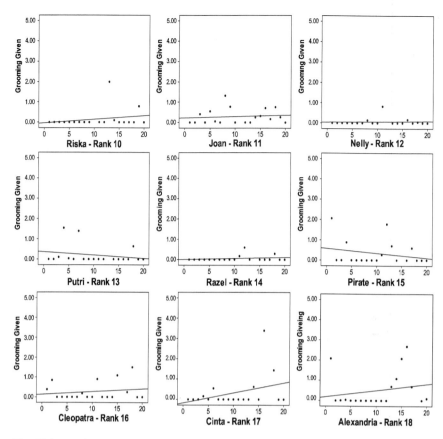

Fig. 18.4 (continued)

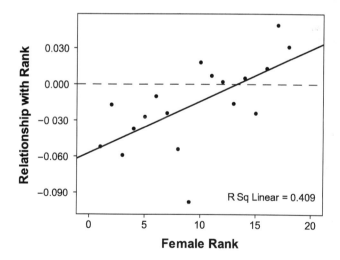

Fig. 18.5 The relationship between an adult female's rank and the slope of relationship between giving grooming and her partner's rank. Females of higher rank directed more of their grooming to higher-ranked females than lower-ranked females did

Although I did not find dominance to be highly related to grooming, further investigation did reveal that dominance and competition did affect grooming patterns in nonlinear ways. First, a paired comparison between up and down-rank grooming within each female showed an imbalance of grooming in the up-rank direction, where higher-ranked females received more grooming from their lower-ranked partners than they reciprocated. This dominance-related effect did not have an equal influence on all females though. Further analysis revealed that top-ranked females received more grooming than they gave from their lower-ranked partners than did mid and low-ranked females when they interacted with lower-ranked partners. Moreover, when top-ranked females were removed from the analysis there was no longer any clear correlation between rank and grooming imbalance, and so this pattern appeared to largely affect the elite-ranked females.

A second result on the role of dominance on grooming was that female long-tailed macaques did not appear to show the consequences of severe competitive exclusion from access to partners. This is because when grooming patterns toward adjacent-ranked partners were analyzed, I could find no significant difference between grooming their up-rank partner or down-rank partner, nor was it clear that they preferred to groom adjacent partners. It appeared that a female's grooming was not restricted by competition to groom mainly females of adjacent rank. Despite this, there were some observable competition effects. A female's rank was related to how much grooming she directed toward the highest-ranked females and therefore females of higher rank groomed at the top of the hierarchy more than lower-ranked females. This indicated that competitive exclusion may limit a female's grooming patterns to some extent, but may not have been severe enough to limit it mainly to adjacent partners. Overall, rank and competition affected grooming in this study, but not as specifically as Seyfarth's (1977) model would predict.

In other studies on long-tailed macaques, there have been tests of Seyfarth's model. Wheatley (1999) studied free-ranging temple monkeys in Bali, Indonesia and found that 73% of grooming episodes were directed up rank, that higher-ranked females received more grooming, and that higher-ranked females showed less constraint in their grooming relationships. He also identified direct competition through displacement of grooming partners, where mid-ranked females occasionally disrupted grooming bouts between a lower-ranked competitor and a higher-ranked partner. Other work has been less clear, and in a captive study it was not possible to clearly identify rank effects on affiliation patterns in a study on two groups (Butovskaya et al. 1995). According to Thierry (2007), long-tailed macaques are graded as a second-level intolerant species, and thus are intermediate to species such as *Macaca mulatta* (Grade 1), *M. sylvanus* (Grade 3) and *M. tonkeana* (Grade 4). They are considered to have a more intolerant social style relative to Grade 3 and 4, and therefore behavioral patterns should be more structured by dominance and kinship (Thierry 1985, 2004, 2007). Mine and other results indicate that dominance does influence grooming patterns in *M. fascicularis*, but not precisely how Seyfarth predicted. Perhaps Seyfarth's model provides a better prediction for highly despotic societies. Long-tailed macaques are Grade 2 and the influence of dominance and competitive exclusion may be less rigid.

Grooming Reciprocity

Reciprocity occurs in grooming relationships in many anthropoids primates (Silk 1982; de Waal and Luttrell 1988; Hemelrijk and Ek 1991; Muroyama 1991; Rowell et al. 1991; O'Brien 1993; Di Bitetti 1997; Barrett et al. 1999, 2000; Henzi and Barrett 1999; Silk et al. 1999; Vervaecke et al. 2000; Leinfelder et al. 2001; Watts 2002; Arnold and Whiten 2003; Manson et al. 2004; de Waal and Brosnan 2006; Mitani 2006). In species where females are the philopatric sex, it is expected that they will develop close bonds with each other (Wrangham 1980), which can be evident by a higher degree of grooming reciprocity (Hemelrijk 2005). However, when there is also high intra-group competition, females are expected to be highly nepotistic and develop a despotic rank order (van Schaik 1983; Sterck et al. 1997) which will structure patterns of affiliation. Long-tailed macaques fit both of these patterns. They show reciprocity because females groom the same females that groom them, and are thus bonded. Despite this, their grooming trade is not well balanced and this may be related to competition and rank structure driven by intra-group competition. Levels of grooming reciprocity are known to fluctuate under differing competitive regimes based on fluctuations in social and ecological conditions (Hemelrijk and Luteijn 1998; Barrett et al. 2002; Henzi et al. 2003; Hemelrijk 2005; Stevens et al. 2005; Barrett and Henzi 2006). Therefore, factors that alter the degree of grooming reciprocity may be related to rank and competition. These include intensity of feeding competition (Barrett et al. 2002), the amount of overt aggression (Barrett et al. 2002; Hemelrijk 2005; Stevens et al. 2005), and the demography of males and females in the group (Hemelrijk and Luteijn 1998; Hemelrijk 2005).

Investigations have shown that within bouts female grooming is time matched in *Papio hamadryas cynocephalus* (Barrett et al. 2000), *Macaca radiata*, and *Cebus capucinus*, but that time-matching accounts for very little of the variation found (Manson et al. 2004). In another study on this group, grooming bouts were not well time-matched and the grooming initiator tended to provide more grooming into the bout (Gumert and Ho 2008). In this study, I evaluated the grooming balance within each dyad and found that grooming was only weakly time-matched, supporting what was previously found. Since grooming is clearly not that well matched in this group, it is possible that grooming balance might be influenced by the power differentials within the group. According to some, grooming should be less balanced where dominance effects are more pronounced because subordinates will groom dominants to develop tolerance and alleviate social competition (Henzi and Barrett 1999; Barrett and Henzi 2001; Barrett et al. 2002). Moreover, Seyfarth (1977) suggested the major reason for such imbalance was the need to obtain high-ranked social allies for coalitionary support. In this group of macaques, rank has been found in this and other studies to be related to grooming imbalances, with lower-ranked females grooming higher-ranked females longer than vice versa (Gumert 2007a; Gumert and Ho 2008). Since long-tailed macaques are an intolerant macaque species (i.e., grade 2 of 4) (Thierry 1986, 2004, 2007) and their social status relies on support from allies (de Waal 1977; Chapais and Gauthier 2004), we might expect social power to influence dyadic grooming patterns by skewing grooming exchange in favor of higher-ranked individuals to gain tolerance and support.

Grooming and Dominance

Recently Schino (2001) conducted a meta-analysis on 14 different species consisting of prosimians, New World, and Old World monkeys. He found that Seyfarth's model applied. Attraction to rank, attraction to kin, and competition over grooming partners could explain grooming patterns across species. Furthermore, Seyfarth's theory on rank attraction has been consistent with the grooming patterns in many species of Old World primates, including *Cercocebus aethiops* (Fairbanks 1980; Seyfarth 1980), *Papio hamadryas hamadryas* (Kummer 1968; Stammbach and Kummer 1982), *Papio anubis* (Buirski et al. 1973), *Papio cynocephalus ursinus* (Seyfarth 1976), *M. mulatta* (Bernstein and Sharpe 1966; Sade 1972b; Chapais 1983), *Macaca fuscata* (Oki and Maeda 1973), *Macaca nemestrina* (Oi 1990), *M. fascicularis* (Wheatley 1999), *M. radiata* (Silk et al. 1981; Silk 1982) *Erythrocebus patas* (Hall and Mayer 1967), and *Pan paniscus* (Vervaecke et al. 2000; Stevens et al. 2005). The model therefore seems to be broadly applicable to numerous species, but not all data, including this study, has clearly supported its predictions, indicating some limitations in the model's use (de Waal and Luttrell 1986; Henzi and Barrett, 1999).

Some species do not show clear rank-related attraction in grooming patterns. For example, dominance had little effect on social grooming patterns in *Macaca tonkeana* (Thierry et al. 1990) and *Macaca assamensis* (Cooper and Bernstein 2000, 2008). In *Cebus* monkeys, grooming is directed down-rank and is less related to rank than in cercopithecine monkeys (O'Brien 1993; Parr et al. 1997; Fragaszy et al. 2004). In addition, some of the same species that supported the model have provided contradictory data in other studies. For example, some studies on *M. mulatta* (Lindburg 1973; Chapais 1983; de Waal and Luttrell 1986) have not clearly supported attraction to high rank in grooming patterns. Furthermore, later studies on *P. cynocephalus ursinus*, the species Seyfarth's drew his original conception of the model from, demonstrated only a small effect of rank on grooming patterns (Silk et al. 1999). Lastly, Schino et al. (2003), after his meta-analysis, found that the degree of reciprocation was not related to rank distance in a captive group of *M. fuscata.*

Rank does affect primate societies, but its influence is not always clear and consistent. Therefore, we need to determine the degree to which the effects of a stable hierarchy on patterns of affiliation are static or dynamic. Are dominance effects consistent within a species or rather does the influence of a social hierarchy more depend on the contextual needs of dominants and subordinates in different social scenarios? Studies on *Papio hamadryas ursinus* have shown no clear long-term rank effects on grooming patterns (Henzi et al. 2003) and rather have found that rank influences on grooming were best explained by fluctuations in dominance gradient based on alterations in competition over food resources (i.e., degree of hierarchical influence over access to resources) (Barrett et al. 2002). Such correlations between dominance gradient and grooming patterns have also been found in *P. paniscus* (Stevens et al. 2005). In another study on *Papio cynocephalus anubis*, rank effects were not clear, and grooming relationships were not linearly related to rank (Sambrook et al. 1995). Rather, it was found that an attractive elite group (i.e., top-ranked individuals) gave and received grooming at a higher level than their lower-ranked conspecifics.

These type of results indicate that rank influences are not static and do not appear to structure the affiliation patterns of a society in any singular consistent pattern across all social conditions or across all individuals. Rather, a social hierarchy appears to have a stochastic influence on affiliation patterns based on situational contexts that alter the competitive ability of the dominants and subordinates in a society. Dominance also does not exert equal influences across each member of a society. These factors all likely contribute to why we see such great variation in the results coming from studies on dominance and affiliation. Moreover, these factors could also explain why in this study I found mixed results on the degree to which dominance structured patterns of affiliation. Consequently, it is clear that both temporal and demographic variation needs to be considered when attempting to investigate the influence of dominance on grooming patterns and other forms of affiliation.

Understanding the Social Function of Grooming

Work has shown that female grooming can be used to obtain access to several social resources under differing conditions. Female monkeys use grooming to access infants from mothers (Henzi and Barrett 2002; Gumert 2007a) and to establish tolerance (Henzi and Barrett 1999; Barrett and Henzi 2001; Gumert and Ho 2008). In addition, patterns of female grooming change with fluctuating conditions of competition over males (Hemelrijk and Luteijn 1998; Hemelrijk 2005). Although rank effects occur in many of these exchanges, these studies show that grooming is not mainly based on rank or attaining support, but rather is used in several cooperative and competitive contexts. If obtaining coalition support is the central function of grooming, than we would expect that females should be adapted to provide grooming during conditions when support is needed most and to direct that grooming to those best fit to give support under such conditions. Grooming has been clearly shown to directly coordinate coalition support (Seyfarth and Cheney 1984; Hemelrijk 1994). Moreover, some evidence suggests chimpanzees (*Pan troglodytes*) and stump-tailed macaques (*Macaca arctoides*) will groom potential alliance partners in times of rank instability or in times prior to potential conflict, such as before feeding on clumped resources (de Waal 1982; Koyama and Dunbar 1996; Koyama 2000). Grooming appears to be used in some species to develop alliance partners that will offer coalition support in times of aggression, but grooming is clearly not limited to this context. Rather it appears to be a generalized medium of social trade where coalition support is just one need that a female macaque could exchange grooming for depending on her conditional needs.

When trying to ascribe a functional explanation to the social aspects of grooming, it may be most appropriate to consider grooming as a way to provide low cost benefits to others in order to achieve a general array of goals related to social exchange, competition, and the development of relationships (Dunbar 1991). Animals that have exchange relationships are argued to benefit by evolving specialized behaviors aimed at directing low cost benefits to others that will increase the likelihood of receiving return acts (Tooby and Cosmides 1996). Evolving such traits would increase the likelihood of gaining from one's social partners, and thus would be advantageous to succeeding and potentially reproducing in one's society. Anthropoid primates particularly, seem to have co-opted (Gould 2002) grooming, a behavior originally adapted for its hygienic purposes, to provide low cost benefits to their partners to establish short and long-term exchange relationships. These exchanges need not be specific to coalition support, but could be utilized in potentially any type of social exchange. As a result, identifying the context of grooming bouts will be critical in future studies attempting to uncover functional aspects of grooming because the social function of grooming varies across conditions.

Overall, grooming reciprocity is an integral component of a female-bonded society. Despite this, not all female grooming is reciprocal and the results of this study and others show that actually most grooming is not reciprocal. Such imbalance is likely because of both exchange relationships and rank effects. No matter what context grooming occurs in there is always the underlying influence of rank. High-ranked individuals have more power and can use their influence to out-compete others and obtain what they need in their societies. Therefore, all exchanges with higher-ranked females, especially the top-ranking elite, will tend to be partly corrupted in terms of the balance of trade. I would speculate here that a large proportion of rank-related grooming is the result of such power imbalance. This forces the subordinate to spend extra effort to appease and establish tolerance from a more powerful partner in order to foster a cooperative relationship. For this reason, Henzi and Barrett's (1999) grooming for tolerance hypothesis should be considered in models of female grooming in cercopithecine primates, which argues that social trade is related to appeasing more powerful females, but is not necessarily related to receiving the benefit of obtaining alliance partners. Rather, we should expect that grooming balance should always be skewed to some extent in the favor of powerful high-ranked females because subordinates will be attempting to lessen power barriers and coordinate cooperative forms of exchange with more powerful partners. Moreover, the strength of relationship between grooming and dominance should vary with the social style of the society as well as the proximate needs and goals of the grooming females.

Acknowledgments Parts of this project were funded by a Fulbright Graduate Research Fellowship from the American-Indonesian Exchange Foundation (AMINEF) and from The Ministry of Education in Singapore (Grant # RG95-07). The Indonesian Institute of Sciences (LIPI) issued a research permit to conduct research in Indonesia (Permit #: 3044/SU/KS/2003), and the Indonesian Department of Forestry (PHKA) authorized permission to enter and reside in Tanjung Puting National Park (Permit #: 1765/IV-SEK/HO/2003). The Institute of Animal Care and Use Committee of the United States at the University of Georgia in Athens approved the methods of animal observation before initiating field research (Animal Research Protocol #: A2005-10167-0). The research was sponsored by the University of Indonesia through Dr. Noviar Andayani from the Faculty of Mathematics and Natural Sciences. I would like to thank Peltanadanson for assisting in the field research and Karthick Ramanathan for compiling data for analysis. I also thank the Rimba Orangutan Ecolodge for supporting the project during my research.

References

Altmann J (1974) Observational study of behavior: sampling methods. Behaviour 49:227–265

Arnold K, Whiten A (2003) Grooming interactions among the chimpanzees of the Budongo Forest, Uganda: tests of five explanatory models. Behaviour 140:519–552

Barrett L, Gaynor D, Henzi SP (2002) A dynamic interaction between aggression and grooming reciprocity among female chacma baboons. Anim Behav 63:1047–1053

Barrett L, Henzi SP (2001) The utility of grooming in baboon troops. In: Noë R, van Hooff JARAM, Hammerstein P (eds) Economics in nature: social dilemmas, mate choice and biological markets. Cambridge University Press, Cambridge, pp 119–145

Barrett L, Henzi SP (2006) Monkeys, markets and minds: biological markets and primate sociality. In: Kappeler PM, van Schaik C (eds) Cooperation in primates and humans: mechanisms and evolution. Springer-Verlag, Berlin, pp 209–232

Barrett L, Henzi SP, Weingrill T, Lycett JE, Hill RA (1999) Market forces predict grooming reciprocity in female baboons. Proc R Soc Lond B 266:665–670

Barrett L, Henzi SP, Weingrill T, Lycett JE, Hill RA (2000) Female baboons do not raise the stakes but they give as good as they get. Anim Behav 59:763–770

Bernstein IS (1991) The correlation between kinship and behaviour in non-human primates. In: Hepper PG (ed) Kin recognition. Cambridge University Press, Cambridge, pp 6–29

Bernstein IS, Sharpe LG (1966) Social roles in a rhesus monkey group. Behaviour 26:91–104

Boccia ML (1987) The physiology of grooming: a test of the tension reduction hypothesis. Am J Primatol 12:330

Boccia ML, Reite M, Laudenslager M (1989) On the physiology of grooming in a pigtail macaque. Physiol Behav 45:667–670

Buirski P, Kellerman H, Plutchik R, Weininger R, Buirski N (1973) A field study of emotions, dominance, and social behavior in a group of baboons (*Papio anubis*). Primates 14:67–78

Butovskaya M, Kozintsev A, Welker C (1995) Grooming and social rank by birth: the case of *Macaca fascicularis*. Folia Primatol 65:30–33

Chapais B (1983) Dominance, relatedness and the structure of female relationships in rhesus monkeys. In: Hinde RA (ed) Primate social relationships: an integrated approach. Sinauer Associates, Sunderland, pp 208–217

Chapais B (1992) The role of alliances in social inheritance of rank among female primates. In: Harcourt AH, de Waal FBM (eds) Coalitions and alliances in humans and other animals. Oxford University Press, Oxford, pp 29–59

Chapais B, Gauthier C (2004) Juveniles outrank higher-born females in groups of long-tailed macaques with minimal kinship. Int J Primatol 25:429–447

Cheney DL (1977) The acquisition of rank and the development of reciprocal alliances among free-ranging immature baboons. Behav Ecol Sociobiol 2:303–318

Cooper MA, Bernstein IS (2000) Social grooming in Assamese macaques (*Macaca assamensis*). Am J Primatol 50:77–85

Cooper MA, Bernstein IS (2008) Evaluating dominance style in Assamese and rhesus macaques. Int J Primatol 29:225–244

Curley JP, Keverne EB (2005) Genes, brains and mammalian social bonds. Trends Ecol Evol 20:561–567

Datta SB (1983a) Patterns of agonistic interference. In: Hinde RA (ed) Primate social relationships: an integrated approach. Sinauer Associates, Sunderland, pp 289–297

Datta SB (1983b) Relative power and the acquisition of rank. In: Hinde RA (ed) Primate social relationships: an integrated approach. Sinauer Associates, Sunderland, pp 93–103

Datta SB (1983c) Relative power and the maintenance of dominance. In: Hinde RA (ed) Primate social relationships: an integrated approach. Sinauer Associates, Sunderland, pp 103–112

de Ruiter JR, Geffen E (1998) Relatedness of matrilines, dispersing males and social groups in long-tailed macaques (Macaca fascicularis). Proc R Soc Lond B 265:79–87

de Vries H (1993) The rowwise correlation between two proximity matrices and the partial rowwise correlation. Psychometrika 58:53–69

de Waal FBM (1977) The organization of agonistic relations within two captive groups of Java-monkeys (Macaca fascicularis). Z Tierpsychol 44:225–282

de Waal FBM (1982) Chimpanzee politics. Harper & Row, New York

de Waal FBM (1989) Dominance 'style' and primate social organization. In: Standen V, Foley RA (eds) Comparative socioecology: the behavioural ecology of humans and other mammals. Blackwell Scientific Publications, Oxford, pp 243–263

de Waal FBM, Brosnan SF (2006) Simple and complex reciprocity in primates. In: van Schaik CP, Kappeler PM (eds) Cooperation in primates and humans: mechanisms and evolution. Springer-Verlag, Berlin, pp 85–106

de Waal FBM, Leimgruber K, Greenberg AR (2008) Giving is self-rewarding for monkeys. Proc Natl Acad Sci USA 105:13685–13689

de Waal FBM, Luttrell LM (1986) The similarity principle underlying social bonding among female rhesus monkeys. Folia Primatol 46:215–234

de Waal FBM, Luttrell L (1988) Mechanisms of social reciprocity in three primate species: symmetrical relationship characteristics or cognition? Ethol Sociobiol 9:101–118

Di Bitetti MS (1997) Evidence for an important social role of allogrooming in a platyrrhine primate. Anim Behav 54:199–211

Drickamer LC (1976) Quantitative observations of grooming behavior in free-ranging Macaca mulatta. Primates 17:323–335

Dunbar R (1991) Functional significance of social grooming in primates. Folia Primatol 57:121–131

Dunbar R, Sharman M (1984) Is social grooming altruistic? Z Tierpsychol 64:163–173

Fabre-Nys C, Meller RE, Keverne EB (1982) Opiate antagonists stimulate affiliative behaviour in monkeys. Pharmacol Biochem Behav 16:653–659

Fairbanks LA (1980) Relationships among adult females in captive vervet monkeys: testing a model of rank related attractiveness. Anim Behav 28:853–859

Fittinghoff NA Jr, Lindburg DG (1980) Riverine refuging in East Bornean Macaca fascicularis. In: Lindburg DG (ed) The macaques: studies in ecology, behavior and evolution. Van Nostrand Reinhold, New York, pp 182–214

Fragaszy D, Visalberghi E, Fedigan L (2004) The complete capuchin. Cambridge University Press, Cambridge

Goosen C (1987) Social grooming in primates. In: Mitchell G, Erwin J (eds) Comparative primate biology: behavior, cognition, and motivation, vol 2b. Alan R. Liss, New York, pp 107–131

Gould SJ (2002) The structure of evolutionary theory. Belknap Press, Cambridge

Gouzoules S, Gouzoules H (1987) Kinship. In: Smuts B, Cheney DL, Seyfarth RM, Wrangham RW, Struhsaker TT (eds) Primate societies. Chicago University Press, Chicago, pp 299–305

Gumert MD (2007a) Grooming-infant handling interchange: the relationship between infant supply and grooming payment in Macaca fascicularis. Int J Primatol 28:1059–1074

Gumert MD (2007b) Payment for sex in a macaque mating market. Anim Behav 74:1655–1667

Gumert MD, Ho MH (2008) The trade balance of grooming and it relationship to tolerance in Indonesian long-tailed macaques (Macaca fascicularis). Primates 49:176–185

Hall KRL, Mayer B (1967) Social interactions in a group of captive patas monkeys. Folia Primatol 5:213–236

Hemelrijk CK (1990a) A matrix partial correlation test used in investigations of reciprocity and other social interaction patterns at group level. J Theor Biol 143:405–420

Hemelrijk CK (1990b) Models of, and tests for, reciprocity, unidirectionality and other interaction patterns at a group level. Anim Behav 39:1013–1029

Hemelrijk CK (1994) Support for being groomed in long-tailed macaques, *Macaca fascicularis*. Anim Behav 48:479–481

Hemelrijk CK (2005) The process-oriented approach to the social behaviour of primates. In: Hemelrijk CK (ed) Self-organization and evolution of social systems. Cambridge University Press, Cambridge, pp 81–107

Hemelrijk CK, Ek A (1991) Reciprocity and interchange of grooming and 'support' in captive chimpanzees. Anim Behav 41:923–935

Hemelrijk CK, Luteijn M (1998) Philopatry, male presence and grooming reciprocation among female primates: a comparative perspective. Behav Ecol Sociobiol 42:207–215

Henzi SP, Barrett L (1999) The value of grooming to female primates. Primates 40:47–59

Henzi SP, Barrett L (2002) Infants as a commodity on a baboon market. Anim Behav 63:915–921

Henzi SP, Barrett L, Gaynor D, Greeff J, Weingrill T, Hill RA (2003) Effect of resource competition on the long-term allocation of grooming by female baboons: evaluating Seyfarth's model. Anim Behav 66:931–938

Hutchins M, Barash DP (1976) Grooming in primates: implications for its utilitarian functions. Primates 17:145–150

Kapsalis E (2004) Matrilineal kinship and primate behavior. In: Chapais B, Berman CM (eds) Kinship and behavior in primates. Oxford University Press, Oxford, pp 153–176

Kawai M (1958) On the ranks system in a natural troop of Japanese monkeys I & II. Primates 1:111–148

Keverne EB, Martensz ND, Tuite B (1989) Beta-endorphin concentrations in cerebrospinal fluid of monkeys are influenced by grooming relationships. Psychoneuroendocrinology 14:155–161

Koyama N (2000) Conflict-reduction mechanisms before feeding. In: Aureli F, de Waal FBM (eds) Natural conflict resolution. University of California Press, Berkeley, pp 130–132

Koyama NF, Dunbar RIM (1996) Anticipation of conflict by chimpanzees. Primates 37:79–86

Kummer H (1968) Social organization of hamadryas baboons: a field study. The University of Chicago Press, Chicago

Kurland JA (1977) Kin selection in the Japanese monkey. Karger, Basel, New York

Leinfelder I, de Vries H, Deleu R, Nelissen M (2001) Rank and grooming reciprocity among females in a mixed-sex group of captive hamadryas baboons. Am J Primatol 55:25–42

Lindburg DG (1973) Grooming behavior as a regulator of social interaction in rhesus monkeys. In: Carpenter CR (ed) Behavioral regulators of behavior in primates. Bucknell University Press, Lewisburg, pp 124–148

Manson JH, David Navarrete C, Silk JB, Perry S (2004) Time-matched grooming in female primates? New analyses from two species. Anim Behav 67:493–500

Matheson MD, Bernstein IS (2000) Grooming, social bonding, and agonistic aiding in rhesus monkeys. Am J Primatol 51:177–186

Mitani JC (2006) Reciprocal exchange in chimpanzee and other primates. In: van Schaik CP, Kappeler PM (eds) Cooperation in primates and humans: mechanisms and evolution. Springer-Verlag, Berlin, pp 107–120

Muroyama Y (1991) Mutual reciprocity of grooming in female Japanese macaques (*Macaca fuscata*). Behaviour 119:161–170

O'Brien TG (1993) Allogrooming behaviour among adult female wedge-capped capuchin monkeys. Anim Behav 46:499–510

Oi T (1990) Patterns of dominance and affiliation in wild pig-tailed macaques (*Macaca nemestrina nemestrina*) in West Sumatra. Int J Primatol 11:339–356

Oki J, Maeda Y (1973) Grooming as a regulator of behavior in Japanese macaques. In: Carpenter CR (ed) Behavioral regulators of behavior in primates. Bucknell University Press, Lewisburg, PA, pp 149–163

Parr LA, Matheson MD, Bernstein IS, de Waal FBM (1997) Grooming down the hierarchy: allogrooming in captive brown capuchin monkeys, *Cebus apella*. Anim Behav 54:361–367

Pereira ME (1995) Development and social dominance among group-living primates. Am J Primatol 37:143–175

Pusey A, Packer C (1987) Dispersal and philopatry. In: Smuts B, Cheney DL, Seyfarth RM, Wrangham RW, Struhsaker TT (eds) Primate societies. University of Chicago Press, Chicago, pp 250–266

Reiss MJ (1984) Kin selection, social grooming and the removal of ectoparasites: a theoretical investigation. Primates 25:185–191

Rowell TE, Wilson C, Cords M (1991) Reciprocity and partner preference in grooming of female blue monkeys. Int J Primatol 12:319–336

Sade DS (1965) Some aspects of parent-offspring and sibling relations in a group of rhesus monkeys, with a discussion of grooming. Am J Phys Anthropol 23:1–18

Sade DS (1972a) A longitudinal study of social behavior of rhesus monkeys. In: Tuttle RH (ed) The functional and evolutionary biology of primates. Aldine, Chicago, pp 378–398

Sade DS (1972b) Sociometrics of *Macaca mulatta*. I: linkages and cliques in grooming matrices. Folia Primatol 18:196–223

Sambrook TD, Whiten A, Strum SC (1995) Priority of access and grooming patterns of females in a large and a small group of olive baboons. Anim Behav 50:1667–1682

Schino G (2001) Grooming, competition and social rank among female primates: a meta-analysis. Anim Behav 62:265–271

Schino G (2007) Grooming and agonistic support: a meta-analysis of primate reciprocal altruism. Behav Ecol 18:115–120

Schino G, Scucchi S, Maestripieri D, Turillazzi PG (1988) Allogrooming as a tension-reduction mechanism: a behavioral approach. Am J Primatol 16:43–50

Schino G, Ventura R, Troisi A (2003) Grooming among female Japanese macaques: distinguishing between reciprocation and interchange. Behav Ecol 14:887–891

Seyfarth RM (1976) Social relationships among adult female baboons. Anim Behav 24:917–938

Seyfarth RM (1977) A model of social grooming among adult female monkeys. J Theor Biol 65:671–698

Seyfarth RM (1980) The distribution of grooming and related behaviours among adult female vervet monkeys. Anim Behav 28:798–813

Seyfarth RM (1983) Grooming and social competition in primates. In: Hinde RA (ed) Primate social relationships: an integrated approach. Blackwell Scientific Publications, Oxford, pp 182–190

Seyfarth RM, Cheney DL (1984) Grooming, alliances and reciprocal altruism in vervet monkeys. Nature 308:541–543

Shutt K, MacLarnon A, Heistermann M, Semple S (2007) Grooming in Barbary macaques: better to give than to receive? Biol Lett 3:231–233

Silk JB (1982) Altruism among female *Macaca radiata*: explanations and analysis of patterns of grooming and coalition formation. Behaviour 79:162–188

Silk JB (2006) Practicing Hamilton's rule: kin selection in primate groups. In: van Schaik CP, Kappeler PM (eds) Cooperation in primates and humans: mechanisms and evolution. Springer-Verlag, Berlin, pp 25–46

Silk JB (2007) Social components of fitness in primate groups. Science 317:1347–1351

Silk JB, Alberts SC, Altmann J (2003) Social bonds of female baboons enhance infant survival. Science 302:1231–1234

Silk JB, Samuels A, Rodman PS (1981) The influence of kinship, rank, and sex on affiliation and aggression between adult female and immature bonnet macaques (*Macaca radiata*). Behaviour 78:111–177

Silk JB, Seyfarth RM, Cheney DL (1999) The structure of social relationships among female savanna baboons in Moremi Reserve, Botswana. Behaviour 136:679–703

Stammbach E, Kummer H (1982) Individual contributions to a dyadic interaction: an analysis of baboon grooming. Anim Behav 30:964–971

Sterck EHM, Watts DP, van Schaik CP (1997) The evolution of female social relationships in nonhuman primates. Behav Ecol Sociobiol 41:291–309

Stevens J, Vervaecke H, de Vries H, Van Elsacker L (2005) The influence of the steepness of dominance hierarchies on reciprocity and interchange in captive groups of bonobos (*Pan paniscus*). Behaviour 142:941–960

Tanaka I, Takefushi H (1993) Elimination of external parasites (lice) is the primary function of grooming in free-ranging Japanese macaques. Anthropol Sci 101:187–193

Terry R (1970) Primate grooming as a tension reduction mechanism. J Psychol 76:129–136

Thierry B (1985) Patterns of agonistic interactions in three species of macaque (*Macaca mulatta, M. fascicularis, M. tonkeana*). Aggress Behav 11:223–233

Thierry B (1986) A comparative study of aggression and response to aggression in three species of macaque. In: Else JG, Lee PC (eds) Primate ontogeny, cognition and social behavior. Cambridge University Press, Cambridge, pp 307–313

Thierry B, Gautheir C, Peignot P (1990) Social grooming in Tonkean macaques (*Macaca tonkeana*). Int J Primatol 11:357–375

Thierry C (1990) Feedback loop between kinship and dominance: the macaque model. J Theor Biol 145:511–521

Thierry B (2004) Social epigenesis. In: Thierry B, Singh M, Kaumanns W (eds) Macaque societies. Cambridge University Press, Cambridge, pp 267–290

Thierry B (2007) Unity in diversity: lessons from macaque societies. Evol Anthropol 16:224–238

Tooby J, Cosmides L (1996) Friendship and the banker's paradox: other pathways to the evolution of adaptations for altruism. Proc Br Acad 88:119–143

van Hooff JARAM (1967) The facial displays of the Catarrhine monkeys and apes. In: Morris D (ed) Primate ethology. Weidenfield, London, pp 7–68

van Noordwijk MA, van Schaik CP (1985) Male migration and rank acquisition in wild long-tailed macaques (*Macaca fascicularis*). Anim Behav 33:849–861

van Noordwijk MA, van Schaik CP (1987) Competition among female long-tailed macaques, *Macaca fascicularis*. Anim Behav 35:577–589

van Noordwijk MA, van Schaik CP (1999) The effects of dominance rank and group size on female lifetime reproductive success in wild long-tailed macaques, *Macaca fascicularis*. Primates 40:105–130

van Schaik CP (1983) Why are diurnal primates living in groups? Behaviour 87:120–144

van Schaik CP, van Amerongen A, van Noordwijk MA (1996) Riverine refuging by wild Sumatran long-tailed macaques (*Macaca fascicularis*). In: Fa JE, Lindburg DG (eds) Evolution and ecology of macaque societies. Cambridge University Press, Cambridge, pp 160–181

Vandenbergh JG (1967) The development of social structure in free-ranging rhesus monkeys. Behaviour 29:179–194

Vervaecke H, De Vries H, Van Elsacker L (2000) The pivotal role of rank in grooming and support behavior in a captive group of bonobos (*Pan paniscus*). Behaviour 137:1463–1485

Walters JR, Seyfarth RM (1987) Conflict and cooperation. In: Smuts BB, Cheney DL, Seyfarth RM, Wrangham RW, Struhsaker TT (eds) Primate societies. University of Chicago Press, Chicago, pp 306–317

Watts DP (2002) Reciprocity and interchange in the social relationships of wild male chimpanzees. Behaviour 139:343–370

Wheatley B (1980) Feeding and ranging of East Bornean *Macaca fascicularis*. In: Lindburg DG (ed) The macaques: studies in ecology, behavior and evolution. Van Nostrand Reinhold, New York, pp 215–246

Wheatley B (1999) The sacred monkeys of Bali. Waveland Press, Inc., Prospect Heights, Illinois

Wrangham RW (1980) An ecological model of female-bonded primate groups. Behaviour 75:262–299

Zamma K (2002) Grooming site preferences determined by lice infection among Japanese macaques in Arashiyama. Primates 43:41–49

Chapter 19
Selamatkan Yaki! Conservation of Sulawesi Crested Black Macaques *Macaca nigra*

Vicky Melfi

Introduction

The Island of Sulawesi is the largest in the Wallacea region, a biodiversity hotspot where Asian and Australasian flora and fauna met and merge. Wallacea is one of 25 regions described as a biodiversity hotspot; designation is attributed to areas with a high degree of endemism and where 70% of primary vegetation has already been lost, namely due to its high diversity of endemic wildlife (Myers et al. 2000). Endemism is greater on the island of Sulawesi than any of the other Indonesian islands (Whitten et al. 1987); estimations suggest that 96/380 (25%) bird species and 79/127 (62%) mammal species, though this rises to 98% if only flightless mammals are considered i.e., bats are excluded, are unique to this island (Holmes and Phillips 1996). Unfortunately, the survival of this biodiversity is under threat as many of the species found in Sulawesi are listed as critically endangered, endangered or threatened by the IUCN red data list; examples include the anoa (*Bubalus depressicournis*), maleo (*Macrocephalon maleo*), Sulawesi forest turtle (*Leucocephalon yuwonoi*), dwarf pygmy goby, (*Pandaka pygmaea*), the "potentially extinct" duck-billed buntingi (*Adrianichthys kruyti*; Harrison and Stiassny 1999), along with a great deal of flora e.g., *Shorea montigena* (IUCN 2008). The importance of wildlife on Sulawesi is so great that it has even been suggested that conservation efforts should be prioritised towards preserving the habitat and species of Sulawesi, ahead of that found on other islands in Indonesia (Wilson et al. 2005). Despite this, the majority of conservation funding and effort is directed towards other islands within Indonesia (Wilson et al. 2005) and it would seem that priority is given to islands noted for their charismatic megafauna. No one can doubt the appeal of elephants, rhinos, tigers, orang-utans and gibbons, but a jewel in the biodiversity crown of Sulawesi, which it seems is often overlooked, is the radiation of seven endemic and extant macaque species.

V. Melfi (✉)
Whitley Wildlife Conservation Trust, Field Conservation and Research Department, Paignton Zoo Environmental Park, Totnes Road, Paignton, Devon, UK TQ4 7EU
e-mail: vicky.melfi@paigntonzoo.org.uk

S. Gursky-Doyen and J. Supriatna (eds.), *Indonesian Primates*,
Developments in Primatology: Progress and Prospects,
DOI 10.1007/978-1-4419-1560-3_19, © Springer Science+Business Media, LLC 2010

Table 19.1 A summary of conservation status and distribution of the seven recognised endemic macaque species found in Sulawesi (Figures of captive animals represent animals living in ISIS member zoos and as such provide an underestimate of the captive populations, as some non-ISIS member zoos also hold these species). Data taken from ISIS 2009

Macaca species	Common name	Distribution	IUCN Red List Category	Captive population (male.female.unknown)
nigra	Black crested	Northeast	Critically endangered	95.139.26
nigrescens	Gorontalo	North	Vulnerable	No
hecki	Heck's	Northwest	Vulnerable	7.11.0
tonkeana	Tonkean	Central	Vulnerable	43.63.25
maura	Moor	Southwest	Endangered	4.2.1
ochreata	Booted	Southeast	Vulnerable	0.1.0
brunnescens	Buton	Buton and Muna Islands		No

Of the seven endemic macaque species on the Island of Sulawesi (Fooden 1969), it is the Sulawesi black crested macaque *Macaca nigra* which is currently most threatened with extinction. This was recently confirmed by their designation in the IUCN red data list as Critically Endangered (Supriatna and Andayani 2008; Table 19.1); *Macaca maura* were listed as Endangered and the five other Sulawesi macaque species were considered Vulnerable (Supriatna and Andayani 2008). As is inevitable, this trend in designation reflects the different degrees to which these wild populations have been studied and thus data availability. Knowledge of cultural differences in the island would also predict that *M. nigra* are at a higher risk of extinction than the other Sulawesi macaque species, as they face different and more dangerous threats to their survival.

Current and Historical Status of *M. nigra*

M. nigra live in the Northern most tip of Sulawesi, in the province of Minahasa. The socio-economic and religious demographic of this area is unusual within Indonesia, and the rest of the island, as 85–90% of the population are Christians and are considered relatively wealthy; though the distribution of wealth has been described as unbalanced (Lee 2000; O'Brien and Kinnaird 2000). The human population in this area has increased substantially in the past few years, which is due to both "natural" population growth and also transmigration within Indonesia. The consumption of *M. nigra* in this area is a long held tradition, the impact of which increases as the population grows and represents a devastating and unique threat to the survival of the species. Added to this, increased development for housing has led to fragmentation of *M. nigra* populations making them locally endemic, which means they like many other tropical species they are at particularly high risk of extinction (Stuhsaker 1975; Terborgh 1992).

Census data are available over the past 30 years which document the decline in the *M. nigra* population size. In the earliest census, *M. nigra* numbers were burgeoning with an estimation of more than 300 animals/km^2 (MacKinnon and MacKinnon 1980).

A decline was first noted less than 20 years later, when only 76 individuals/km^2 were estimated (Sugardjito et al. 1989). A further 10 years later and estimations of the population were as low as 23.5 animals/km^2; though higher numbers were recorded on the distant island of Bacan, where about 66.7 animals/km^2 were estimated to reside (Rosenbaum et al. 1998). These census data evidenced a >90% reduction in *M. nigra* population in barely less than 20 years; the decline is less (>75%) if data from Bacan are considered though still catastrophic. Worse still, it is likely that the current numbers of *M. nigra* are much bleaker than portrayed in the summary of these census results. Data on which previous *M. nigra* population densities have been estimated were gathered from only a few areas within their home range of Minahasa. The majority of census data were collected at one site, the Tangkoko Nature Reserve (Tangkoko); the designation of this reserve has changed over time including the neighbouring volcanoes and environs of Batuangus and/or DuaSudara, but the Tangkoko volcano and its surrounds form the core of the reserve. It is likely that the data collected at Tangkoko does not reflect fairly, and indeed probably provides an overestimate of, the likely population of *M. nigra* elsewhere within their range. As a nature reserve Tangkoko is a protected under law and is unique within the region as it is has long been the site of scientific research and ecotourism (O'Brien and Kinnaird 2000). As a consequence, it is home to habituated groups of *M. nigra* and also "wild" groups which are familiar with people who do not necessarily hunt them. *M. nigra* outside of Tangkoko therefore appear to flee much more readily from people and are thus hard to observe, in contrast to *M. nigra* within Tangkoko which appear more robust to the approach of people. Therefore, it seems likely that *M. nigra* counts in Tangkoko could represent total macaque numbers for groups, which all gather unabashed in the observer's line of sight. It is easy therefore to see how estimates of *M. nigra* populations which are made from extrapolating counts made at Tangkoko might lead to overestimations. The presence of research activity and tourism at Tangkoko also affords the area a greater degree of protection from hunting and farming than other areas within Minahasa.

A recent census of 22 locations throughout Minahasa, resulted in very few sightings of *M. nigra* outside of Tangkoko (Table 19.2; Melfi et al. 2007). Surveys of villagers living near to forested areas considered that there were few *M. nigra* living outside the nature reserves (Feistner 2001) and in some locations, it was considered that hunting by local people had stripped areas of all wildlife (Melfi et al. in preparation). The census conducted by Melfi et al. (2007) included many sites but was limited as the distance travelled at each of these sites was short and, therefore, population estimates were not calculated on the basis of these data.

Why Are *M. nigra* Numbers Declining?

There has been an undoubted and dramatic decline in *M. nigra* population heralding the immediate need for conservation action to ensure the long term preservation of this species. In order to reduce, halt, or reverse the current predicament which faces *M. nigra*, it is necessary to establish what pressures are placing them in danger of extinction.

Table 19.2 A summary of the data collected during three census surveys of *M. nigra* which included the total distance covered and/or the number of trials followed and the population estimates calculated following Sugardjito et al. (1989)

Area	Location	Transect distance and observation freq and popul est.		
		Sugardjito et al. (1989)	WCS (1999–2000)	Melfi et al. (2007)
Northern tip	Tangkoko Duasudara Batuangus	124.5 km (+++/++)	99.5 km 21 (++)	15.3 km 6 (++)
Northern tip	Mt. Wiau			4.5 km Ø
Northern tip	Mt. Klabat	19.5 km Ø		7.8 km Ø
Northern tip	Likupang-Wori			☒
Northern tip	Mt. Tumpa			3.3 km Ø
Tatawiran complex	Mahawu			3.9 km Ø
Tatawiran complex	Lokon (Tinno instead)			1 km Ø
Tatawiran complex	Tatawiran			4.32 km Ø
Tatawiran complex	Manembo-nembo Senduk Wawona	6.5 km (++)	10.8 km 3 (++)	☒ 10.7 km Ø
	Kumu			5.7 km 1 (+)
Lembean range	Kombi			2.2 km Ø
Lembean range	Eris			☒
Lembean range	Kakas			8.4 km Ø
Inland Minahasa	Mt. Tampusu			3.9 km Ø
Inland Minahasa	Mt. Kawatak			6.6 km Ø
Motoling landscape	Eluson	20 km Ø (+)		☒
Motoling landscape	Motoling			☒
Ambang range	Mt. Ambang (Sinsingon)	22.5 km(+)	63.15 km 5 (+)	14.52 km Ø
Ambang range	Mt. Sinonsayang			6.6 km Ø

(continued)

Table 19.2 *(continued)*

Area	Location	Transect distance and observation freq and popul est.		
		Sugardjito et al. (1989)	WCS (1999–2000)	Melfi et al. (2007)
Ambang range	Mt. Bunbungon	ø		4.1 km 1 (+)
Pasaan landscape	Belang			☒
Pasaan landscape	Ratatotok			2.4 km Ø
Pasaan landscape	Kotabunan			12 km Ø
Crossing S borders	Pinolosian	Ø		11.8 km 1 (+)
Crossing S borders	Lolak			9 km Ø
Islands	Manado tua			5.5 km Ø
Total		7 locations 193.5 km	3 locations 173.5 km	22 locations 150.14 km

* indicates animals were observed. +++ indicates that there are >50 *M. nigra*/km²; ++ indicates that there are 10–40 *M. nigra*/km²; + indicates that there are <10 *M. nigra*/km²
* Lee *et al.*, (2000); ** Lee *et al.*, (1999), *** Riley *et al.*, (2000).

The pressures which threaten the future survival of *M. nigra* can be viewed as similar to those which threaten many other wild primate populations and the result of anthropogenic activities, either direct or indirect. Local human populations have overwhelming increased, as have agri-business (especially mining), non-sustainable farming practises are implemented (slash and burn; cash crops; transition to monoculture) all of which have brought the survival of *M. nigra* populations in conflict with the needs of local people as they compete for land. In addition, hunting of *M. nigra* has been implemented since people settled in Sulawesi, over 30,000 years ago, but it is only in recent times, since the early 1970s that the impact of these activities has become unsustainable (O'Brien and Kinnaird 2000).

Threats: Habitat Loss and Fragmentation

A major factor devastating *M. nigra* populations is effective habitat loss, and it's associated negative ramifications. Due to radiation, *M. nigra* already inhabit a small range but this has declined further into small fragments which do not appear to have any obvious, or safe, corridors linking them (pers. comm. J. Tasirin).

The immediate impact of this habitat restriction is that the carrying capacity available to *M. nigra* has declined. Though the genetic integrity of the current population is not known, it is readily acknowledged that small populations are especially vulnerable to the deleterious results of inbreeding (Frankham et al. 1986). The potential for inbreeding is high as the small fragmented populations are isolated from each other making migration of animals between troops highly unlikely, due to distance and physical obstacles. It is possible that these small fragmented populations may be augmented with animals and thus new genetic material. This occurs on an ad hoc basis when pet *M. nigra* are released, or escape back into the wild. The value to augmenting small populations, especially in this haphazard way is not necessarily beneficial, as it is associated with increased risk of anthrozoonotic disease transfer and the genetic contribution of a single animal in small population is incredibly limited (Jones-Engel et al. 2004; Lees and Wilcken 2009).

The conservation value of fragmented habitats depends on their ecological integrity and thus their ability to support biodiversity. Both physical (e.g., size and shape; Laurance and Bierregaard 1997) and biological (e.g., flora and fauna density and diversity; Laurance et al. 2002) variables will affect ecological integrity of fragments, however, many of the smaller fragments of land which currently exist within the home range of *M. nigra* do not represent viable habitat for the species (pers. comm. John Tasirin). Chapman et al. (2003) noted that habitats where fragmentation occurs were usually unprotected and thus especially vulnerable from anthropogenic factors which further threaten the preservation of the habitat and species within; this certainly seems to hold true throughout Minahasa where many forest fragments are heavily used by local villagers (see below; Melfi et al. in preparation).

Logging is ubiquitous to forested areas in Minahasa. The activity of logging is known to have an immediate deleterious impact reducing primate populations; e.g. Bennett and Dahaban (1995) reported reductions of 30–73% in *Hylobates muelleri* and *Presbytis* spp. populations resulting from logging. But there are also long term negative repercussions of logging, as highlighted by Chapman et al. (2000), who noted that decades after logging populations of *Cercopithecus mitis* and *C. ascanius* continued to decline. By contrast, some species appear to favour logged conditions (Plumptre and Reynolds 1994; Chapman et al. 2000). This contradiction may be explained by different life history strategies which primates adopt, making them more or less adaptable to change. Data suggest that *M. nigra* are robust and can adapt to very different environments, evidenced by a similarity in the activity budgets of wild and zoo-housed conspecifics which are exposed to dramatically different environmental variables (Melfi and Feistner 2002). The interplay between different environmental factors does have a significant impact on *M. nigra* behaviour however, as observed by O'Brien and Kinnaird (1997) in a study at Tangkoko; the larger troop had a small territory but access to more abundant food supplies, compared to a smaller group which travelled further in an area with poorer food abundance. Logging doesn't merely strip habitats of valuable trees, but reduces the quality of the habitat, leads to disturbance and enhances accessibility into prime habitat. Forest fragmentation and logging can bring with it, therefore, other deleterious consequences, for example, it facilitates hunting through the creation of increased access to prime habitat (Robinson and Bennett 2000).

Threats: Hunting

Hunting has been identified as one of the main threats to the survival of *M. nigra* (Lee 2000). Hunting macaques for food is particularly unique to Minahasa, within Sulawesi, where they are considered a traditional ceremonial food. This is highlighted by the number of macaques on sale at traditional markets which appears to rise before Christian festivals e.g., Christmas (Clayton and Milner-Gulland 2000). As noted previously, many Minahasans are Christian and so unlike people in much of Indonesia, and the rest of the Island of Sulawesi, they are not restricted in their food selection by religious food taboos. Though the principal reason for hunting *M. nigra* is for consumption, some animals are kept as pets; but the latter appears largely to be a precursor to being eaten or sold as food (*pers. obs.*).

Hunting pressure has increased substantially in recent years and attributed to several factors: (1) disappearance of tropical forest, so hunting pressure is concentrated into smaller areas; (2) increased human population density, leading to more people which hunt; (3) increased number of people living near forests and increased level of hunting commercialism, resulting in less subsistence hunting but instead more people who hunt for a supplementary income and to supply traditional market demand; (4) disappearance of traditional hunting practises and progression of hunting technologies, taboos and restrictions on hunting have been overlooked and there is a rise in the use of air rifles/snares which are more efficient and indiscriminate methods of hunting; (5) increased access into tropical forests, which enables more people to hunt (including loggers and non-local people), who are able to more readily buy new hunting technologies and sell bush meat on to dealers; and finally (6) increased income, which in Asia appears to increase demand for bush meat (Riley et al. 2002).

O'Brien and Kinnaird (2000) suggested that the uneven distribution of wealth which exists in Minahasa has exacerbated two different types of hunting, "subsistence", where hunting *M. nigra* meets the nutritional needs of hunters and their families, versus, "supplementary", where much of the bush meat ends up in traditional markets and is sold to wealthy urban dwellers. In a survey of people living next to two protected areas in Minahasa (Manembonembo Nature Reserve and Gunung Ambang Nature Reserve), Lee (2000) established that most people ate bush meat (92 and 96% of those surveyed, respectively). He suggested that hunting was not undertaken to meet economic or nutritional needs (subsistence) in these families, but instead was considered a supplementary activity which was undertaken "as long as there were tasks to be done in the garden, snares and traps were set" (in this instance, the garden refers to local forests). Essentially, if no additional effort was required to go into the forest to set traps, because people were there already, they would also set traps. This passive approach to hunting means that it is inexpensive in terms of time and resources and, therefore, does not need to be socioeconomically viable. Furthermore, data collected showed the amount of bush meat consumed did not parallel family size; Lee (2000) argued the converse might be assumed to be true if indeed bush meat were simply fulfilling nutritional requirements.

It appears that the majority of *M. nigra* hunted are caught to fulfil the demand of consumers at traditional markets. The potential profit from bush meat is higher at traditional markets, where the dealer to the market can earn up to five times that paid to hunters. There are several traditional markets in Minahasa and the number of dealers has increased steadily in the last few years. For example, Clayton and Milner-Gulland (2000) report that between 1948 and 1970, there was one pig meat dealer, but by 1996, there were 30.

A recent survey of people in 16 villages located close to or in potential *M. nigra* habitat, throughout Minahasa, established that hunting *M. nigra* was still prevalent (Melfi et al. in preparation). High levels of hunting were further evidenced by a large number of traps which were found in 13/22 sites throughout Minahasa (Melfi et al. 2007). Many sites where traps were absent, were considered by local people to be devoid of wildlife and thus hunting was not considered a valuable use of time (Melfi et al. in preparation). Monitoring of traditional markets has been taking place for several years and it has been noted that the number of macaques seen has reduced, but species of other macaques are now observed (pers. comm. John Tasirin); Clayton and Milner-Gulland (2000) observed *M. nigra*, *M. hecki* and *M. nigrescens* in traditional markets. It would seem that as numbers of *M. nigra* decline, the capture and transport of other Sulawesi macaque species from further South in the island are increasing to meet the Minahasan demand for macaque meat. Both O'Brien and Kinnaird (2000) and Lee (2000) have demonstrated through the use of population growth models that rates of *M. nigra* hunting are not sustainable. Quite simply, there are more animals being removed from the forests that can be replaced through population growth, which has lead to local extirpation in several other species in Minahasa too (e.g., anoa, Burton et al. 2005; babirusa, Clayton and Milner-Gulland 2000).

Threats: Health

In areas of Africa, parasite-related infection has been reported to kill about seven times more primates, compared to the numbers lost to hunting (Wallis and Rick 1999). A key determining factor in the reduction of great ape populations has also been attributed to pathogens and their negative impact on health (Leendertz et al. 2006). Poor health, whether due to parasitic infections and/or disease are thus considered to severely jeopardise the long-term survival of primate species (Gillespie and Chapman 2006). Survival of the host is compromised as can be their reproductive rate and success, through direct pathological effects or indirectly by dramatically reducing the host's condition and well being (Beckage 1997). The relationship between parasite and host can however, be complex and does not always have negative ramifications for the host, but is contingent on other factors, for example, food availability. Food scarcity can lead to stress which manifests itself as immunosuppression and supports a rise in parasite burden which can not be comfortably accommodated by the host (Chapman et al. 2006).

There are two main routes by which the health of wild primate populations can be compromised by parasitic infection: direct transmission of pathogens from people to primates or their habitats; or anthropogenic activities which indirectly compromise health and result in immunocompromised animals which are more susceptible to infestation.

Direct transmission between people and primates occurs because of the great many similarities between the two, which means both are susceptible to similar pathogens resulting in potential zoonotic and anthrozoonotic transmission and infection (Brack 1987). Therefore, increased interactions between people and primates serve to exacerbate the potential to transfer pathogens. Although human–primate interactions are not a new phenomenon, the frequency, variety and intensity of encounters which occur has increased dramatically in recent years, in part due to an increase in human population and the activities they pursue (Phillips et al. 2004; Travis et al. 2006). *M. nigra* come into contact with people when they are kept as pets, or when people (locals, tourists and/or researchers) go into their habitat. Data collected by Jones-Engel et al. (2001) showed that these contacts have resulted in some wild and pet *M. nigra* harbouring antibodies for human diseases, e.g., measles, influenza and parainfluenza. In addition, Jones-Engel et al. (2004) documented seven protozoa and three nemtodes in the several macaque species found in Sulawesi which were housed as pets. In a recent study which opportunistically collected faecal samples voided by *M. nigra* found that proximity to humans was significantly positively correlated with endoparasite burden; thus burdens were higher in pets, than habituated *M. nigra* and finally wild *M. nigra* had the lowest burdens (Jonas and Melfi submitted). The determining factor and mechanism which underlies this relationship is still being investigated, though potential causal factors may be one or a combination of: direct transmission from people or their livestock to *M. nigra*; restricted home range of pets (as tethered) and thus rapid re-infestation; or immunocompetence resulting from disturbance by people at close proximity. This pattern between proximity to people and parasite burden is not surprising, as disturbance resulting from anthropogenic activities has been seen previously to greatly increase gastro-intestinal microbes (Gillespie and Chapman 2006).

Mitigating Threats Which Endanger the Survival of *M. nigra*

Population decline in *M. nigra* has resulted from multi-factorial causes, so it is difficult to identify with certainty the impact of particular pressures upon their population (Rosenbaum et al. 1998). Attempts to reduce or eliminate these causes (threats) are needed to secure the future survival of *M. nigra*, which despite the gloomy outlook portrayed does have some happy endings! Proactive initiatives are currently being implemented in an attempt to safeguard this population from the inevitable extinction which will occur soon if circumstances do not change. The Selamatkan Yaki programme, though in its infancy, aims to integrate resources, knowledge and skills available *in situ* and *ex situ* in a focussed attempt to reduce threats which endanger *M. nigra*. Although the conservation of this species will be determined by the actions and activities of people within Minahasa, and more broadly Indonesia, integration of efforts will undoubtedly improve the likelihood that this species can be successfully conserved.

Looking towards the future, and considering the threats outlined above, key areas which need to be addressed in an attempt to conserve *M. nigra* are notably an increase in: law enforcement, protection of *M. nigra* habitats, awareness and conservation education, and management of tourism in Tangkoko. An improvement in law enforcement and protection of *M. nigra* habitats would have a direct impact reducing the number of animals lost to hunting or through competition for land (O'Brien and Kinnaird 2000; Lee 2000). The contribution conservation education and tourism could play in supporting *M. nigra* conservation is more indirect, but potentially an area which could have far greater importance and, potentially change hearts and minds and thus the need for enforcement. .

Mitigation: Education and Awareness

The provision of conservation education and increasing awareness of environmental issues and species extinction has been implemented by various NGO's previously. In the past: Durrell Wildlife Conservation Trust (at the time, Jersey Wildlife Preservation Trust) initiated a public awareness campaign, through the provision of materials and slide shows at villages, in 1996 and again in 2001; Wildlife Conservation Society (WCS) – Sulawesi programme have been actively running education programmes for a variety of audiences throughout Minahasa for about 20 years, which have included running school workshops, providing public access to scientific and natural history materials, and initiating media coverage of conservation topics (pers. comm. John Tasirin); and more recently, education and training opportunities have been provided in the village of Batu putih, the gateway village to Tangkoko, via researchers studying in the reserve as part of the *M. nigra* project (http://www.macaca-nigra.org/). The efficacy of these initiatives at reducing threats to *M. nigra* survival have not been monitored empirically, however, comparisons of perceptions held of *M. nigra* and hunting activity, reported above, give the impression that changing attitudes and activities which threaten the continued existence of this species is going to continue to be a challenge.

Mitigation: Ecotourism

An area which could have a dramatic impact on changing attitudes to *M. nigra* and thus their conservation is the development of ecotourism. Various activities have been provided under the banner of ecotourism, the definition and interpretation of which can be highly variable. The four basic tenets which are essential in eco-tourist pursuits are that they should be, (1) nature-based, (2) provided with minimal-impact on the environment and social community (3) provide environmental education and

(4) actively contribute to conservation (Buckley 2009). Alternatively, tourism activities should aim to meet the "triple-bottom-line", where there is a net environmental, social (local community) and financial gain (Buckley 2003a). Sadly, a review by Kiss (2004) noted that very few community based ecotourism projects achieve these outcomes, but instead result in "little change in existing local land and resource-use practises, (and) provide only a modest supplement to local livelihoods". More specifically, ecotourism involving primates can compromise their health (through disease transmission) and increase their vulnerability to hunting pressures (through reduced fear and habituation to people) (e.g. Goldberg et al. 2007; Buckley 2003b). To date there is only one study which has monitored the impact of tourism on *M. nigra* populations and their habitats. Data collected by Kinnaird and O'Brien (1996) documented the rapid expansion of tourists visiting Tangkoko. Unfortunately, they noted tourists had a deleterious impact on *M. nigra* behaviour and the presence of tourists was not achieving social, environmental or financial benefits either; it was suggested that this was due to lack of comprehensive planning. Despite the potential costs of tourism, it does appear that to date the benefits afforded by these activities in Tangkoko has outweighed the costs and been of great benefit to the preservation of *M. nigra* and their habitats; which is possibly the final viable population of *M. nigra*. If managed properly, the adoption of a triple bottom line approach to tourism in this area could work with conservation and have social benefits.

Hope for the Future

And, if the worst happens! There is a relatively stable *ex situ* population of *M. nigra* the majority of which are maintained in European Zoos as a part of a European Captive Breeding Programme (Melfi 2009). Within Minahasa, there are also captive animals, comprising individuals held within sanctuaries or as pets. However, whilethe threats which endanger this species' survival are still current, namely hunting, reintroduction attempts would be tantamount to supplementing the hunting market. Even when we consider how complex it is to evaluate reintroduction attempts (Seddon 1999), rates of success are still considered low (Wolf et al. 1996). So really we have to avoid the worst case scenario and put all efforts into the preservation of *M. nigra* habitats and reduction of human–animal conflict, through habitat protection, greater law enforcement, and reduced hunting. These efforts will do more than save *M. nigra* from extinction as the preservation of their habitats will also save some of the richest and unique biodiversity seen globally.

Acknowledgements I would like to acknowledge the continued help and support of Dr John Tasirin and the WCS – Sulawesi programme staff. Data collected by Melfi et al. (2007) were facilitated and funded by: German Primate Research Centre, Lembaga Ilmu Pengetahuan Indonesia, Pekan Kebudayaan Aceh, Paignton Zoo Environmental Park, Universities Federation for Animal Welfare, Universitas Sam Ratulangi, Wildlife Conservation Society – Sulawesi Programme, Whitley Wildlife Conservation Trust, and Zoological Society of London.

References

Beckage NE (1997) Parasites and pathogens: effects on host hormones and behavior. Chapman and Hall, New York

Bennett EL, Dahaban Z (1995) Wildlife responses to disturbances in Sarawak and their implications for forest management. In: Primack RB, Lovejoy TE (eds) Ecology, conservation, and management of Southeast Asian rainforests. Yale University Press, New Haven, pp 66–86

Brack M (1987) Agents transmissible from simians to man. Springer, Berlin

Buckley R (2003a) Environmental inputs and outputs in ecotourism: geotourism with a positive triple bottom line? J Ecotourism 2:76–82

Buckley R (2003b) Case studies in ecotourism. CAB International, Wallingford, UK

Buckley R (2009) Ecotourism: principles and practices. CAB International, Oxford

Burton JA, Hedges S, Mustari AH (2005) The taxonomic status, distribution and conservation of the lowland anoa *Bubalus depressicornis* and mountain anoa *Bubalus quarlesi*. Mammal Rev 35(1):25–50

Chapman CA, Balcomb SR, Gillespie TR, Skorupa JP, Struhsaker TT (2000) Long-term effects of logging on African primate communities: a 28 year comparison from Kibale National Park, Uganda. Conserv Biol 14(1):207–217

Chapman CA, Lawes MJ, Naughton-Treves L, Gillespie TR (2003) Primate survival in community-owned forest fragments: are metapopulation models useful amidst intensive use? In: Marsh LK (ed) Primates in fragments: ecology and conservation. Kluwer/Plenum, New York, pp 63–78

Chapman CA, Wasserman MD, Gillespie TR, Speirs ML, Lawes MJ, Saj TL, Ziegler TE (2006) Do food availability, parasitism, and stress have synergistic effects on red colobus populations living in forest fragments? Am J Phys Anthropol 131:525–534

Clayton L, Milner-Gulland EJ (2000) The trade in wildlife in North Sulawesi, Indonesia. In: Robinson JG, Bennett EL (eds) Hunting for sustainability in tropical forest. Columbia University Press, New York, pp 473–496

Feistner ATC (2001) Conservation of Sulawesi crested black macaques *Macaca nigra* in Indonesia. In: Norcup S (ed) The third Sulawesi crested black macaque Macaca nigra European studbook. Durrell Wildlife Conservation Trust, Jersey

Fooden J (1969) Taxonomy and evolution of the monkeys of Celebes (Primate: Cercopithecidae). Bibliotheca Primatologica No. 10, S. Karger, Basel

Frankham R, Hemmer H, Ryder O, Cothran E, Soulé M, Murray N, Synder M (1986) Selection of captive populations. Zoo Biol 5:127–138

Gillespie TR, Chapman CA (2006) Prediction of parasite infection dynamics in primate metapopulations based on attributes of forest fragmentation. Conserv Biol 20(2):441–448

Goldberg TL, Gillespie TR, Rwego IB, Wheeler E, Estoff EL, Chapman CA (2007) Patterns of gastrointestinal bacterial exchange between chimpanzees and humans involved in research and tourism in western Uganda. Biol Conserv 135(4):511–517

Harrison IJ, Stiassny MLJ (1999) The quiet crisis: a preliminary listing of the freshwater fishes of the World that are extinct or "missing in action". In: MacPhee RDE (ed) Extinctions in near time: causes, contexts, and consequences. Kluwer/Plenum, New York, pp 271–332

Holmes D, Phillips K (1996) The birds of Sulawesi. Oxford University Press, Oxford

International Species Inventory System (ISIS) (2009) International Species Information System (ISIS), Eagan, MN

IUCN (2008) IUCN Red List of Threatened Species. <www.iucnredlist.org>. Downloaded on 06 March 2009

Jonas A, Melfi V (submitted) Does the level of human contact affect parasite burden in *Macaca nigra?* Anim Welf

Jones-Engel L, Engel GA, Schillaci MA, Babo R, Froehlich J (2001) Detection of antibodies to selected human pathogens among wild and pet macaques (*Macaca tonkeana*) in Sulawesi, Indonesia. Am J Primatol 54(3):171–178

Jones-Engel L, Engel GA, Schillaci MA, Kyes K, Froehlich J, Paputungan U, Kyes RC (2004) Prevalence of enteric parasites in pet macaques in Sulawesi, Indonesia. Am J Primatol 62:71–82

Kinnaird MF, O'Brien T (1996) Ecotourism in the Tangkoko DuaSudara nature reserve: opening Pandora's box? Oryx 30(1):65–73

Kiss A (2004) Is community-based ecotourism a good use of biodiversity conservation funds? Trends Ecol Evol 19(5):232–236

Laurance WF, Bierregaard RO Jr (1997) Tropical forest remnants: ecology, management, and conservation of fragmented communities. University of Chicago Press, Chicago

Laurance WF, Lovejoy TE, Vasconcelos HL, Bruna EM, Didham RK, Stouffer PC, Gascon C, Bierregaard RO, Laurance SG, Sampaio E (2002) Ecosystem decay of Amazonian forest fragments: a 22-year investigation. Conserv Biol 16:605–618

Lee RJ (2000) Impact of subsistence hunting in North Sulawesi, Indonesia and Conservation option. In: Robinson JG, Bennett EL (eds) Hunting for sustainability in tropical forest. Columbia University Press, New York, pp 455–472

Lee RJ, Riley J, Suyatno N (1999) Biological surveys and management recommendations: Manembonembo nature reserve. A report to the Department of Forestry (PKA). Wildlife Conservation Society

Lee RJ, Riley J, Teguh H (2000) Biological surveys and management recommendations: Gunung Ambang nature reserve. A report to the Department of Forestry (PKA). Wildlife Conservation Society

Leendertz FH, Pauli G, Maetz-Rensing K, Boardman W, Nunn C, Ellerbrok H, Jensen SA, Junglen S, Boesch C (2006) Pathogens as drivers of population declines: the importance of systematic monitoring in great apes and other threatened mammals. Biol Conserv 131:325–337

Lees CM, Wilcken J (2009) Sustaining the Ark: the challenges faced by zoos in maintaining viable populations. Int Zoo Yearb 43(1):6–18

MacKinnon J, MacKinnon K (1980) Cagar Alam Gunung Tangkoko-DuaSudara, Sulawesi Utara Management Plan 1981-1986. Bogor, Indonesia

Melfi VA (2009) The 10th European Studbook for Sulawesi crested black macaques. Paignton Zoo Environmental Park, Totnes Road, Devon, TQ4 7EU

Melfi VA, Feistner ATC (2002) A comparison of the activity budgets of wild and captive Sulawesi crested black macaques (*Macaca nigra*). Anim Welf 11:213–222

Melfi VA, Tasirin J, Jonas A, Yosep A, Sabintoe B, Houssaye F, Jago N, Kambey R (2007) Impact of disturbance of wild *Macaca nigra* populations and habitats in Sulawesi. Report to LIPI

Melfi VA, Tasirin J, Jonas A, Yosep A, Sabintoe B, Houssaye F, Jago N, Kambey R (in preparation) Attitudes and activities of villagers to wildlife in Northern Sulawesi, Indonesia

Myers NRA, Mittermeier CG, Mittermeier GAB, da Fonseca, Kent J (2000) Biodiversity hotspots for conservation priorities. Nature 403:853–858

O'Brien TG, Kinnaird MF (1997) Behavior, diet, and movements of the Sulawesi crested black macaque (*Macaca nigra*). Int J Primatol 18(3):321–351

O'Brien T, Kinnaird M (2000) Differential vulnerability of large birds and mammals to hunting in North Sulawesi, Indonesia, and the outlook for the future. In: Robinson JG, Bennett EL (eds) Hunting for sustainability in tropical forest. Columbia University Press, New York, pp 199–213

Phillips KA, Haas ME, Grafton BW, Yrivarren M (2004) Survey of the gastrointestinal parasites of the primate community at Tambopata National Reserve, Peru. J Zool 264:149–151

Plumptre AJ, Reynolds V (1994) The impact of selective logging on the primate populations in the Budongo Forest Reserve, Uganda. J Appl Ecol 31:631–641

Riley J, Wangko M, Nusalawo M, Soleman T (2000) Monitoring and patrolling. In: Tangkoko Duasudara Nature Reserve. *A Report to Department of Forestry Department* (PKA). The Wildlife Conservation Trust

Riley J, Siwu S, Antono B, Keintjem J, Manurung M, Pontonowu S (2002) Monitoring and control of the trade in protected wildlife in North Sulawesi, Indonesia. A report to Forestry Department (PKA). The Wildlife Conservation Society

Robinson JG, Bennett EL (2000) Hunting for the Snark. In: Bennett EL, Robinson JG (eds) Hunting for sustainability in tropical forests. Columbia University Press, New York, pp 1–13

Rosenbaum B, O'Brien T, Kinnaird M, Supriatna J (1998) Population densities of Sulawesi crested black (*Macaca nigra*) on Bacan and Sulawesi, Indonesia: effects of habitat disturbance and hunting. Am J Primatol 44:89–106

Seddon PJ (1999) Persistence without intervention: assessing success in wildlife reintroductions. Trends Ecol Evol 14(12):503

Stuhsaker TT (1975) The red colobus monkey. University of Chicago Press, Chicago

Sugardjito J, Southwick CH, Supriatna J, Kohlhass A, Baker S, Erwin J, Froehlich J, Lerche N (1989) Population survey of macaques in Northern Sulawesi. Am J Primatol 18:285–301

Supriatna J, Andayani N (2008) *Macaca nigra*. In: IUCN 2008. 2008 IUCN Red List of Threatened Species. www.iucnredlist.org. Downloaded on 06 March 2009

Terborgh J (1992) Diversity and the tropical rain forest. Scientific American Library, New York

Travis DA, Hungerford L, Engel GA, Jones-Engel L (2006) Disease risk analysis: a tool for primate conservation planning and decision making. Am J Primatol 68:855–867

Wallis J, Rick LD (1999) Primate conservation: the prevention of disease transmission. Int J Primatol 20(6):803–826

Whitten A, Mustafa M, Henderson G (1987) The ecology of Sulawesi. Oxford University Press, Oxford

Wilson KA, McBride MF, Bode M, Possingham HP (2005) Prioritizing global conservation efforts. Nature 10:1038

Wolf CM, Griffith B, Reed C, Temple SA (1996) Avian and mammalian translocations: update and reanalysis of 1987 survey data. Conserv Biol 10(4):1142–1154

Part III
Indonesia's Prosimians

Chapter 20
The Function of Scentmarking
in Spectral Tarsiers

Sharon Gursky-Doyen

Introduction

One of the primary trends characterizing primates as distinct from other mammals is their increasing dependence on vision and the concomitant reduced dependence on their sense of olfaction (Fleagle 1998; Martin 1990). Although most primates rely heavily on their sense of vision, their olfactory sense still plays an important role in their day to day activities (Charles-Dominique 1977; Harcourt 1981; Clark 1982; Epple et al. 1988; Dugmore and Evans 1990; Fornasieri and Roeder 1992; Harrington 1977). This is especially true of prosimian primates that have undergone the least reduction in their olfactory apparatus relative to the other primates (Martin 1990; Fleagle 1998).

Prosimian primates are known to deposit scent around their territory. Two major hypotheses have been proposed to explain the function of scentmarking behavior in primates and other mammals. First, it has been hypothesized that scentmarking serves to mark off territorial boundaries. For example, in *Lemur catta*, the majority of the scent marks were in a narrow band within the area of overlap that coincided with the positions of inter-troop confrontations (Mertl-Millhollen 1988). Scentmarks in *Lemur catta* thus appear to demarcate territorial boundaries and not the completed home range boundaries. This hypothesis has also been proposed for a variety of other animals including pronghorn antelope (Gilbert 1973), hyenas (Gorman 1990), oribi (Gosling 1981), African dwarf mongoose (Rasa 1973), aardwolf (Richardson 1990), and European badger (Roper et al. 1986).

It has also been proposed that scentmarking enables group members to monitor female reproductive condition, thereby serving a mate defense function. Male cotton-top tamarins *Saguinus oedipus* are capable of discerning the chemical signals of ovulation as are meadow voles (Ferkin et al. 1995; Ziegler et al. 1993).

S. Gursky-Doyen (✉)
Department of Anthropology, Texas A&M University, MS 4352,
College Station, TX 77843, USA
e-mail: gursky@tamu.edu

S. Gursky-Doyen and J. Supriatna (eds.), *Indonesian Primates*,
Developments in Primatology: Progress and Prospects,
DOI 10.1007/978-1-4419-1560-3_20, © Springer Science+Business Media, LLC 2010

This hypothesis has been proposed to account for the ability of males to locate sexually receptive females in many of the nocturnal solitary foraging prosimians (Charles-Dominique et al. 1980; Doyle and Martin 1979; Tattersall and Sussman 1977).

Spectral tarsiers have a number of scent marking glands that they use to deposit scent throughout their range. These include (1) the ano-genital gland, (2) the epigastric gland, and (3) the circum-oral gland (Niemitz 1984). They are also known to scentmark by depositing small droplets of urine on substrates. Although previous studies have described the form of scent marking behavior in semi-wild (caged in their natural habitat) (Niemitz 1984), and wild spectral tarsiers (MacKinnon and MacKinnon 1980), no quantitative attempts have been made to identify the function of this behavior.

If the function of spectral tarsier scent marking is to communicate information about the female's reproductive condition, then it is predicted that (1) scentmarks will be randomly distributed throughout the territory and not restricted to the territorial borders and (2) scentmark frequency will increase during the mating season compared to the nonmating season. If spectral tarsier scentmark in order to defend their territory, then it is predicted that (1) both males and females will scentmark equally; (2) scentmarks will not be randomly distributed throughout the group's territory, but will be restricted to the territory borders, particularly at the areas of overlap; and (3) scent marking behavior will increase in frequency during territorial disputes.

Methods

Study Site

Sulawesi, Indonesia, formerly known as Celebes, is a four-armed island located east of Borneo and northwest of Australia and New Guinea (longitude 125 14' east and latitude 1 34'). This study was conducted at Tangkoko Nature Reserve on the easternmost tip of the northern arm of the island. The reserve, which is approximately 3,000 ha, exhibits a full range of forest types, including beach formation forest, lowland forests, sub-montane forests to mossy cloud forests on the summits of Dua Saudara and the Tangkoko Crater (MacKinnon and MacKinnon 1980; Whitten et al. 1987). The reserve is far from pristine due to heavy selective logging and encroaching gardens along its borders. The forest canopy is very discontinuous and contains a high proportion of *Ficus* trees (Gursky 1997, 1998). Rainfall averaged approximately 2,300 m annually (World Wildlife Fund 1980; Kinnaird and O'Brien 1993; Gursky 1997). Resource availability, as measures according to insect biomass (Brower et al. 1990) ranged from 6.9 to 11.1 g between April 1994 and June 1995 (Gursky 2000). Additional details concerning the habitat type at Tangkoko Nature Reserve can be found in Gursky (1997).

Study Species

The spectral tarsier is a small nocturnal primate found exclusively on the island of Sulawesi in Indonesia. Although most groups exhibit a monogamous social system, a few exhibit a polygynous social system (Gursky 1994, 1995). They are highly insectivorous eating a wide variety of insects (Gursky 2000). Spectral tarsiers have a 191 day gestation period that is followed by a 78 day period of lactation. The mean interbirth interval is 12.7 months. Births are seasonal, with most occurring in April–May and a few occurring in November–December. Infants are not continuously transported by the mother or other group members following birth. Rather, they exhibit a cache and carry infant caretaking strategy. Infants are transported in the mother's mouth and then parked on branches while the mother forages nearby. Spectral tarsiers at Tangkoko Nature Reserve primarily utilized strangling *Ficus* trees for the sleeping sites. Most ranges contained 1–3 sleeping sites, but one site was preferentially used.

Capture and Attachment

The following procedures were used to locate individuals. Prior to dawn, my field assistant and I would stand on the periphery of a one hectare plot. Plots were chosen randomly (following a block design) within one square kilometer of the trail system. As the tarsiers returned to their sleeping sites, or at their sleeping sites, they gave loud vocal calls for three to five minutes that could be heard from 300 to 400 m (MacKinnon and MacKinnon 1980; Niemitz 1984). All groups that were heard vocalizing were then followed to their sleeping sites. My field assistant and I then returned to the sleeping site prior to dusk to set up several mist nets in the vicinity of the sleeping site (Bibby et al. 1992). The mist nets were continually monitored for captured tarsiers. Upon capture, individuals were placed in a cloth bag and weighed with a portable scale providing an accuracy of ±1 g. An SM1 radio collar (manufactured by AVM Instrument Co., Ltd., Livermore, CA) weighing either 3.5 or 7.0 g (depending on the size of the battery in the radio collar) with a groove-loop was attached to the tarsier's neck by a simple folding of the thermoplastic band.

Data Collection

A radio receiver using 151 MHz frequency and a three element collapsible Yagi antenna were used to determine the location of each individual radio frequency. Each night, a single spectral tarsier individual was followed with the aid of moonlight and flashlights. Initially, an Indonesian student and I conducted focal follows together until approximately 99% of our data were the same. At this point, we began to conduct independent focal follows. Once each month thereafter, we conducted an inter-observer reliability test to determine if we were still consistent in our data recording: It was >98% during each inter-observer reliability test. A total of 16 individuals from seven groups were trapped in mist nets. Thirteen adult individuals were radio-collared and observed using focal follows between April 1994 and June 1995. The focal individual's behavior was recorded at 5 min intervals (Altmann 1974). The following behaviors were recorded: foraging, feeding, resting, traveling, and socializing (i.e., scent mark, allo-grooming, playing, and vocalizing). Definitions of all behaviors recorded are presented in Gursky (1997). Using these techniques, approximately 485 h of behavioral observation data were collected on the five adult males ($n = 138$ nights) and 600 h of behavioral observation data on the eight adult females ($n = 304$ nights).

In addition, all occurrences of scent marking and territorial disputes were collected continuously, ad libitum. The location of all scent marking and territorial disputes was marked using reflective flagging tape. The actual location was measured with the aid of a compass and tape measure with reference to the 50 m trail system in the study area and then plotted.

Home range data were collected at 15 min intervals. Each location was marked with flagging tape, which noted the time, the individual and the date. The next day, all flagging tapes were re-located. The actual location was measured with the aid of a compass and tape measure with reference to the 50 m trail system in the study area. On the basis of these locational data points, the actual home range size was calculated using minimum concave polygons (Kenward 1987; White and Garrott 1987).

Results

During this study, a total of 3,427 scentmarks were observed being deposited. This includes scentmarks deposited during scan samples as well as ad libitum during focal follows. A mean of 7.75 scentmarks were deposited each night (SD = 17.46). Nearly one third (32%) of all scentmarks were made with urine ($n = 1,092$), 28% ($n = 943$) were made with the epigastric gland, 25% ($n = 879$) were made with the ano-genital gland, and 15% ($n = 513$) were made with the circumoral gland. During the course of the study, males deposited 1,443 scentmarks. Males averaged 10.46 (SD = 7.77) scentmarks per night and ranged from 0 to as many as 41 per night of observation. Females deposited 1,984 scentmarks per night averaging only 6.53 (SD = 4.35) scentmarks and ranged from 0 to 18 per night of observation. Male and female spectral tarsiers differed statistically in the frequency that they scent marked ($X^2 = 19.12$, $P = 0.0001$, df = 1). The only other major difference in scent marking behavior between males and females was that males consistently sniffed and licked the markings left by other individuals ($n = 562$), whereas this

Table 20.1 The area of the range where male and female spectral tarsiers sniffed scentmarks they came across during their nightly travels

Sex	# Scentmarks sniffed	# Scentmarks sniffed in territory center	# Scentmarks sniffed in territory periphery	# Scentmarks sniffed in areas of overlap
Male	562	173	184	205
Female	126	21	66	39

behavior was rarely exhibited by the female ($n = 126$) ($X^2 = 166.95$, $P = 0.0001$, df $= 1$). Females were more likely to sniff scent markings deposited along the edge of the territory, whereas male spectral tarsier's sniffed scentmarkings deposited in all quadrats of the territory (Table 20.1).

On nights when there was a territorial dispute, approximately 8.49 (SD $= 3.06$) scentmarks were deposited per night. However, on nights when there was no territorial dispute, the average number of scentmarks was only 6.85 scentmarks per night (SD $= 4.67$). The frequency that individuals scentmarked on nights when there was a territorial dispute was statistically greater than the frequency that individuals scentmarked on nights when there were no territorial disputes ($X^2 = 4.43$, $P = 0.0353$, df $= 242$).

Spectral tarsiers did not randomly deposit their scentmarks throughout the group's territory (Table 20.2). Group territories ranged in size between 2.12 and 4.05 ha and averaged 2.9 ha (Gursky 1998). Areas of territory overlap ranged from 7 to 20% and averaged 16% of each group's territory. Although the majority of each territory's quadrats were located within the center of the group's territory, only a small proportion of all scentmarks (3.8–14.6%) were placed within the central area of the territory. The average number of scentmarks in these centrally located quadrats ranged from 0.17 to 0.41 scentmarks per quadrat and averaged 0.24 (Table 20.3). Quadrats along the territory's periphery received substantially greater numbers of scentmarks (10.8–21.6%), despite the fact that they constituted a smaller proportion of the territory's quadrats. The average number of scentmarks in territory periphery quadrats ranged from 0.98 to 4.84 averaging 2.93 scentmarks per quadrat along the periphery. Even though the areas of overlap represented only a relatively minor proportion of the group's territory, the majority of the scentmarks were placed in quadrats that overlapped more than one group's range. The proportion of all scentmarks that were deposited in areas of overlap ranged from 54% to as much as 75% of the group's scentmarks. The proportion of scentmarks in areas of territory overlap ranged from 2.2 to 13.5 and averaged 7.5 scentmarks per quadrat. The frequency that individuals scentmarked in quadrats that overlapped two group's ranges was statistically greater than the frequency that individuals scentmarked in quadrats that were non-overlapping ($X^2 = 22.11$, $P = 0.0001$, df $= 1$).

There was a non-significant seasonal component to spectral tarsier scentmarking behavior. Figure 20.1 illustrates the frequency spectral tarsier's scentmarked each month. The average number of scentmarks deposited by females during the

Table 20.2 The number of scentmarks deposited by each spectral tarsier group in the territory center, periphery and areas of overlap

Group	# Scentmarks	# Quadrats in territory	# Scentmarks in territory center	# Quadrats in territory center	# Scentmarks in territory periphery	# Quadrats in territory periphery	# Scentmarks in areas of overlap	# Quadrats in areas of overlap
C600	478	291	43	208	174	67	261	16
E650	381	212	56	143	91	45	234	24
F600	837	283	32	201	237	59	568	23
G850	356	345	51	271	102	56	203	18
G1000	229	228	22	138	50	62	157	28
J700	772	405	84	317	106	66	582	22
M600	374	268	53	218	85	21	236	29

Table 20.3 The average number of scentmarks (SM) deposited by each spectral tarsier group in each area of the territory

Group	Average # of SM in all territory center quadrats	Average # of SM in territory periphery quadrats	Average # of SM in territory overlap quadrats
C600	0.21	2.64	12.43
E650	0.41	2.53	5.71
F600	0.17	4.84	13.21
G850	0.21	2.91	2.94
G1000	0.17	0.98	3.41
J700	0.29	2.59	7.97
M600	0.25	4.05	7.15

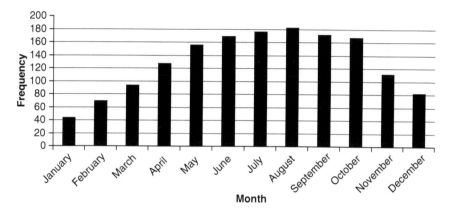

Fig. 20.1 The frequency male and female spectral tarsiers scentmarked each month from April 1994 through June 1995

mating season was 116.5 (SD=27.9) per month. The average number of scentmarks deposited by females during the non-mating season was 142.5 (SD=19.1) per month. There was no statistical difference between scentmarking behavior during the mating season (April–May & November–December) compared to the non-mating season ($X^2=-0.10$, $P=0.9208$, df=13 females; $X^2=-1.34$, $P=0.2045$, df=13 males). There was also no increase in scentmark sniffing frequency when comparing the mating season with the non-mating season ($X^2=1.322$, $P=0.2090$, df=13 females; $X^2=0.506$, $P=0.6215$, df=13 males) (Fig. 20.2). However, there was a statistical difference between scentmarking behavior during the wet (November–April) and the dry season (May–October). The average number of scentmarks deposited during the wet season was only 87.7 (SD=39.7) per month whereas the average number of scentmarks deposited each month during the dry season was 170.2 (SD=25.4) per month. The frequency of scentmarking increased statistically during the dry season compared to the wet season ($X^2=63.65$, $P=0.0001$, df=1).

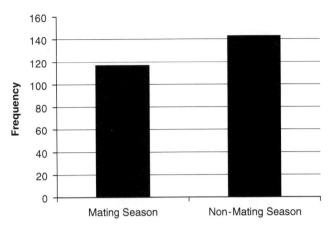

Fig. 20.2 The mean frequency spectral tarsiers sniffed scentmarks during the mating and non-mating season

Discussion

My observations reject the hypothesis that the primary function of scentmarking behavior by spectral tarsiers is to monitor female reproductive condition. There was no statistical increase in the frequency of scentmarking behavior during the mating seasons compared to the non-mating season. Because the mating season is concurrent with the birth season (due to this species' 6 month gestation and postpartum estrus) (Wright et al. 1986, 1988), this result also means that there was no increase in the frequency of scentmarking during the birth season relative to the non-birth season. Similarly, the scentmarks were not distributed throughout all quadrats of the territory. Instead, the results of this study support the hypothesis that the primary function of scentmarking behavior by spectral tarsiers is territorial defense. As predicted, the majority of the scentmarks were deposited along territorial borders, particularly areas of overlap. Scentmarks were not randomly distributed throughout all quadrats within the group's territory. As predicted, the number of scentmarks increased on nights, when there was a territorial dispute relative to nights when there was no territorial dispute. Lastly, the majority of scentmarking was performed during the dry season, when resources were more limited and not the wet season when resources were more abundant (Gursky 2000). Together, these data supports a territorial function for scentmarking behavior in spectral tarsiers.

However, in contrast to the predictions of the second hypothesis, male and female spectral tarsiers differed in the frequency with which they each scentmarked. Male spectral tarsiers scentmarked nearly two times as frequently as females. The lack of concordance for this prediction does not in itself provide enough evidence to refute this hypothesis. Rather, the lack of concordance between the prediction and the data may reflect the fact that the male is responsible for the majority of the territorial defense.

The territorial defense function for scentmarking in spectral tarsiers is not surprising given the presence of other territorial behaviors exhibited by this species (MacKinnon and MacKinnon 1980; Niemitz 1984; Gursky 1997). Spectral tarsiers emit vocal duets each morning and occasionally in the evening. These vocal duets by male, female pair are more appropriately called family choruses since all members of the immediate family group except infants participate in the vocalizations. The family choruses are believed to have a territorial function whereby territorial owners are warned that the territory is occupied and will be defended against intruders.

Spectral tarsiers also exhibit extensive territorial behavior when encountering another spectral tarsier group. Every time individuals of neighboring groups were encountered, a vocal battle with individuals lunging and retreating occurred until members of one group retreated back to their own territory. Interestingly though, previous studies (Gursky 2000) demonstrated that there was a decrease in the number of within group encounters during the dry season. This suggests that the increase in scentmarking behavior is used during the dry season to maintain exclusive access to the territory, but without the cost of an intergroup encounter.

Conclusions

The results of this study provide support for the territorial defense function of scentmarking in spectral tarsiers. As predicted by the territorial defense function hypothesis, the majority of the scentmarks were deposited in areas of overlap. In addition, the number of scentmarks increased on nights when there was a territorial dispute, and during the dry season. There was no statistical increase in the frequency of scentmarking behavior during the mating seasons and the scentmarks were not distributed throughout all quadrats of the territory refuting the mate defense function hypothesis.

Acknowledgments The author acknowledges that this research would not have been possible without the permission and assistance of the following organizations and people: LIPI (The Indonesian Institute of Sciences), SOSPOL, POLRI, PHPA (Manado, Bitung, Tangkoko, and Jakarta), Romon Palette, Yoppy Muskita (WWF), Jatna Supriatna, the University of Indonesia-Depok, and Tigor P.N. (UNas). Thanks go to my field assistants for their help in collecting the data (Nestor, Petros, Celsius, Frans, Ben, Nolde, Nellman, and Uri). Funding for this research was provided by: Wenner Gren Foundation, National Science Foundation Grant SBR-9507703, Douroucouli Foundation, Chicago Zoological Society, L.S.B. Leakey Foundation, Primate Conservation Incorporated, and Sigma Xi.

References

Altmann J (1974) Observational study of behavior: sampling methods. Behaviour 49:227–267
Bibby R, Southwood T, Cairns P (1992) Techniques for estimating population density in birds. Academic, New York

Brower J, Zar J, von Ende C (1990) Field and laboratory methods for general ecology. Wm. C. Brown Publishers, Iowa

Charles-Dominique P (1977) Ecology and behavior of nocturnal primates. Columbia University Press, New York

Charles-Dominique P, Cooper H, Hladik A, Hladik C, Pages E, Pariente G, Petter-Rousseaux A, Petter J, Schilling A (1980) Nocturnal Malagasy primates. Academic, New York

Clark A (1982) Scentmarks as social signals in *Galago crassicaudatus*. I. Sex and reproductive status as factors in signals and responses. J Chem Ecol 8(8):1133–1151

Doyle G, Martin R (1979) The study of prosimian behavior. Academic, New York

Dugmore S, Evans C (1990) Discrimination of conspecific chemosignals by female ringtailed lemurs, *Lemur catta*. In: MacDonald D, Muller-Schwarze D, Natynczuk S (eds) Chemical signals in vertebrates. Oxford University Press, Oxford, pp 361–366

Epple G, Kuderling I, Belcher A (1988) Some communicatory functions of scentmarking in the cotton-top tamarin (*Saguinus oedipus oedipus*). J Chem Ecol 14:503–515

Ferkin M, Sorokin E, Johnston R, Lee C (1995) Attractiveness of scent varies with protein content of the diet in meadow voles. Anim Behav 53:133–141

Fleagle J (1998) Primate adaptation and evolution. Academic, New York

Fornasieri I, Roeder J (1992) Marking behaviour in two Lemur species (*L. fulvus* and *L. macaco*): relation to social status, reproduction, aggression and environmental change. Folia Primatol 59:11137–11148

Gilbert B (1973) Scentmarking and territoriality in pronghorn (*Antilope americana*) in Yellowstone National Park. Mammalia 37:25–33

Gorman M (1990) Scentmarking strategies in mammals. Rev Suisse Zool 97:3–29

Gosling L (1981) Demarcation in a gerenuk territory: an economic approach. Z Tierpsychol 56:305–322

Gursky SL (1994) Infant care in the spectral tarsier, *Tarsius spectrum*: a preliminary analysis. Int J Primatol 15(6):843–853

Gursky SL (1995) Group size and composition in the spectral tarsier, *Tarsius spectrum*: implications for social organization. Trop Biodivers 3(1):57–62

Gursky S (1997) Modeling maternal time budgets: the impact of lactation and gestation on the behavior of the spectral tarsier, *Tarsius spectrum*. Ph.D. dissertation, Doctoral Program in Anthropological Sciences, SUNY-Stony Brook

Gursky SL (1998) The conservation status of the spectral tarsier, Tarsius spectrum, in Sulawesi Indonesia. Folia Primatologica 69:191–203

Gursky SL (2000) Effect of seasonality on the behavior of an insectivorous primate. Int J Primatol 21:477–495

Harcourt C (1981) The function of urine washing in *Galago senegalensis*. Z Tierpsychol 55:119–128

Harrington J (1977) Discrimination between males and females by scent in *Lemur fulvus*. In: Tattersall I, Sussman R (eds) Lemur biology. Plenum, New York, pp 259–279

Kenward R (1987) Wildlife radio tagging. Academic, New York

Kinnaird M, O'Brien T (1993) Species list of trees found within Tangkoko Nature Reserve. Unpublished MS

MacKinnon J, MacKinnon K (1980) The behavior of wild spectral tarsiers. Int J Primatol 1:361–379

Martin RD (1990) Primate origins and evolution. Chapman & Hall, London

Mertl-Millhollen A (1988) Olfactory demarcation of territorial but not home range boundaries by *Lemur catta*. Folia Primatol 50:175–187

Niemitz C (1984) The biology of tarsiers. Gustav Fischer, Stuttgart

Rasa O (1973) Marking behaviour and its social significance in the African dwarf mongoose, *Helogale undulata rufula*. Z Tierpsychol 32:293–318

Richardson P (1990) Scentmarking and territoriality in the aardwolf. In: MacDonald D, Muller-Schwarze D, Natynczuk S (eds) Chemical signals in vertebrates. Oxford University Press, Oxford, pp 378–387

Roper T, Shepherdson D, Davies J (1986) Scentmarking with faeces and anal secretion in the European badger (*Meles meles*): seasonal and spatial characteristics of latrine use in relation to territoriality. Behaviour 97:94–117

Tattersall I, Sussman R (1977) Lemur biology. Plenum, New York

White G, Garrott R (1987) Analysis of wildlife radiotracking data. Academic, New York

Whitten T, Mustafa M, Henderson G (1987) The ecology of Sulawesi. Gadjah Mada University Press, Yogyakarta

World Wildlife Fund (1980) Cagar Alam Gunung Tangkoko Dua Saudara Sulawesi Utara Management Plan 1981–1986. Bogor, Indonesia

Wright PC, Izard M, Simons E (1986) Reproductive cycles in *Tarsius bancanus*. Am J Primatol 11:207–215

Wright PC, Toyama L, Simons E (1988) Courtship and copulation in *Tarsius bancanus*. Folia Primatol 46:142–148

Ziegler T, Epple G, Snowdon C, Porter T, Belcher A, Kuderling I (1993) Detection of the chemical signals of ovulation in the cotton-top tamarin, *Saguinus oedipus*. Anim Behav 5:313–322

Chapter 21
The Population Ecology of Dian's Tarsier

Stefan Merker

Introduction

Over the past two decades, the previously hotly debated but little studied tarsiers have received increased attention from primatologists. Befitting a growing number of recognized taxa and our greatly improved knowledge on behavioral, ecological, and genetic characteristics of tarsiers, it seems appropriate to summarize what we have learnt so far on these enigmatic primates. Recently, Gursky (2007) has done so in a comprehensive volume on north Sulawesi's spectral tarsiers. In a greatly condensed form, I review here what we know on the other well-studied Sulawesi endemic: Dian's tarsier, the predominant species of the central region of the island (Nietsch 1999).

Taxonomic History

Owing to a number of field studies within the past 20 years, Dian's tarsier became one of the best-known Sulawesi primate species. The correct name of this taxon, however, is still disputed. Niemitz et al. (1991) described *Tarsius dianae* from the type locality Kamarora on the basis of acoustic, anatomical, and morphological characteristics. Shekelle et al. (1997) pointed out a likely nomenclatural conflict between *T. dianae* and a taxon described by Miller and Hollister (1921) as *T. fuscus dentatus*, a central Sulawesi subspecies of *T. fuscus* – which, in turn, was later synonymized with *T. tarsier* (Brandon-Jones et al. 2004). Not only does the grayer than usual coloration of *T. f. dentatus*' pelage match the appearance of *T. dianae*, but the well-studied duet vocalization of the latter species is also found throughout

S. Merker (✉)
Department of Ecology and Evolution, Goethe University Frankfurt, Siesmayerstr. 70-72, D-60323 Frankfurt am Main, Germany;
e-mail: smerker@bio.uni-frankfurt.de; tarsius@gmx.net

S. Gursky-Doyen and J. Supriatna (eds.), *Indonesian Primates*, Developments in Primatology: Progress and Prospects, DOI 10.1007/978-1-4419-1560-3_21, © Springer Science+Business Media, LLC 2010

much of central Sulawesi including the type locality for *T. f. dentatus*, Labua(n) Sore (Shekelle et al. 1997). Following Shekelle's and other authors' suggestions, the IUCN 2008 Red List advocates replacing the name *T. dianae* with its probable senior synonym *T. dentatus*, acknowledging its status as a full species (Shekelle and Merker 2008). The common name Dian's tarsier, although not matching the Latin name *T. dentatus*, has been preserved for the sake of recognition of this well-known species.

Distribution

Dian's tarsier is endemic to the central core of the Indonesian island Sulawesi. The northern boundary of its range has been localized through recordings of vocalizations in Kebun Kopi (Nietsch 1999), recordings and captures at Marantale (Shekelle et al. 1997), and my own surveys in the years 2001 and 2008 (Merker unpublished). The western range limit roughly coincides with the western boundary of Lore-Lindu National Park (Merker and Groves 2006). To the east, the species appears to be occupying most of the eastern part of the Central Sulawesi province probably reaching as far as Luwuk (Brandon-Jones et al. 2004; Shekelle and Leksono 2004; Shekelle 2008). As to the southern range limit, very little is known. Shekelle and Leksono (2004) hypothesize that ranges of Sulawesi macaques and toads (Evans et al. 2003) may coincide with ranges of tarsier taxa, and thus we may ultimately find Dian's tarsier's southern boundary somewhere in the area of the Malili lake system.

Vocalizations

Niemitz et al. (1991) described the conspicuous duet song of Dian's tarsier as a species-diagnostic character. The loud, sex-specific duet vocalizations are performed by adults and subadults, almost every morning around dawn and – rarer though – at dusk. The timing of the highest call rate correlates closely with sunrise, and therefore it changes slightly over the course of the year (Merker 1999). Irregularly, tarsier duet calls can be heard at other times during their activity period. As also observed in spectral tarsiers (Gursky 2007), intergroup encounters near territorial boundaries often result in loud and extended vocalizations. Recordings of Dian's tarsier duet calls used in playback experiments elicited vocal and behavioral response in conspecific wild (Niemitz et al. 1991; Shekelle et al. 1997) and caged animals (Nietsch and Kopp 1998), while individuals of other tarsier species generally failed to respond to these playbacks. The unique duet song is likely to be a mate recognition system enabling tarsiers to easily discriminate between conspecific and alien individuals (Niemitz et al. 1991; Nietsch and Kopp 1998; Nietsch 2003; Shekelle 2003, 2008).

Group Composition

Group sizes of *T. dentatus* range from 2 to 7 individuals (Merker 1999, 2003, 2006; Merker et al. 2004, 2005) and are thus similar in size and composition to groups of northern Sulawesi's spectral tarsier (e.g., MacKinnon and MacKinnon 1980; Nietsch 1993; Gursky 1995, 2000, 2007). Almost always, groups were found to comprise one adult male, between one and three adult females, and their offspring. The usually higher number of adult females than adult males in a group could mean that female offspring leave their natal group at a later stage than adolescent males. In a recent population genetic study on *T. lariang*, a species occurring parapatrically to *T. dentatus*, very similar group compositions were revealed (Driller et al. 2009).

Group sizes depend on habitat type and degree of human land use (Merker 2003; Merker et al. 2005): the largest groups were found in primary forest (4.7 ± 0.8 individuals/group, $n=6$ groups) and in slightly disturbed forest with small-scale timber harvesting (5.2 ± 1.9 ind./group, $n=6$); smaller groups characterized moderately disturbed forest interspersed with agroforestry plots, signs of selective logging, and rattan harvesting (3.4 ± 1.1 ind./group, $n=5$) and also heavily affected mixed-species plantations (3.2 ± 1.1 ind./group, $n=5$). More detailed descriptions of these study plots were given by Merker (2003) and Merker et al. (2005). The generally larger groups in the lesser disturbed habitats comprised slightly more adult females and more youngsters than groups in the disturbed plots. Again, all of these included only one adult male (Merker 2003; Merker et al. 2005).

Merker (2003) simultaneously radio-tracked males and females of Dian's tarsier groups and found distances between the mated individuals to be highly variable over the course of the night. Usually spending daylight hours in a common sleeping tree, males and females of a group most often separated for foraging. This confirms the perception of tarsiers as solitary foragers (MacKinnon and MacKinnon 1980; Bearder 1987; Gursky 2000, 2002). In contrast to findings of MacKinnon and MacKinnon (1980) and Gursky (2002, 2007) for *T. spectrum*, males and females of mated pairs of *T. dentatus* did not usually keep close contact during most of the night. The source of the discrepancy is not known, but could include different prey densities, predation pressures, or individual social context. Interindividual distances as illustrated in Fig. 21.1 have been obtained from two pairs of Dian's tarsiers. Pair 1 (M1-F1) comprised two individuals that were judged to be relatively old (Merker 2003). Thus, there is a reasonably high probability that this pair-bond had already been established for a longer period of time. These animals probably met not more than once during the night of radio-tracking and finally chose different sleeping sites. In contrast, pair 5 (M5-F5) consisted of rather young individuals (Merker 2003). They probably encountered each other three times over the course of one night (Fig 21.1). At dawn, they chose a common sleeping site quickly bridging a distance of more than 100 m from each other. These differing observations might certainly be coincidental, yet they might also be interpreted as an effort of the young pair to strengthen the newly established pair-bond. Merker (2003) radio-tracked four additional mated pairs and found similarly low encounter frequencies for all of them.

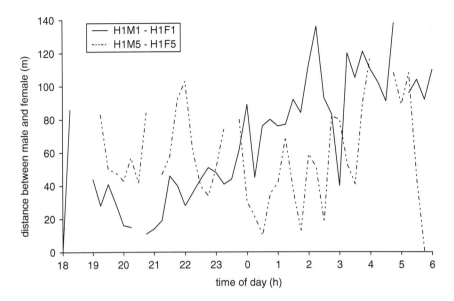

Fig. 21.1 Distance between males and females of two mated pairs of *T. dentatus* in undisturbed forest of Lore-Lindu National Park over the course of one night. Gaps mark missing positional data (after Merker 2003)

Sleeping Tree Choice

The choice of sleeping sites by Dian's tarsiers is highly variable and heavily dependent on habitat type (Merker 2003). In primary forest where strangler figs (*Ficus spp.*) are abundant, tarsiers almost exclusively use crevices and holes in such trees as places to spend daylight hours. Lianas and adjacent trees were frequently used as pathways to reach these sites. Fig trees are preferred whenever available. In secondary forest and mixed-species plantations, tarsiers usually seek shelter in bamboo stands and dense shrubbery (Merker 2003). Niemitz et al. (1991) stated that observed groups (in primary forest) "consistently failed to return to the sleeping site which they had left the evening before" and acknowledged this to be clearly different to *T. spectrum*, the species then thought to be ubiquitous on Sulawesi. Based on Tremble et al.'s (1993) and my own long-term observations of Dian's tarsiers at the type locality as well as other locations in central Sulawesi, this statement can be refuted. Dian's tarsiers regularly conclude their sequence of morning choruses before actually returning to the sleeping site. Thus, potential predators – and human observers – might be tricked into assuming the wrong sleeping tree. Radio-tracking studies on this species confirmed the continued though not exclusive use of the same sleeping sites over long time periods (Merker 2003, 2006). Only when disturbed, e.g., by loggers, rattan harvesters, or tarsier researchers, the animals frequently retreat to alternate shelters – of which there seem to be between one and three for each group. Most of these can be found in close proximity to the home range boundary (Merker 2003, 2006).

Population Densities

Table 21.1 gives an overview of population density estimates for various tarsier species. Densities of Dian's tarsiers vary depending on altitude and habitat type (Gursky 1998; Merker and Mühlenberg 2000; Merker 2003; Merker et al. 2004, 2005; Yustian et al. 2008). Using quadrat census methods, Gursky (1998) found 22 tarsiers per square kilometer of primary forest. Merker and colleagues localized sleeping trees and converted their distance to population densities. Merker and Mühlenberg's (2000) estimation of 136 groups/km² in primary forest was recalculated by Merker et al. (2004) to amount to 105 groups/km² based on a more sophisticated formula to convert sleeping tree distance to population density. In a subsequent study, 57 groups or 268 individuals were found per square kilometer of undisturbed forest (Merker 2003; Merker et al. 2004, 2005; Yustian et al. 2008). Considering that these numbers all stem from pristine forest, biologists might be wondering about the variation in reported densities. One explanation for this lies in the application of different methodologies to quantify tarsier abundance, another one is related to different types of primary forest at varying altitudes and slopes. Microclimate, forest stratification, and species assemblage may well have significant effects on resource availability for tarsiers.

Population densities of *T. dentatus* in secondary habitats range from 14 groups or 45 individuals/km² in a heavily disturbed mixed-species plantation (Merker 2003; Merker et al. 2004, 2005) to 250 individuals/km² in secondary forest (Gursky 1998), or 120 groups/km² (approximately 400 individuals/km²) in old-growth forest

Table 21.1 Published population densities of tarsiers. Each table row marks a separate study

Species	Groups/km²	Individuals/km²	Source
T. dentatus	14–57	45–268	Merker 2003, Merker et al. 2004, 2005, Yustian et al. 2008
T. dentatus	56–156 (*43–120)		Merker and Mühlenberg 2000 (*re calculated, Merker et al. 2004)
T. dentatus		22–250	Gursky 1998
T. spectrum		83–87	Gursky 2007
T. spectrum	56	156	Gursky 1998
T. spectrum		300–1,000	MacKinnon and MacKinnon 1980
T. syrichta		57 adults	Neri-Arboleda et al. 2002
T. syrichta		100–300	Lagapa 1993
T. bancanus		19–47	Yustian 2007
T. bancanus		14–20	Crompton and Andau 1987
T. bancanus		<80	Niemitz 1979

interspersed with small plantations (Merker et al. 2004, recalculated from Merker and Mühlenberg 2000). Whereas in some studies, tarsiers were found to be more abundant in secondary than in primary habitats (Gursky 1998; Merker and Mühlenberg 2000 for *T. dentatus*; MacKinnon and MacKinnon 1980 for *T. spectrum*), other data suggest highest densities in pristine forest (Merker 2003; Merker et al. 2004, 2005). Densities vary strongly – as we have seen even among undisturbed forests – and therefore, the question whether tarsiers are more abundant in primary or human-altered habitats cannot be conclusively answered for now.

It is noteworthy that Dian's tarsier densities are similar to those of northern Sulawesi's spectral tarsier yet somewhat higher than those of *T. bancanus* from Belitung and Borneo and of the Philippine *T. syrichta* (Table 21.1). Merker (2003) suggests the structural difference of forests east and west of the Wallace line to be one possible reason for different tarsier densities. Whereas abundant *Ficus* trees in Sulawesi provide shelter for tarsiers and an all-season food supply for their insect prey (Whitmore 1984; Kinnaird et al. 1999), gregarious flowering of the abundant dipterocarps in Oriental forests affects seasonal food availability for animals (Whitmore 1984). Home ranges of Western and Philippine tarsiers are generally larger than those of their Sulawesi congenerics (see below) and consequently, population densities of these territorial primates are different as well. Nietsch (1993) assumed mainly social factors to be underlying the variations in habitat use of Western and Sulawesi tarsiers.

Population densities of tarsiers are similar to those of nocturnal African and Malagasy strepsirrhine primates that are comparable in size and that occupy analogous ecological niches (reviewed in Merker 2003). Tropical forests of different types and on different continents often maintain populations of small, mainly or partly insectivorous, primates that number up to several hundred individuals per square kilometer.

Ranging Patterns

Tremble et al. (1993) were the first to estimate home range sizes of Dian's tarsiers. At this species' type locality, Kamarora, these authors radio-tracked four individuals (one adult female and three subadult males) and calculated their home ranges as 0.5–0.8 ha. Based on present knowledge from long-term studies at the same location (see below), these numbers seem too low. They were probably affected by heavy radio transmitters (11.5 g) or an insufficient duration of tracking (Merker 2003, 2006). Merker and colleagues subsequently radio-tracked 30 adult Dian's tarsiers in the area of Kamarora (Merker 2003, 2006; Merker et al. 2005; Merker and Yustian 2008). These included six females in each of four study plots along a gradient of human disturbance and six additional males only in undisturbed forest. At the time of this study, none of the females had an infant or was obviously gravid. Home ranges of the males (1.77 ha) were slightly larger than those of their mates (1.58 ha, Figs. 21.2 and 21.3). Mean female range sizes varied significantly between habitats, with 1.08 ha in intermediately disturbed forest and 1.81 ha in a heavily affected mixed-species plantation (Figs. 21.3 and 21.4, for a detailed description of disturbance

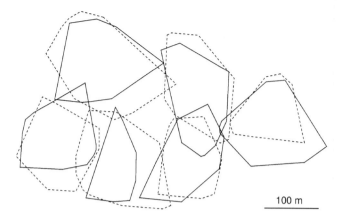

Fig. 21.2   Home ranges of six adjacent mated pairs of *T. dentatus* in undisturbed forest of Lore-Lindu National Park. *Solid line*=female, *dashed line*=male (Merker 2006)

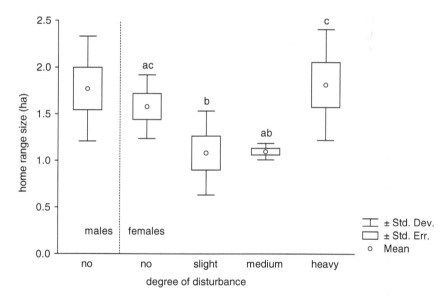

Fig. 21.3   Home range sizes of *T. dentatus* in four study plots along a gradient of human disturbance in Lore-Lindu National Park ($n=6$ for each category). Different letters denote significant differences between plots (ANOVA, least significant difference test, $P<0.05$) (Merker 2006)

parameters see Merker 2003). Tarsiers used the least space in forest that was slightly or intermediately disturbed by human land use. Animals generally range and defend smaller areas when resources are plentiful (Bolen and Robinson 1995). The moderately disturbed study plots were characterized by a high insect abundance caused by small gaps in the canopy, denser undergrowth, and small forest-gardens (Merker 2006). In such forests, an optimized net energy uptake by tarsiers effectively counteracts the potentially adverse effects of small-scale land use. Judging from home range

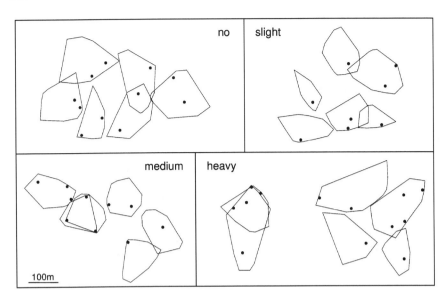

Fig. 21.4 Home ranges of female *T. dentatus* in four study plots along a gradient of human disturbance in Lore-Lindu National Park (degree of disturbance shown). Individuals of all neighboring groups are represented (no and slight disturbance: six females of six groups; medium and heavy disturbance: six females of five groups). *Black dots* mark frequently used sleeping sites (after Merker 2006)

sizes, such areas can be considered high-quality habitat for *T. dentatus*. In primary forest, prey abundance was slightly lower, and the density of tarsier locomotor supports was significantly lower than in moderately disturbed habitats (Merker 2003, 2006; Merker et al. 2005). Both factors probably contribute to shaping medium-sized home ranges in undisturbed forest. Frequent human interference, scarce sleeping trees, low substrate abundance, and – owing to the application of pesticides to protect cash crops – a low insect density in the mixed-species plantation resulted in relatively large tarsier home ranges in this habitat (Merker 2003, 2006; Merker et al. 2005).

The distance traveled by Dian's tarsiers during the course of the night varies only slightly between habitat types. In undisturbed forest, males covered an average of 905 m ($n=6$) and females traveled 945 m ($n=6$) per night. In the slightly and intermediately disturbed plots, mean female path lengths were calculated as 1081 m ($n=6$) and 1030 m ($n=6$), respectively. The longest nightly distances were covered by females in heavily disturbed forest (1,263 m, $n=6$) (Merker 2003, 2006; Merker et al. 2005). Given that path lengths in pristine forest are smaller and home ranges are larger than in moderately disturbed habitats; it is reasonable to assume that in the latter, the animals use a greater fraction of their home range per single night than in primary forest. In fact, Merker (2003, 2006) showed that Dian's tarsiers, in his undisturbed study plot, utilized less than one-third of their total range per night, while the animals in moderately disturbed areas traveled almost one half of their home range during a 12-h activity period.

Dian's Tarsiers in Plantations

The clearance of old-growth rainforest to make room for cash-crop plantations plays a major role in Indonesia's recent history, and such plantations are now a conspicuous part of Sulawesi's landscape. With the loss of their original habitat, it became imperative to know how well tarsiers adapt to human land use and what is needed to prevent this genus from extinction. Merker and Yustian (2008) studied the habitat use of Dian's tarsier in such a mixed-species plantation outside of the natural forest. Radio-tracking the animals, these authors compared the proportion of particular vegetation types within tarsier home ranges with the proportionate use of these structures by the animals. Tarsiers avoided fallows, maize fields, and – unsurprisingly – rice paddies, but strongly selected dense shrubbery for shelter and traveling. The animals neither avoided nor specifically selected young secondary forest, bamboo groves, *Imperata cylindrica* grassland, or cocoa plantations. Coffee or cocoa cultivations, however, are only used by tarsiers if dense shrubbery, forests remnants, or bamboo stands are nearby to provide shelter for the day. As shown above, tarsier population density in this plantation is lower, and home ranges and path lengths are bigger than in lesser disturbed habitats (Merker 2003, 2006; Merker et al. 2005). Nevertheless, such areas can play a vital role in tarsier metapopulation dynamics or as stepping stones and corridors between larger forest patches. Merker and Yustian (2008) suggest to keep or create a mosaic of natural structures (e.g., shrubs or bamboo stands) among cash crops to sustain tarsier populations.

Population Genetics

Microsatellite markers to study the population biology of Dian's tarsier have recently been characterized (Merker et al. 2007a). An ongoing genetic study, including mtDNA as well as Y-chromosomal and microsatellite markers, focuses on the contact zone between *T. dentatus* and the parapatric *T. lariang*. Preliminary findings suggest occasional unidirectional gene flow between these species (Merker et al. 2007b). Dian's tarsier samples from Kamarora and Marantale were involved in analyses of 12S rRNA to infer phylogenetic relationships among north and central Sulawesi tarsiers (Shekelle 2003; Shekelle et al. 2008).

Conservation Status

Massive forest destruction, ethnic conflict, and political upheaval in the province of Central Sulawesi led Gursky et al. (2008) to advocate upgrading the conservation status of *T. dentatus* to Vulnerable. Indeed, for the recently updated IUCN Red List, the species was assessed as Vulnerable A4c (Shekelle and Merker 2008). During long-term studies at the type locality, Dian's tarsier populations were found to decrease – mainly due to clearcut or selective logging (Merker 2003; Merker et al. 2004). While the species occurs in at least two national parks (Lore-Lindu and Morowali), Shekelle and Merker (2008) pointed out that even within these conservation areas, only little good-quality lowland forest habitat is left. Dian's tarsier is, however, still widely distributed, occurs in lowland as well as lower montane forest, and is quite well-adapted to widespread types of human land use (Merker 2003). Despite the negative population trend, *T. dentatus* is thus not in imminent danger of extinction. Enriching the ubiquitous cash-crop plantations with shelter opportunities for tarsiers is an affordable and easy-to-manage way of providing room for these animals. A common misbelief of farmers about tarsiers, however, is that these primates feed on their cultivated crops (Leksono et al. 1997). Striving to overcome such public misconceptions about this fascinating Sulawesi endemic is probably the most urgent and most effective conservation action we could possibly take!

References

Bearder SK (1987) Lorises, bushbabies, and tarsiers: diverse societies in solitary foragers. In: Smuts B, Cheney D, Seyfarth R, Wrangham R, Struhsaker T (eds) Primate societies. University of Chicago Press, Chicago, pp 11–24

Bolen EG, Robinson WL (1995) Wildlife ecology and management. Prentice Hall, Englewood Cliffs, NJ

Brandon-Jones D, Eudey AA, Geissmann T, Groves CP, Melnick DJ, Morales JC, Shekelle M, Stewart CB (2004) Asian primate classification. Int J Primatol 25:97–164

Crompton RH, Andau PM (1987) Ranging, activity rhythms, and sociality in free-ranging *Tarsius bancanus*: a preliminary report. Int J Primatol 8:43–71

Driller C, Perwitasari-Farajallah D, Zischler H, Merker S (2009) The social system of Lariang tarsiers (*Tarsius lariang*) as revealed by genetic analyses. Int J Primatol 30:267–281

Evans BJ, Supriatna J, Andayani N, Setiadi MI, Cannatella DC, Melnick DJ (2003) Monkeys and toads define areas of endemism on Sulawesi. Evolution 57:1436–1443

Gursky S (1995) Group size and composition in the spectral tarsier, *Tarsius spectrum*: implications for social organization. Trop Biodiversity 3:57–62

Gursky S (1998) Conservation status of the spectral tarsier *Tarsius spectrum*: population density and home range size. Folia Primatol 69(suppl 1):191–203

Gursky S (2000) Sociality in the spectral tarsier, *Tarsius spectrum*. Am J Primatol 51:89–101

Gursky S (2002) The behavioral ecology of the spectral tarsier, *Tarsius spectrum*. Evol Anthropol 11:226–234

Gursky S (2007) The spectral tarsier. Prentice Hall, New Jersey

Gursky S, Shekelle M, Nietsch A (2008) The conservation status of Indonesia's tarsiers. In: Shekelle M, Maryanto I, Groves C, Schulze H, Fitch-Snyder H (eds) Primates of the oriental night. LIPI Press, Jakarta, pp 105–114

Kinnaird MF, O'Brien TG, Suryadi S (1999) The importance of figs to Sulawesi's imperiled wildlife. Trop Biodiversity 6:5–18

Lagapa EPG (1993) Population density estimate and habitat analysis of the Philippine tarsier (Tarsius syrichta, Linnaeus) in Bohol. Bachelors thesis, University of the Philippines, Los Banos

Leksono SM, Masala Y, Shekelle M (1997) Tarsiers and agriculture: thoughts on an integrated management plan. Sulawesi Primate Newslett 4:11–13

MacKinnon J, MacKinnon K (1980) The behavior of wild spectral tarsiers. Int J Primatol 1:361–379

Merker S (1999) Der Einfluß traditioneller Landnutzungsformen auf die Populationsdichten des Koboldmakis *Tarsius dianae* im Regenwald Sulawesis. MSc thesis, Georg-August-Universität Göttingen, Germany

Merker S (2003) Vom Aussterben bedroht oder anpassungsfähig? – Der Koboldmaki *Tarsius dianae* in den Regenwäldern Sulawesis. PhD dissertation, Georg-August-Universität Göttingen, Germany. Available online at http://webdoc.sub.gwdg.de/diss/2003/merker

Merker S (2006) Habitat-specific ranging patterns of Dian's tarsiers (*Tarsius dianae*) as revealed by radiotracking. Am J Primatol 68:111–125

Merker S, Groves CP (2006) *Tarsius lariang*: a new primate species from western central Sulawesi. Int J Primatol 27:465–485

Merker S, Mühlenberg M (2000) Traditional land-use and tarsiers – human influences on population densities of *Tarsius dianae*. Folia Primatol 71:426–428

Merker S, Yustian I (2008) Habitat use analysis of Dian's tarsier (*Tarsius dianae*) in a mixed-species plantation in Sulawesi, Indonesia. Primates 49:161–164

Merker S, Yustian I, Mühlenberg M (2004) Losing ground but still doing well – *Tarsius dianae* in human-altered rainforests of central Sulawesi, Indonesia. In: Gerold G, Fremerey M, Guhardja E (eds) Land use, nature conservation and the stability of rainforest margins in Southeast Asia. Springer, Heidelberg, pp 299–311

Merker S, Yustian I, Mühlenberg M (2005) Responding to forest degradation: altered habitat use by Dian's tarsier *Tarsius dianae* in Sulawesi, Indonesia. Oryx 39:189–195

Merker S, Driller C, Perwitasari-Farajallah D, Zahner R, Zischler H (2007a) Isolation and characterization of 12 microsatellite loci for population studies of Sulawesi tarsiers (*Tarsius spp.*). Mol Ecol Notes 7:1216–1218

Merker S, Driller C, Perwitasari-Farajallah D, Zischler H (2007b). Hybridisation in tarsiers. Abstracts, International Congress Prosimians 2007, Ithala, South Africa

Miller GS, Hollister N (1921) Twenty new mammals collected by H. C. Raven in Celebes. Proc Biol Soc Wash 34:93–104

Neri-Arboleda I, Stott P, Arboleda NP (2002) Home ranges, spatial movements and habitat associations of the Philippine tarsier (*Tarsius syrichta*) in Corella, Bohol. J Zool 257:387–402

Niemitz C (1979) Outline of the behavior of *Tarsius bancanus*. In: Doyle GA, Martin RD (eds) The study of Prosimian behavior. Academic, New York, pp 631–660

Niemitz C, Nietsch A, Warter S, Rumpler Y (1991) *Tarsius dianae*: a new primate species from central Sulawesi (Indonesia). Folia Primatol 56:105–116

Nietsch A (1993) Beiträge zur Biologie von *Tarsius spectrum* in Sulawesi. PhD-dissertation, Freie Universität Berlin, Germany

Nietsch A (1999) Duet vocalizations among different populations of Sulawesi tarsiers. Int J Primatol 20:567–583

Nietsch A (2003) Outline of the vocal behavior of *Tarsius spectrum*: call features, associated behaviors, and biological functions. In: Wright PC, Simons EL, Gursky S (eds) Tarsiers – past, present, and future. Rutgers University Press, New Brunswick, pp 196–220

Nietsch A, Kopp ML (1998) Role of vocalization in species differentiation of Sulawesi tarsiers. Folia Primatol 69(suppl. 1):371–378

Shekelle M (2003) Taxonomy and biogeography of Eastern Tarsiers. PhD-dissertation, Washington University, Saint Louis, USA

Shekelle M (2008) Distribution of tarsier acoustic forms, north and central Sulawesi: with notes on the primary taxonomy of Sulawesi's tarsiers. In: Shekelle M, Maryanto I, Groves C, Schulze H, Fitch-Snyder H (eds) Primates of the Oriental Night. LIPI Press, Jakarta, pp 35–50

Shekelle M, Leksono SM (2004) Rencana konservasi di Pulau Sulawesi: dengan menggunakan *Tarsius* sebagai flagship spesies [Indon.: Conservation strategy in Sulawesi Island using *Tarsius* as flagship species]. Biota 9:1–10

Shekelle M, Merker S (2008) *Tarsius dentatus*. In: IUCN 2008. 2008 IUCN Red List of Threatened Species. www.iucnredlist.org. Downloaded on 11 Dec 2008

Shekelle M, Leksono SM, Ichwan LLS, Masala Y (1997) The natural history of the tarsiers of North and Central Sulawesi. Sulawesi Primate Newslett 4(2):4–11

Shekelle M, Morales JC, Niemitz C, Ichwan LLS, Melnick D (2008) Distribution of tarsier haplotypes for some parts of northern and central Sulawesi. In: Shekelle M, Maryanto I, Groves C, Schulze H, Fitch-Snyder H (eds) Primates of the oriental night. LIPI Press, Jakarta, pp 51–69

Tremble M, Muskita Y, Supriatna J (1993) Field observations of *Tarsius dianae* at Lore Lindu National Park, Central Sulawesi, Indonesia. Trop Biodiversity 1:67–76

Whitmore TC (1984) Tropical rain forests of the far east. Oxford University Press, Oxford

Yustian I (2007) Ecology and conservation status of *Tarsius bancanus saltator* on Belitung Island, Indonesia. PhD dissertation, Georg-August-Universität Göttingen, Germany

Yustian I, Merker S, Supriatna J, Andayani N (2008) Relative population density of *Tarsius dianae* in man-influenced habitats of Lore Lindu National Park, Central Sulawesi, Indonesia. Asian Primates J 1(1):10–16

Chapter 22
Using Facial Markings to Unmask Diversity: The Slow Lorises (Primates: Lorisidae: *Nycticebus* spp.) of Indonesia

K.A.I. Nekaris and Rachel Munds

Introduction

The slow lorises (*Nycticebus*) are the only strepsirrhine primates found in Indonesia (Nekaris and Bearder 2007). In addition to features such as a toothcomb and moist nose, these small nocturnal primates were given their name based on their trademark steady, stealthy, and fluid locomotion. Morphologically incapable of leaping (Sellers 1996), slow lorises rather slither through the treetops, and if startled, they may freeze or even cover their face, resulting in one of their many Indonesian names, *malu malu* or "the shy one" (Supriatna and Wahyono 2000). Alternatively, they can fleetingly but silently escape, resulting in the name *buah angin* or "wind monkey" in Acehnese (Nekaris and Nijman 2007a). One of two genera of nocturnal primates found in Indonesia (the other being *Tarsius*), slow lorises are a unique part of Indonesian primate communities, and are widely spread on at least 27 of Indonesia's islands, including Borneo, Sumatra, and Java (Table 22.1) (Nijman and Nekaris in press). Despite this, studies of Indonesian slow lorises are in their infancy.

Lack of studies of slow lorises seems to have derived not only from their nocturnal habits but also from a belief that slow lorises were unspeciose and common, resulting in a conservation status of Least Concern (Meijaard et al. 2005). These beliefs are being upturned, making studies of the behavior and taxonomy of slow lorises imperative to their conservation. Indonesia is notable for its loris diversity, with at least three taxa (*N. coucang, N. javanicus, N. menagensis*) recognized on genetic and morphological bases (Roos 2003; Chen et al. 2006; Groves and Maryanto 2008); all are considered Vulnerable or Endangered (IUCN 2009).

Habitat loss is a serious threat to all Indonesia's primates, and slow lorises are no exception. Owing to the paucity of population data from the ground, Thorn et al. (2009) used ecological niche modeling to elucidate the current conservation status

K.A.I. Nekaris (✉) and R. Munds
Nocturnal Primate Research Group, School of Social Sciences and Law,
Oxford Brookes University, Oxford, OX3 0BP, UK
e-mail: anekaris@brookes.ac.uk

S. Gursky-Doyen and J. Supriatna (eds.), *Indonesian Primates*,
Developments in Primatology: Progress and Prospects,
DOI 10.1007/978-1-4419-1560-3_22, © Springer Science+Business Media, LLC 2010

Table 22.1 Taxonomy, body weight range (from unpublished records and museum labels), conservation status and distribution of Indonesian slow lorises

Species	Common name	Body weight (grams)	Conservation status	Distribution
N. coucang	greater or Sunda slow loris	480–710	VU A2cd	Sumatra, Bunguran, Riau archipelago Tebingtinggi
N. javanicus	Javan slow loris	565–900	EN A2cd	Java, Panaitan
N. menagensis	Philippine or Bornean slow loris	265–325	VU A2cd	Banggi Bangka, Belitung, Borneo; Karimata; Labuan; Sulu archipelago

of Indonesian lorises by predicting the likely remaining loris habitats throughout Sumatra, Java, and Borneo. They found that Javan lorises, in particular, are threatened with habitat loss, followed closely behind by Sumatran lorises.

Trade, too, has been highlighted for many years as a factor seriously underpinning the ability of many Indonesia's primates to persist (Nijman 2005, 2009). The omnipresence of slow lorises as amongst the most common protected mammals in Indonesia's many bird markets (Shepherd et al. 2005) has also been a factor used by researchers to suggest that they are plentiful in the wild (Meijaard et al. 2005). Detailed studies from other parts of Asia, and new data emerging from Indonesia itself, suggest, however, that it is more the inability of a loris to escape from expert hunters combined with the opportunity for easy financial reward that leads to abundance in markets (Ratajszczak 1998; Collins and Nekaris 2008). The stark impact of pet trade on Javan slow lorises has lead to their inclusion on Conservation International's biennial list of the "Top 25 Most Endangered Primates" (Nekaris et al. 2009). Although we still know little about Indonesian loris life history, parameters of closely related species are slow even among the primates (Rasmussen and Izard 1988), with a gestation period of about 6 months (Fitch-Snyder and Ehrlich 2003); combined with typical litters of one or two infants that require 3–6 months for weaning, their extremely slow life history does not lend well to this level of off-take.

The large number of animals coming through pet markets has had the side effect of offering scientists a glimpse of Indonesian loris diversity (Nekaris and Jaffe 2007). Earlier taxonomists recognized greater diversity in the Sundaland region (Osman Hill 1953), and ongoing research is investigating the validity of previously proposed taxa (Ravosa 1998; Nekaris and Jaffe 2007; Groves and Maryanto 2008). In recent times, numerous cryptic species have been revealed amongst other nocturnal primates (e.g., galagos – Bearder 1999; tarsiers – Merker and Groves 2006; lemurs – Thalmann and Geissmann 2005); it would be unsurprising to find hitherto unappreciated diversity amongst slow lorises (Groves 1971, 1998; Schwartz and Beutel 1995).

In this chapter, we had two aims. In the first half, we review studies of the behavior and ecology of wild Indonesian slow lorises. Understanding the behavior and ecology of lorises hinges upon resolving their taxonomic diversity (Chen et al. 2006). Variation in external characters, including facial masks, has improved our knowledge in discerning species within many nocturnal primates (Musser and Dagasto 1987; Ford 1994; Bearder 1999; Rasoloarison et al. 2000; Defler 2003). Several factors have been implicated as selecting for species-specific "facial masks": species-specific recognition devices (Bearder et al. 1995), individual recognition within species (Barash 1974) and predator deterrents (Newman et al. 2005). Nekaris and Jaffe (2007) showed facial masks distinguished Indonesian loris species, but they did not specifically define characters of the mask. Furthermore, the animals they examined were from trade and thus their localities were not known. In this chapter, we further explore the utility of the facial mask to determine loris taxa, based on a sample of living lorises and museum specimens measured from known localities. In particular, we examine if any characters statistically distinguish the three recognized species, and if the face mask provides any evidence for further diversity.

Methods

For the behavioral overview, we compiled data on wild Indonesian slow lorises from the literature and report novel data from our own observations, summarized by taxon. We examined the facemasks of Indonesian slow lorises from photographs and museum specimens from known localities, either given by field workers or taken from museum tags. We used SPSS v.14.0 for the analysis, applying appropriate nonparametric and descriptive statistics. We coded each face mask for 12 discrete characters; measures were either presence/absence or on an ordinal scale. Characters fell into four general groups: circumocular eye patch characters ($n=6$), ear characters ($n=2$), nose skin color ($n=1$), and crown characters ($n=3$). These are patch top distinctly pointed, rounded, or diffused into crown; patch middle (midline of eye) barely visible around the edge of the eye, a distinct band around the eye, or broad extending to cheek; patch bottom ended as a small line just below the eye, as a wider band on top of the zygomatic, or as a broad band below the zygomatic; presence/absence of an additional distinct black rim around the eye; width and shape of the interocular stripe (narrow or wide; rectangle, hourglass, diamond) (Fig. 22.1); the size of the preauricular hair that differed in color from the circumocular patch (narrow/absent, medium, wide, Fig. 22.2); ears were naked, distinctly furred but pressed to the head (appearing hidden), distinctly furred but erect on the head, or furred with additional tufts; nasal skin color – pink, black, or pink/ black mix; crown patch small with distinct pointed forks, large with rounded forks, or diffuse; and the general color of the crown and facial markings – brownish, blackish, reddish, or yellowish.

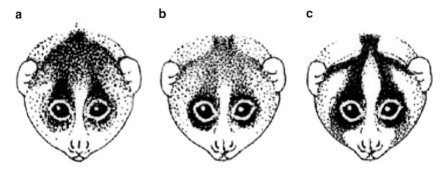

Fig. 22.1 Characteristics of the loris facial mask: (**a**) shows a mask with an hourglass interocular stripe, where the top of the patch is rounded into a large crown with rounded forks. (**b**) shows a mask with a rectangular interocular stripe, with diffuse patches and crown, and black rim around the eye. (**c**) shows a diamond-shaped interocular stripe, with distinct pointed forks leading to a small crown; eye patch extends beneath the zygomatic

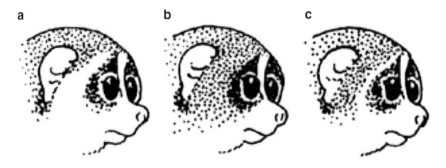

Fig. 22.2 The width of differently colored preauricular hair: (**a**) wide, (**b**) narrow/absent, (**c**) medium

Results

Studies from the Wild

N. coucang

Also found in peninsular Malaysia, Thailand, and Singapore, in Indonesia, *N. coucang* is restricted to the island of Sumatra and some smaller islands. Presence of a brown and a red variant on Sumatra may ultimately require taxonomic revision (see below). Nekaris and Nijman (2007a) spent several weeks conducting surveys of the red variant of this taxon in Aceh, Sumatra. Of five sites surveyed, lorises were found at two sites within the Ulu Masen forest complex: SP Limon and Jantho. At SP Limon, in 13.8 km of transects, the only evidence for lorises was a decaying animal found on a path. Residents described the recent removal by a specialist loris dealer of a large number of lorises from the area in conjunction with deforestation of a large area of forest for agriculture, perhaps explaining the paucity of encounters. The Jantho site was a two-day walk from human settlements, and yielded an encounter rate of 0.39 lorises/km, with 22 km covered. The average distance of lorises from the line was 7.5 m. Animals were patchily distributed, with areas of high concentration, followed by 1–5 km without a single loris sighting.

Pairs of lorises, or lorises separated by less than 30 m, were encountered five times. Three of these included a mother and her male infant. The infant was seen "parked", as well as being carried ventrally by its mother. He was caught and measured; weighing 115 g, his age was determined to be about 3 months old (Zimmermann 1989), suggesting birth in February. Animals emitted a high-pitched whistle, which was heard only three times over five nights. Habitat analysis of the area of high loris density revealed that median tree height was 8 m, with most sightings of lorises occurring at 5 m, on trees that averaged 10.2 cm DBH. Lorises were seen to catch unidentified insects three times, and were also seen to consume the bland fruit of *ranuk dong*, a tree belonging to the Piperaceae (Nekaris and Nijman 2007a). In the nearby Gunung Leuser ecosystem, *N. coucang* is also known to be a prey item of Sumatran orangutans (*Pongo abelii*) (van Schaik et al. 2003). Wild-born animals measured at the Schmutzer Primate Center had an average neck circumference of 145 mm (±sd 10.1, $n = 9$), a figure important for future radio-collar studies.

In southern Sumatra in Lampung Province, seven wild-born brown lorises, only recently caught and confiscated, were observed in captivity for two months before being reintroduced to the wild (Collins and Nekaris 2008). Although captive lorises accepted fruit (particularly the native duku – *Lansium domesticum*), they preferred insects and live birds (yellow-vented bulbuls, *Pycnonotus goiavier*), the latter of which was shared with other lorises (Streicher et al. in review). They also gouged the timber of their enclosure 441 times, favoring *sengon* (Fabacea: *Paraserianthes falcataria*) (Nekaris et al. in press). Animals regularly emitted affiliative calls ("krik") (c.f. Schulze and Meier 1995), and also counter-called with a loud call strongly reminiscent of a crow's caw. Lorises allogroomed, played, and fed together, the behaviors that continued after their release. Upon release, animals moved quickly on small branches and maintained group cohesion for the single night during which observations were made (Collins and Nekaris 2008). Wild-born animals had an average neck circumference of 146 mm (±sd 14.9, $n = 16$).

Nycticebus coucang was the subject, too, of a long-term study in peninsular Malaysia. Although the Malaysian form may in fact be a distinct species, we summarize key aspects of its behavior and ecology. In a study lasting more than two years, Wiens et al. (2006) found that nectar and gum comprised more than 70% of the diet, with fruit and insects playing a limited role. Indeed, lorises seem to have a number of morphological specializations for exudativory (Nekaris et al. in press). One to three animals interacted in overlapping home ranges ranging from about 10 to 25 ha in size, and social sleeping occurred. The general social organization was uni-male, uni-female (Wiens and Zitzmann 2003). Mating, however, was promiscuous with multiple males pursuing a single estrous female (Elliot and Elliot 1967; Wiens 2002), a behavior seen also in slender lorises (Nekaris 2003).

N. javanicus

Supriatna and Wahyono (2000) provided the first opportunistic observations of Javan slow lorises. They found them in primary and secondary forest, bamboo forest, mangrove forest, and plantations, with a preference for chocolate plantations. They observed slow lorises consuming fruit, lizards, eggs, and chocolate seeds. Nijman and van Balen (1998) confirmed presence of lorises in the Dieng Mountains. Surveys in Gunung Gede Pangrango National Park, West Java, revealed low densities of lorises, with encounter rates ranging from 0.02 to 0.20 animals/km (Arisona, pers. comm.; Nekaris et al. 2008). Another survey in the nearby Mt. Salak National Park found few lorises (0.03 animals/km), and only in areas where human distur-bance was minimal (Collins 2007; Munds et al. 2008). Lorises were encountered alone or in pairs, and occurred at heights from 1.5 to 9.5 m.

A pilot study by Winarti (2008) revealed further aspects of the ecology of Javan lorises. In Ciamis and Tasikmalaya regencies, West Java, she found that lorises were able to persist in mixed-crop home gardens with high levels of human distur-bance. They did not use nest holes, but slept curled into a ball on branches in tangles of rope bamboo; multiple animal sleeping groups were observed. She observed them actively gouging gum from Fabaceae: *Albizia*. They moved at heights of 3–22 m, and were also observed crossing open spaces on the ground in their dis-turbed habitat. Wild-born animals measured at the Schmutzer Primate Center had an average neck circumference of 136 mm (±sd 12.1, $n=6$).

N. menagensis

Nycticebus menagensis is the least studied of Indonesia's lorises. In 1971, seeking this species in Kinabalu National Park, Sabah, Jenkins (1971) described it as present, but rarely seen. A first attempt to study this species in more detail at the Sabangau National Park, Central Kalimantan, yielded only 12 sightings in 75 days (Nekaris et al. 2008). The median distance of a loris from the transect line was about 13 m and all were seen at heights of 15–20 m in the trees. Lorises were encountered singly, mother and offspring, or in adult trios. Of two trios spotted, both were on fruiting trees: *Calophyllum hosei* and *Syzygium cf. nigricans*. Another survey at Wehea, East Kalimantan yielded similar results, with only one loris encountered in more than 30 km (0.02 animals/km). This animal was seen at 30 m height (Munds et al. 2008). Other attempts to find *N. menagensis* have proved equally futile. In 46,000 trap nights in Kinabalu National Mark, Wells et al. (2004) caught only one animal, albeit thrice. Duckworth (1997) was unable to record loris presence in Similajau National Park, Sarawak after 77 h of nocturnal walks. Baker (2008 pers. comm.) was able to locate only three lorises (one pair and one single adult) at Danum Valley Research Station, Sabah, during more than 60 h of night walks. In her on-going study in the Lower Kinabatangan Wildlife Sanctuary (LKWS), Sabah, Malaysia, Munds has found *N. menagensis* to be in relatively low densities. The primary study site, Danau Girang

Field Centre, is a riparian secondary forest within the LKWS. Of 35 night surveys, only three lorises have been spotted, on average at 20 m height. All were alone and traveling between two trees by lianas or vines. All lorises were sighted at least 100 m away from any of the major rivers that surround the field centre. One loris, spotted at 5:30 a.m., moved along a branch toward a 15 m high thicket of vines and leaves. The sighting may indicate that such thickets provide sleeping sites for Bornean lorises. On the basis of its craniodental morphology, Ravosa (1998) proposed that this species might be more insectivorous than its congeners.

Facial Masks

We collected data from 106 individual lorises *N. coucang* (brown), $n = 16$; *N. coucang* (red), $n = 16$; *N. menagensis* (Borneo and offlying islands, Sulu archipelago), $n = 29$; *N. menagensis* (Bangka), $n = 3$; *N. javanicus* (short-furred form), $n = 26$; *N. javanicus* (long-furred form, $n = 16$). Table 22.2 summarizes 12 characters of the facial masks, all of which were significantly different among taxa. Certain characteristics always distinguish the three species (Fig. 22.3). For example, *N. coucang* has medium width preauricular hair, *N. menagensis* always has circumocular patches that end just below the eye, and *N. javanicus* always has a diamond interocular stripe.

Other features are suggestive of additional species or subspecies. Sumatran *N. coucang* occurs in two color phases: red and brown. The red form is further distinguished by significant presence of rounded forks leading to a distinctly shaped crown, and a dark rim around the eye. Figure 22.4 reveals that the majority of red specimens in our sample are restricted to Northern Sumatra, with brown specimens restricted to the southern 2/3 of the island. Although similar to other *N. menagensis* in our sample in many respects, facemask color, width of preauricular hair, and furred ears give lorises from Bangka a strikingly distinct appearance. Other than their long-silky fur, the only characters distinguishing the long-furred Javan lorises from those with short fur were a tendency for black facial markings and a pink nose. Figure 22.4 reveals substantial overlap between these two forms; the current analysis, however, does not allow for altitude to be taken into account.

Discussion

Here, we have reviewed the handful of data available to us on Indonesian lorises in the wild. We reaffirm earlier work showing that lorises occur in sparse numbers in their natural habitat (Nekaris et al. 2008). Indeed, Indonesian lorises appear to occur at even lower densities than mainland *Nycticebus* (Duckworth 1997). This rarity combined with high volume in the pet trade highlight the conservation

Table 22.2 Percent that each character was present among the individuals in our sample; all characters were statistically significantly different ($p < 0.001$) when tested with a chi-square cross tabulation, even when *menagensis* from Bangka was removed from the sample

Character	coucang (brown) $n = 16$	coucang (red) $n = 16$	menagensis $n = 29$	menagensis (Bangka) $n = 3$	javanicus (short fur) $n = 26$	javanicus (long fur) $n = 16$
Circumocular patch						
Top	P: 94% R: 6%	P: 6% R: 94%	D: 100%	P: 100%	P: 100%	P: 100%
Middle	B: 94% N: 6%	B: 94% N: 6%	N: 100%	N: 100%	B: 100%	B:100%
Bottom	AZ: 94% UE: 6%	AZ: 94% UE: 6%	UE: 100%	UE:100%	UZ: 100%	UZ: 100%
Rim	A: 94% P: 6%	P:100%	A:10% P:90%	P:100%	A:100%	A:100%
Interocular stripe						
Width	N: 100%	N: 100%	N: 100%	N: 100%	W:100%	W:100%
Shape	H: 75% R: 25%	H: 94% R: 6%	R: 67%	H: 33% R: 67%	D:100%	D:100%
Nose color	P:25% M:75%	B:6% M:94%	P:45% B:21% M:35%	B:100%	P:27% M: 73%	P:94% M:6%
Ear characters						
Preauricular hair	M:94% W:6%	M:100%	N:31% M:69%	M:100%	W:100%	W:100%
Ears furred?	F:100%	F:94% T:6%	H:100%	F:100%	T:100%	T:100%
Crown						
Shape	RF:100%	DF: 6% RF: 94%	DF: 7% D: 93%	DF:43% RF: 33% D: 26%	DF: 100%	DF: 100%
Forks	D: 100%	D: 100%	B: 100%	D: 100%	D: 100%	D: 100%
Color	R: 25% Br: 75%	R: 100%	R: 14% Br: 3% Y: 83%	R: 100%	R: 12% Br: 85% Bl: 4%	R: 6% Br: 6% Bl: 88%

Top – *P* pointed, *R* rounded, *D* diffuse; Middle – *B* broad, *N* Narrow; Bottom – *UE* under eye, *AZ* above zygomatic, *UZ* under zygomatic; Rim – *P* present, *A* absent; Width – *N* narrow, *W* wide; Shape – *H* hourglass, *R* rectangle, *D* diamond; Nose color – *P* pink, *B* black, *M* mixed; Preauricular hair – *N* narrow, *M* medium, *W* wide; Ears – *H* hidden, *F* furred, *T* tufted; Shape – *DF* distinct forked, *RF* rounded forked, *D* diffuse; Forks – *D* distinct, *B* blended; Color – *R* red, *Br* brown, *Y* yellow, *Bl* black.

plight faced by lorises. In Indonesia, lorises are protected under Decree of Agriculture Ministry No. 66 of 1973, the Government Regulation No. 7 of 1999 concerning the Protection of Wild Flora and Fauna and Act No. 5 of 1999 concerning Biodiversity Conservation (Streicher et al. 2008). Furthermore, Indonesia supported the 2007 up-listing of *Nycticebus* to CITES I (Nekaris and Nijman 2007b). Despite this protection, enforcement is challenging, and penalties issued in the rare instances when lorises are confiscated are meager (Shepherd et al. 2005).

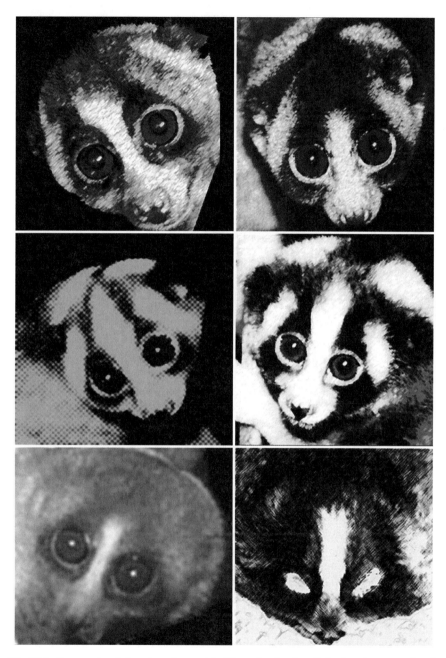

Fig. 22.3 Face masks of six subspecies used in this study, showing distinctive characters of each taxon. Clockwise from upper left – *N. coucang* (*brown*), *N. coucang* (*red*), *N. javanicus* (long-furred form), *N. menagensis* (Bangka form; museum specimen), *N. menagensis* (Borneo), *N. javanicus* (short-furred form)

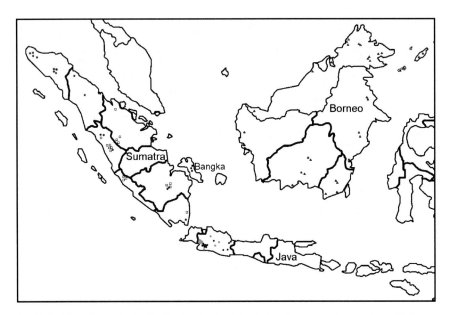

Fig. 22.4 Map of slow loris distribution in Sundaland showing points for species with known localities used in this analysis. In Sumatra, the dark spots demarcate the red form, with open squares indicating brown form localities. In Java, the darker points are the short-furred form, with the lighter points representing the long-furred form. Triangles indicate *N. menagensis,* with the Bangka form denoted with dark squares

Our literature review also shows that quantitative data are sorely lacking on their diet, ranging patterns, social behavior, and habitat preferences. These data are vital to managing lorises in the wild, determining reintroduction programs, as well as to improve captive management (Fitch-Snyder and Schulze 2001). Fitch-Snyder et al. (2008) outline extensive guidelines on keeping lorises in captivity, and Streicher et al. (2008) provide details on following IUCN protocol for reintroducing wild animals. All of these authors reiterate, however, the fundamental importance of understanding loris taxonomy. Currently, the reintroduction of the wrong species into the wrong area only increases taxonomic and conservation havoc (Schulze and Groves 2004).

Our data on facial mask characteristics solidify support for classification of Indonesian lorises into at least three species. Based on craniometric evidence, Ravosa (1998) and Groves and Maryanto (2008) found that *N. javanicus* was highly distinguishable by several characters, including its larger size. Pelage differences have been noted before (Groves 1998), and our study yields quantifiable evidence that *N. javanicus* is easily distinguished from other Indonesian lorises by its facial mask. Nekaris and Jaffe (2007) pointed out two forms may be present on Java, with a key feature being significantly longer fur length. Our study only slightly distinguished a long-coated form from Java. Whether or not it can be classified as a distinct taxon remains to be seen. Data on altitudinal variation are sorely needed;

it may follow a coat pattern under the influence of similar ecological factors as those dictating variation in ebony langurs (*Trachypithecus auratus*) (Nijman 2000).

Long considered a subspecies of *N. coucang,* two independent genetic studies have distinguished *N. menagensis* as a taxon (Roos 2003; Chen et al. 2006). In our analyses, pelage characters also clearly distinguished Bornean lorises from their congeners. Lorises from Bangka are clearly united with *N. menagensis* by distinctive cranial morphology, including the persistent absence of the second upper incisor (Groves 1971; Groves and Maryanto 2008). Unique facial characters and striking pelage warrant additional review of the taxonomy of these lorises.

Although Sumatran lorises shared a number of traits distinguishing *N. coucang* from *N. menagensis* and *N. javanicus,* our analysis points toward a red form found throughout Sumatra and a brown one restricted to the southern two-thirds. Northern Sumatra is known as a faunal transition area. Certain species, including tapirs (*Tapirus indicus*), agile gibbons (*Hylobates agilis*) and banded langurs (*Presbytis melalophos),* are only found south of this boundary. It might be reasonable to assume that similar selective pressures may have resulted in two taxa of slow loris. In the case of nocturnal mammals, where morphology is only a small part of the story in distinguishing cryptic species (Bearder et al. 1995), genetic studies and more current locality data from the wild are required to elucidate Sumatran loris taxonomy.

Another area to explore is *why* these facial characters are so distinct. Bearder (1999) showed that galago species can be easily discriminated by a suite of facial characters and suggested that face masks may be part of complex system of species recognition. Face masks clearly aid in individual recognition; in the field, both galagos and slender lorises can identify individuals by vision from a distance of 20–50 m away (Bearder et al. 2006). A striking face mask may also serve as a form of aposematism to make a species look larger or threatening (Newman et al. 2005). This hypothesis would be particularly interesting to test for slow lorises; appearing larger would be a valuable antipredator benefit to these otherwise slow and relatively helpless primates (Nekaris et al. 2007). Although still an area under study, the purportedly toxic bite of slow lorises has even been known to kill humans (Hagey et al. 2007); aposematic face masks could also serve as a warning to potential predators.

In conclusion, we clearly have much to learn about Indonesia's lorises. Long thought to be common, conservation of these primates is of utmost importance, as is an understanding of the complex biology of these unique strepsirrhines.

Acknowledgments We thank S. Gursky and J. Supriatna for inviting us to contribute to this volume. We are grateful to the following individuals for access to and help with their zoological collections: H. van Grouw (Naturalis Leiden), D. Hill and P. Jenkins (Natural History Museum London), W. Stanley (Field Museum of Natural History Chciago), A. Rol and V. Nijman (Zoological Museum Amsterdam), and M. Nowak-Kemp (Natural History Museum Oxford). Others who have and are continuing to assist with this project include: J. Ariosona, S. Bearder, M. Bruford, B. Goossens, F. den Haas, F. Jalil, A. Knight, K. L. Sanchez, and I. Winarti. H. Schulze provided the loris figures, and the map was adapted from one produced by J. Thorn. We thank K. Wells for the photo of the ever elusive *N. menagensis.* C. Groves, H. Schulze, and V. Nijman provided valuable comments on the manuscript. Funding was provided by the Systematics Research Fund of the Linnaean Society, Primate Conservation Inc, International Animal Rescue Indonesia, the Royal Society, and Oxford Brookes University Research Strategy Fund. This research received support from the SYNTHESYS Project, which is financed by European Community Research Infrastructure Action under the FP6 "Structuring the European Research Area" Programme (NL-TAF-3491).

References

Bearder SK (1999) Physical and social diversity among nocturnal primates: a new view based on long term research. Primates 40:267–282

Bearder SK, Honess PE, Ambrose L (1995) Species diversity among galagos with special reference to mate recognition. In: Alterman L, Doyle G, Izard MK (eds) Creatures of the dark: the nocturnal prosimians. Plenum, New York, pp 331–352

Bearder SK, Nekaris KAI, Curtis DJ (2006) A re-evaluation of the role of vision in the activity and communication of nocturnal primates. Folia Primatol 77(1–2):50–71

Barash DP (1974) Neighbor recognition in two 'solitary' carnivores: the raccoon (*Procyon lotor*) and the red fox (*Vulpes fulva*). Science 185(4153):794–496

Chen JH, Pan D, Groves CP (2006) Molecular phylogeny of *Nycticebus* inferred from mitochondrial genes. Int J Primatol 27(4):1187–1200

Collins R (2007) Preliminary study of behaviour and population densities of Nycticebus coucang and *N. javanicus* in Sumatra and Java, Indonesia. MSc Thesis, Oxford Brookes University, Oxford

Collins R, Nekaris KAI (2008) Release of greater slow lorises, confiscated from the pet trade, to Batutegi Protected Forest, Sumatra, Indonesia. In: Soorae PS (ed) Global re-introduction perspectives. IUCN Reintroduction Specialist Group, Abu Dhabi, pp 192–195

Defler TR (2003) Primate of Colombia. Conservation International, Bogota, Colombia

Duckworth JW (1997) Mammals in Similajau National Park, Sarawak, in 1995. Sarawak Mus J 51:171–192

Elliot O, Elliot M (1967) Field notes on the slow loris. Malaya J Mammal 48:497–498

Fitch-Snyder H, Ehrlich A (2003) Mother-infant interactions in slow lorises (*Nycticebus bengalensis*) and pygmy lorises (*Nycticebus pygmaeus*). Folia Primatol 74(5–6):259–271

Fitch-Snyder H, Schulze H (2001) Husbandry manual for Asian Lorisines (Nycticebus and Loris). Center for Reproduction of Endangered Species (CRES) Zoological Society of San Diego, San Diego

Fitch-Snyder H, Schulze H, Streicher U (2008) Enclosure design for captive slow and pygmy lorises. In: Shekelle M, Maryanto I, Groves C, Schulze H, Fitch-Snyder H (eds) Primates of the oriental night. LIPI, Jakarta, pp 123–135

Ford SM (1994) Taxonomy and distribution of the owl monkey. In: Baer JF, Weller RE, Kakoma I (eds) Aotus: the owl monkey. Academic, San Diego, pp 1–57

Groves CP (1971) Systematics of the genus *Nycticebus*. In: Biegert J, Leutenegger W (eds) Taxonomy, anatomy, and reproduction, vol 1, Proceedings of the third international congress of primatology. S Karger, Basel, pp 44–53

Groves CP (1998) Systematics of tarsiers and lorises. Primates 39:13–27

Groves CP, Maryanto I (2008) Craniometry of slow lorises (genus *Nycticebus*) of insular Southeast Asia. In: Shekelle M, Maryanto I, Groves C, Schulze H, Fitch-Snyder H (eds) Primates of the oriental night. LIPI, Jakarta, pp 115–122

Hagey LR, Fry BG, Fitch-Snyder H (2007) Talking defensively, a dual use for the brachial gland exudate of slow and pygmy lorises. In: Gursky SL, Nekaris KAI (eds) Primate anti-predator strategies. Springer, New York, pp 253–272

IUCN (2009) IUCN red list of threatened species. Version 2009.1. www.iucnredlist.org. Accessed 07 June 2009

Jenkins DV (1971) Animal life of Kinabalu National Park. Malay Nat J 24:177–183

Meijaard E, Sheil D, Nasi R (2005) Life after logging: reconciling wildlife conservation and production forestry in Indonesian Borneo. CIFOR and UNESCO, Bogor

Merker S, Groves C (2006) *Tarsius lariang*: a new primate species from western central Sulawesi. Int J Primatol 27(2):465–485

Munds RA, Collins R, Nijman V, Nekaris KAI (2008) Abundance estimates of three slow loris taxa in Sumatra (*N. coucang*), Java (*N. javanicus*) and Borneo (*N. menagensis*). Primate Eye 96:902

Musser GG, Dagasto M (1987) The identity of *Tarsius pumilus*, a pygmy species endemic to the montane mossy forests of central Sulawesi. Am Mus Novit 2867:1–53

Nekaris KAI (2003) Observations on mating, birthing and parental care in three taxa of slender loris in India and Sri Lanka (*Loris tardigradus* and *Loris lydekkerianus*). Folia Primatol 74:312–336

Nekaris KAI, Bearder SK (2007) The strepsirrhine primates of Asia and Mainland Africa: diversity shrouded in darkness. In: Campbell C, Fuentes A, MacKinnon K, Panger M, Bearder SK (eds) Primates in Perspective. Oxford University Press, Oxford, pp 24–45

Nekaris KAI, Jaffe S (2007) Unexpected diversity of slow lorises (*Nycticebus* spp.) in the Javan pet trade: implications for slow loris taxonomy. Contrib Zool 76:187–196

Nekaris KAI, Nijman V (2007a) Survey on the abundance and conservation of Sumatran slow lorises (*Nycticebus coucang hilleri*) in Aceh, Northern Sumatra. In: Proceeding of the European Federation of Primatology, Prague, Charles University, Prague, p 47

Nekaris KAI, Nijman V (2007b) CITES proposal highlights rarity of Asian nocturnal primates (Lorisidae: *Nycticebus*). Folia Primatol 78(4):211–214

Nekaris KAI, Blackham GV, Nijman V (2008) Implications of low encounter rates in five nocturnal species (*Nycticebus* spp.). Biodivers Conserv 17(4):733–747

Nekaris KAI, Pimley ER, Ablard K (2007) Anti-predator behaviour of lorises and pottos. In: Gursky SL, Nekaris KAI (eds) Primate anti-predator strategies. Springer, New York, pp 220–238

Nekaris KAI, Collins RL, Starr C, Navarro-Montes A (in press-a). Comparative ecology of exudate feeding by Asian slow lorises (*Nyctecebus*). In: Nash L, Burrows A (eds) The evolution of exudativory. Springer, New York

Nekaris KAI, Sanchez KL, Thorn JS, Winarti I, Nijman V (2009) Javan slow loris. In Primates in peril: the world's top 25 most endangered primates 2008–2010. IUCN/SSC Pst, International Primatological Society, Conservation International, Arlington, VA, pp 44–46

Newman C, Buesching CD, Wolff JO (2005) The function of facial masks in "midguild" carnivores. Oikos 108(3):623–633

Nijman V (2009) An assessment of trade in orang-utans and gibbons on Sumatra, Indonesia. Traffic Southeast Asia, Petaling Jaya

Nijman V (2005) Hanging in the balance: an assessment of trade in orang-utans and gibbons on Kalimantan, Indonesia. Traffic Southeast Asia, Petaling Jaya

Nijman V (2000) Geographic distribution of ebony leaf monkey *Trachypithecus auratus* (E. Geoffroy Saint-Hilaire, 1812) (Mammalia: Primates: Cercopithecidae). Contrib Zool 69(3):157–177

Nijman V, Nekaris KAI (in press) Distribution patterns of slow lorises and tarsiers in insular Southeast Asia – interspecific competition or something else. Int J Primatol

Nijman V, van Balen S (1998) A faunal survey of the Dieng mountains, Central Java, Indonesia: distribution and conservation of endemic primate taxa. Oryx 32:145–156

Osman Hill WC (1953) Primates. Comparative anatomy and taxonomy. I. Strepsirhini. Edinburgh University Press, Edinburgh

Rasoloarison RM, Goodman SM, Ganzhorn JU (2000) Taxonomic revision of mouse lemurs (*Microcebus*) in the western portions of Madagascar. Int J Primatol 21(6):963–1019

Rasmussen DT, Izard MK (1988) Scaling of growth and life-history traits relative to body size, brain size and metabolic rate in lorises and galagos (Lorisidae, Primates). Am J Phys Anthropol 75:357–367

Ratajszczak R (1998) Taxonomy, distribution and status of the lesser slow loris *Nycticebus pygmaeus* and their implications for captive management. Folia Primatol 69:171–174

Ravosa MJ (1998) Cranial allometry and geographic variation in slow lorises (*Nycticebus*). Am J Primatol 45:225–243

Roos C (2003) Molekulare Phylogenie der Halbaffen, Schlankaffen, und Gibbons. Unpubl. Ph.D Thesis, Technische Universität München

Schulze H, Groves CP (2004) Asian lorises: taxonomic problems caused by illegal trade. In: Nadler T, Streicher U, Thang Long H (eds) Conservation of primates in Vietnam. Frankfurt Zoological Society, Frankfurt, pp 33–36

Schulze H, Meier B (1995) Behaviour of captive *Loris tardigradus nordicus*: a qualitative description including some information about morphological bases of behavior. In: Alterman L, Doyle G, Izard MK (eds) Creatures of the dark: the nocturnal prosimians. Plenum Publishing, New York, pp 171–192

Schwartz JH, Beutel JC (1995) Species diversity in lorisids: a preliminary analysis of *Arctocebus, Perodicticus* and *Nycticebus*. In: Alterman L, Doyle G, Izard MK (eds) Creatures of the dark: the nocturnal Prosimians. Plenum Publishing, New York, pp 171–192

Sellers W (1996) A biomechanical investigation into the absence of leaping in the locomotor repertoire of the slender loris (*Loris tardigradus*). Folia Primatol 67:1–14

Shepherd CR, Sukumaran J, Wich SA (2005) Open season: an analysis of the pet trade in Medan, North Sumatra, 1997–2001. TRAFFIC Southeast Asia, Kuala Lumpur

Streicher U, Schulze H, Fitch-Snyder H (2008) Confiscation, rehabilitation and placement of slow lorises – recommendations to improve the handling of confiscated slow lorises *Nycticebus coucang*. In: Shekelle M, Maryanto I, Groves C, Schulze H, Fitch-Snyder H (eds) Primates of the oriental night. LIPI, Jakarta, pp 137–145

Streicher U, Collins R, Navarro-Montes A, Nekaris KAI (in review). Observations on the feeding preferences of slow lorises (*Nycticebus pygmaeus, N. javanicus, N. coucang*) rescued from the trade. In: Masters J, Crompton R, Genin F (eds.) Leaping ahead. New York: Springer.

Supriatna J, Wahyono EH (2000) Panduan lapangan primata Indonesia. Yayasan Obor Indonesia, Jakarta, p 332

Thalmann U, Geissmann T (2005) New species of woolly lemur *Avahi* (Primates: Lemuriformes) in Bemaraha (central western Madagascar). Am J Primatol 67(3):371–376

Thorn JS, Nijman V, Smith D, Nekaris KAI (2009) Ecological niche modelling as a technique for assessing threats and setting conservation priorities for Asian slow lorises (Primates: *Nycticebus*). Divers Distrib 15(2):289–298

van Schaik CP, Ancrenaz M, Borgen G (2003) Orang-utan cultures and the evolution of material culture. Science 299:102–105

Wells K, Pfeiffer M, Lakim MB (2004) Use of arboreal and terrestrial space by a small mammal community in a tropical rain forest in Borneo. Malaysia J Biog 31:641–652

Wiens F (2002) Behavior and ecology of wild slow lorises (*Nycticebus coucang*): social organisation, infant care system and diet. Dissertation, Bayreuth University

Wiens F, Zitzmann A (2003) Social structure of the solitary slow loris *Nycticebus coucang* (Lorisidae). J Zool 261(1):35–46

Wiens F, Zitzmann A, Hussein NA (2006) Fast food for slow lorises: is low metabolism related to secondary compounds in high-energy plant diet? J Mammal 87(4):790–798

Winarti I (2008) Field research on Javan slow loris' population in Sukakerta Ciamis and Kawungsari Tasikmalaya,West Java, Indonesia. Report to IAR Indonesia. Ciapus, Bogor, Indonesia, p 7

Zimmermann E (1989) Reproduction, physical growth and behavioural development in slow loris (*Nycticebus coucang*, Lorisidae). Hum Evol 4(2–3):171–179

Chapter 23
Conclusions

Throughout this volume, we have attempted to present the most up-to-date information on the behavior, ecology, and conservation status of Indonesia's primates. Despite the variety of topics including grooming, scent marking, culture, communication, group composition, ranging behavior, sexual conflict, predator recognition, male-male affiliations, as well as human nonhuman primate commensalism that is covered in this book, a common theme that is evident in all of these papers is CONSERVATION. As scientists working in Indonesia, all of us recognize that without conservation, there would no longer be any primates or any habitat for the primates to inhabit. We recognize that if we want to continue studying Indonesia's primates, we must assist Indonesia in conserving them and the forest they rely on. Thus, the majority of scientists studying Indonesia's primates integrate various conservation actions within their theoretical research designs. These actions often include giving talks at the local schools and religious institutions, making and handing out posters to local agencies and schools, training park guards and local students in basic biological field techniques, as well writing reports for various government agencies on the changing state of threats to the populations we are studying. Nonetheless, despite the best efforts, according to the report, "*Primates in Peril: The World's 25 Most Endangered Primates: 2008–2010*", four of the world's most endangered primates (16%) are from Indonesia. They are *Tarsius tumpara* (the Siau Island tarsier); *Nycticebus javanicus* (the Javan slow loris); *Simias concolor* (Pig tailed langur); and *Pongo abelii* (the Sumatran orangutan). Three of these species are represented in this volume, presenting the newest information on these species. There is no paper in this volume on *Tarsius tumpara* because the first description of this population occurred during the publication of this volume. According to the IUCN/WWF report, the *Tarsius tumpara* population is limited to "low thousands" and is listed as Critically Endangered and faces an imminent threat of extinction. Shekelle and Salim (*in press*) used GIS data and field surveys to identify specific threats to this species. They include a very small geographic range of 125 km², an even smaller area of occupancy (19.4 km²), and a high density of humans (311 people per km²) that habitually hunt and eat tarsiers for snack food. In addition, there are no protected areas within its range.

Nycticebus javanicus is found only on the Indonesian island of Java. This species is mostly threatened by the massive trade as pets and for commerce in traditional medicines.

S. Gursky-Doyen and J. Supriatna (eds.), *Indonesian Primates*,
Developments in Primatology: Progress and Prospects,
DOI 10.1007/978-1-4419-1560-3_23, © Springer Science+Business Media, LLC 2010

Easy to catch due to their slow locomotion, the number of lorises in animal markets is greater than the ability of these slow-reproducing primates to recover their population numbers (Shepherd et al. 2004). This has resulted in an extremely rapid population decline which has been compounded by widespread forest loss. At present, less than 10% of the original forest remains in Java and only 17% of the potential distribution of *N. javanicus* is currently within the protected area network of Java. Surveys showed that this species occurs at a density of 0.02 to 0.20 animals per km, meaning 5–10 km must be walked to see a single loris (Nekaris and Nijman 2008; Winarti 2008).

Simias concolor is restricted in its distribution to the Pagai Islands. New estimates of the amount of forest cover remaining on the Pagai Islands (about 826 km²) have been calculated using Google Earth Pro composite satellite imagery (Paciulli and Viola 2009). The forest cover coupled with primate density data (Paciulli 2004) indicate that there are approximately 3,347 simakobus on the Pagai Islands. The main threats to this species are forest loss due to human encroachment, commercial logging and product extraction, conversion to cash crop, oil palm plantations as well as traditional hunting. *Simias concolor* seems to be particularly sensitive to logging, having a population density of 5 individuals/km² in unlogged Pagai forests compared with a mere 2.5 individuals/km² in Pagai forests that were logged 20 years earlier (Paciulli 2004).

Pongo abelii the Sumatran orangutan faces an immediate threat of extinction and is listed as Critically Endangered on the IUCN Red List of Threatened Species. The species is endemic to the island of Sumatra, Indonesia, but its distribution on the island is restricted to the remaining lowland forests of the two most northerly provinces of the island, Nanggroe Aceh Darussalam (NAD) and North Sumatra. Based on nest density surveys and satellite imagery, the remaining population numbers approximately 6,500 individuals. The primary threat to the Sumatran orangutan is habitat conversion and fragmentation. Logging often leads to the conversion of forests for agriculture or oil palm plantations. It has been suggested that between 1990 and 2000, 301,420 ha, or 13% of the original forest cover were lost in North Sumatra Province alone (Gaveau et al. 2007). On a more positive note, the Indonesian government developed a National Strategy and Action Plan for Orangutan Conservation 2007–2017 (DitJen PHKA 2007), and the Government of NAD has also imposed a moratorium on all logging in the Province. Nevertheless, as with so many plans and laws, they must be strictly followed and enforced or they will have little or no influence on the status quo.

Clearly, Indonesia's primates are still threatened by potential extinction. Yet despite the threat of extinction, it is still a very exciting time to study primates in Indonesia as new species are continually being discovered throughout the archipelago and species that were previously thought to be extinct, such as the pygmy tarsier *Tarsius pumilus* being rediscovered. Many of these new primates have very restricted distributions (one of the reasons they were not discovered sooner) and, some are known only from their type localities. With more localities throughout the Indonesian archipelago being explored, it is probable that many more species will be discovered and identified.

References

Gaveau D, Wandono H, and Setiabudi F (2007) Three decades of deforestation in southwest Sumatra: Have protected areas halted forest loss and logging, and promoted re-growth? Biological Conservation 134:495–504

Nekaris KAI and Nijman V (2008) Survey on the abundance and conservation of Sumatran slow lorises (*Nycticebus coucang hilleri*) in Aceh, Northern Sumatra. *Folia Primatologica* 79:365. Abstract.

Paciulli LM (2004) The Effects of Logging, Hunting, and Vegetation on the Densities of the Pagai, Mentawai Island Primates. State University of New York at Stony Brook, Stony Brook, NY.

Paciulli LM and Viola J (2009) Population estimates of Mentawai primates on the Pagai Islands, Mentawai, West Sumatra, Indonesia. *American Journal of Physical Anthropology* Suppl. 43. In press. Abstract

Shepherd CR, Sukumaran J, Wich SA (2004) Open season: an analysis of the pet trade in Medan, North Sumatra, 1997–2001. TRAFFIC Southeast Asia. Kuala Lumpur

Winarti I (2008) Field research on Javan slow loris' population in Sukakerta Ciamis and Kawungsari Tasikmalaya,West Java, Indonesia. Report to International Animal Rescue Indonesia (IARI), Ciapus, Bogor, Indonesia. 7pp

Index